T0297839

CAMBRIDGE LIBRARY COLLECTION

Books of enduring scholarly value

Life Sciences

Until the nineteenth century, the various subjects now known as the life sciences were regarded either as arcane studies which had little impact on ordinary daily life, or as a genteel hobby for the leisured classes. The increasing academic rigour and systematisation brought to the study of botany, zoology and other disciplines, and their adoption in university curricula, are reflected in the books reissued in this series.

A History of British Fossil Mammals, and Birds

Richard Owen (1804–92) was a controversial and influential palaeontologist and anatomist. During his medical studies in Edinburgh and London, he grew interested in anatomical research and, after qualifying as a surgeon, became assistant conservator in the museum of the Royal College of Surgeons, and then superintendent of natural history in the British Museum. He became an authority on comparative anatomy and palaeontology, coining the term 'dinosaur' and founding the Natural History Museum. He was also a critic of Darwin's theory of evolution by natural selection, and engaged in a long and bitter argument with Thomas Huxley, known as 'Darwin's bulldog' for his belligerent support of the theory. Published in 1846, this is Owen's comparative anatomical analysis of the fossils of British birds and mammals. It compares living species with extinct ones, and explains the characteristics that help identification, using 237 woodcut illustrations to show the traits of different species.

Cambridge University Press has long been a pioneer in the reissuing of out-of-print titles from its own backlist, producing digital reprints of books that are still sought after by scholars and students but could not be reprinted economically using traditional technology. The Cambridge Library Collection extends this activity to a wider range of books which are still of importance to researchers and professionals, either for the source material they contain, or as landmarks in the history of their academic discipline.

Drawing from the world-renowned collections in the Cambridge University Library, and guided by the advice of experts in each subject area, Cambridge University Press is using state-of-the-art scanning machines in its own Printing House to capture the content of each book selected for inclusion. The files are processed to give a consistently clear, crisp image, and the books finished to the high quality standard for which the Press is recognised around the world. The latest print-on-demand technology ensures that the books will remain available indefinitely, and that orders for single or multiple copies can quickly be supplied.

The Cambridge Library Collection will bring back to life books of enduring scholarly value (including out-of-copyright works originally issued by other publishers) across a wide range of disciplines in the humanities and social sciences and in science and technology.

A History of British Fossil Mammals, and Birds

Richard Owen

CAMBRIDGE UNIVERSITY PRESS

Cambridge, New York, Melbourne, Madrid, Cape Town,
Singapore, São Paolo, Delhi, Tokyo, Mexico City

Published in the United States of America by Cambridge University Press, New York

www.cambridge.org
Information on this title: www.cambridge.org/9781108038164

This edition first published 1846
This digitally printed version 2011

ISBN 978-1-108-03816-4 Paperback

A HISTORY OF BRITISH FOSSIL MAMMALS, AND BIRDS.

BY

RICHARD OWEN, F.R.S., F.G.S. ETC.

HUNTERIAN PROFESSOR AND CONSERVATOR OF THE MUSEUM OF THE ROYAL COLLEGE OF SURGEONS OF ENGLAND.

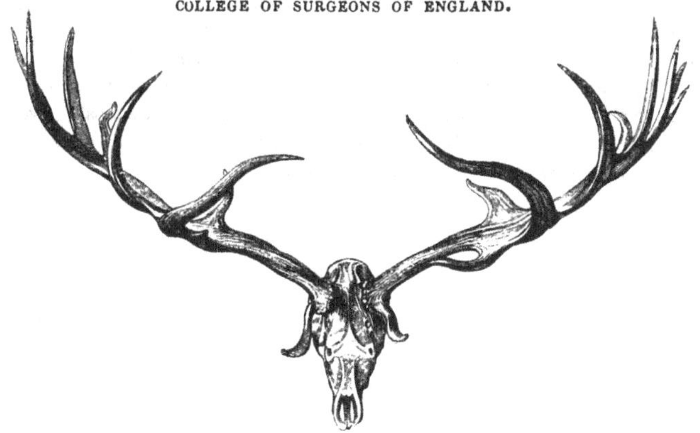

ILLUSTRATED BY 237 WOODCUTS.

LONDON:
JOHN VAN VOORST, PATERNOSTER ROW.
M.DCCC.XLVI.

"There is in the world no kind of knowledge, whereby any part of Truth is seen, but we justly account it precious; yea, that Principal Truth, in comparison whereof all other knowledge is vile, may receive from it some kind of light; whether it be that Egyptian and Chaldean wisdom Mathematical, wherewith Moses and Daniel were furnished; or that Natural, Moral, and Civil wisdom, wherein Solomon excelled all men; or that Rational and Oratorial wisdom of the Grecians, which the Apostle St. Paul brought from Tarsus; or that Judaical, which he learned in Jerusalem, sitting at the feet of Gamaliel: To detract from the dignity thereof, were to injure even God himself, who being that Light which none can approach unto, hath sent out these lights whereof we are capable, even as so many sparkles resembling the bright Fountain from which they rise."—*Hooker's Ecclesiastical Polity*, Book iii. fol. 1705, p. 137.

TO

SIR RODERICK IMPEY MURCHISON, G.C.SS.,

F.R.S., MEM. IMP. ACAD. SCI. ST. PETERSBURGH, &c., &c.

PRESIDENT AND TRUSTEE OF THE BRITISH ASSOCIATION FOR THE ADVANCEMENT OF SCIENCE.

DEAR SIR RODERICK,

I avail myself with much pleasure of the permission to dedicate this work to You, as President of the British Association, because it owes its origin to a course of enquiry suggested by the Geological Section, in which your own labours have been so exemplary; and, also, as an acknowledgment of the kind interest you have manifested in its progress.

I am, yours most sincerely,

RICHARD OWEN.

February, 1846.

PREFACE.

I propose in the present work to describe the most characteristic and remarkable Fossil Remains of Mammals and Birds that have hitherto been found in the British Islands; to deduce therefrom, by Physiological comparisons, the living habits of the extinct species, to trace out their Zoological affinities, and to indicate their Geological relations.

The special researches which have enabled me to fulfil in any degree these intentions, were begun by the desire, and have been carried on chiefly by the liberal aid, of the British Association; and this work may be regarded as one of the fruits of the principle of the combination of individual efforts towards the advancement of science, which is embodied in the Association.

In adopting the octavo size and wood-cut illustrations, I have been actuated by the desire to coöperate with some esteemed friends and fellow-cultivators of Zoology in carrying out Mr. Van Voorst's laudable design of publishing, in a uniform series of works, a complete Zoology of the British Islands. That this scheme ought to embrace a history of the past as well as of the present races of British Animals, is as obvious as the necessity of knowing the whole in order fully to comprehend a part; and will be especially manifest by the light thrown by the remains of our ancient Mammalia upon the origin and relations of the small remnant of the indigenous members of that class which still exists in the British Isles.

To those who coöperate in the progress of Palæontology by collecting and preserving the Fossil Remains of Mammals and Birds, I trust that the present work will be found useful as an aid in determining their acquisitions. I have heard the wish for such a work expressed by many Collectors, to whom the great works on general Palæontology, Osteology, and Geology, including figures and descriptions of British Fossils,—as, for example, the 'Ossemens Fossiles' of Cuvier, the 'Ostéographie' of Professor De Blainville, the 'Reliquiæ Diluvianæ' of Dr. Buckland, and the 'Organic Remains' of Parkinson,—were with difficulty, if at all, accessible; not to speak of Memoirs in the Transactions of British and Foreign Societies, in which, heretofore, the descriptions and figures of some of the most interesting British Fossil Mammals and Birds could alone be found.

The present summary will by no means, indeed, preclude the necessity of studying the valuable works above cited, in order to gain a full knowledge of the nature of our extinct animals; but I am not without hope that it may frequently give such an indication of the value and rarity of a newly-discovered fossil, as may induce greater pains and care in its preservation, and thereby tend to accelerate the progress of our knowledge of the ancient Fauna of Great Britain.

The Treatises and Memoirs cited at the head of each section in this 'History,' will demonstrate how great and valuable a proportion of the information therein systematically set forth has been derived from the labours of my predecessors in this field of enquiry. I most gratefully acknowledge these indispensable sources of knowledge;

but I can assure the reader that I have seldom cited them in reference to British Fossils, without a previous examination of the original specimens.

Another source of information, quite indispensable in the composition of a work like the present, is due to the labours of the Field-Geologist and Collector of Fossil Remains; and to the Curators of Public Museums, who impart so much valuable information by the oral elucidation and the systematic display of the treasures confided to their care.

To some of our most eminent Geologists I am under deep personal obligation for the warm interest they have manifested in the success of my researches.

Dr. Buckland has not only given me the free use of the Mammalian Fossils with which he has so richly stored the Geological Museum at Oxford, but he has also, with his wonted liberality, supplied me with drawings and unpublished proof impressions of the Fossil Bones and Teeth from British caves which have been discovered or explored by him since the publication of the 'Reliquiæ Diluvianæ.'

I gratefully acknowledge the same liberality on the part of the eloquent Lecturer on Geology in the University of Cambridge in affording me the use of the specimens in the Woodwardian Museum, which owes so vast an augmentation of its means of instruction to Professor Sedgewick's liberal management and superintendence.

To Charles König, K.H., and to his able assistant Mr. Waterhouse, I am indebted for the kind facilities afforded me in the examination of the Mammalian and Avian remains in the Mineralogical Department of the British Museum, which has been enriched by some of the rare or unique originals from the cave of Kent's Hole, figured in

Dr. Buckland's drawings and engravings above referred to. These specimens were obtained at the sale of the collection of the late Rev. Mr. Mac Enery, a zealous and successful explorer of that rich depository of the remains of Extinct Mammalia. I owe similar acknowledgments to Mr. Lonsdale and to Mr. Woodward, for their obliging attention during my study of the fossils in the Museum of the Geological Society of London; and to Sir H. de la Beche and Professor Edward Forbes, through whose kindness I have profited by the important and rapidly advancing Museum of Economic Geology.

I have derived much information from those indispensable aids to the progress of British Natural History, the Local Museums, now established in most of our provincial cities and towns; and I beg particularly to express my obligations to Professor Phillips, during my study of the fossils in the Museum at York; and to the Directors and Curators of the Museums in Bristol, Newcastle, Birmingham, Manchester, Hull, Falmouth, Stamford, Saffron Walden, and Lancaster.

The private museums and collections of Mammalian Fossils, for free access to which, and for the loan of specimens described and figured in the present work, I here return most grateful acknowledgments, are those of the Marchioness of Hastings, the Earl of Enniskillen, Lord Braybrooke, Sir Philip de M. Grey Egerton, Bart., the Hon. R. Neville, Mr. Ball, Mr. Bowerbank, Mr. Brown of Stanway, Mr. Colchester of Ipswich, Mr. Dixon of Worthing, Mr. Fitch of Norwich, Mr. John Wickham Flower, the Rev. Darwin Fox, Mrs. Gibson of Stratford, Mr. Green of Bacton, Miss Gurney of Northrepps near Cromer, the Rev. F. Lyte of Torquay, Mr. Lyell, Mr. Pratt, Mr. Richardson, Mr. Stone of Garlick Hill, Mr. Stutchbury of Bristol, Mr. Saull, Mr. Wetherell of Highgate, and Mr. Wigham of Norwich.

I gratefully acknowledge the kind and valuable information which Mr. Lyell has imparted to me respecting the geological position and relations of the matrix of many of the fossils described in this work; and I owe much to my accomplished friend Mr. Broderip for his careful revision of the pages before they went to press. I have cited only his important Memoir on the Stonesfield Marsupials in the body of the work (p. 61), but I should be wanting in acknowledgment of the valuable information which, in common with a large proportion of the reading public, I have derived from Mr. Broderip's writings, were I not to state that I have especially profited, in regard to many of the subjects treated of in the following pages, by the rich and judiciously selected mass of information on extinct animals, which he has appended to his most valuable Zoological contributions to the 'Penny Cyclopædia.'

In the Wood Engravings of the fossils selected for illustration, the skilful artist Mr. W. Bagg, of 63 Gower Street, has ably coöperated in fulfilling the objects of this work, by his intelligence in seizing, and care in expressing, their most characteristic features.

INTRODUCTION.

In the endeavour to complete the Natural History of any class of animals, the mind seeks to penetrate the mystery of its origin, and by tracing its mutations in time past, to comprehend more clearly its actual condition, and gain an insight into its probable destiny in time to come.

But the researches by which such knowledge is to be attained are far from being complete. In many countries the fossil remains of former races of animals have been neither found nor sought; where the quest has commenced, it dates but a few years back; and in our own Island, the geology and fossils of which have been as thoroughly investigated as in any other equal portion of the earth, much may yet remain, even as regards the usually conspicuous and easily recognizable fossils of the highly organised animals which form the subject of the present work, to recompense the toil of the Collector and the skill of the Interpreter. Nevertheless, the evidence already elicited from that part of the earth which, after many changes, now constitutes the British Islands, seems to afford a sufficient basis for the following outline of the Ancient History of its Mammalian Fauna.

We discern the earliest trace of warm-blooded, air-breathing, viviparous quadrupeds at that remote period when the deposition of the Oolitic group of limestones had commenced. The massive evidence of the operations of the old ocean, from which those rocks were gradually precipitated, extends across England, from Yorkshire on the north-east to Dorsetshire on the south-west, with an average breadth of nearly thirty miles; and from some land which formed the shore of this arm of sea, were washed down the remains of small Insectivorous and probably Marsupial quadrupeds, distinct in genus and species from any now known in the world. With these small Mammals there occur elytra of beetles, and debris of *Cycadeæ* and other terrestrial plants. The character of some of the vegetable fossils and of the associated shells, as the *Trigoniæ* for example, and the great abundance in the oolitic ocean of fishes, whose nearest living analogue is the Port-Jackson Shark (*Cestracion*), recall many of the characteristic features of actual organic life in Australia. In contemplating, however, the frail and scanty but precious evidence of the ancient oolitic Insectivora, we naturally ask, could this link of the Mammalian chain of Being have existed detached and insulated? Were there then no representatives of carnivorous Thylacines and Dasyures to enjoy life at the expense of the little quick-breeding Phascolotheres and Amphitheres? We can scarcely resist the latent conviction of such an association, notwithstanding the absence of direct proof, since we find so many indications of coeval conditions, apparently favourable for the development of all forms of organic life: and it is plain, from the scarce and fragmentary parts of the skeletons of the hitherto discovered Stonesfield Mammalia, that many circumstances concurred to destroy or conceal such evidences.

The non-discovery of the remains of marine Mammalia is more conclusive as to their non-existence. Had Whales, Grampuses, Porpoises, or Manatees existed in the oolitic ocean, it is highly improbable that every trace of their bones and teeth should have escaped notice; especially when the remains of the *Cetiosauri* and other Reptilian inhabitants of those ancient seas are so abundant.

From the remote period in which the remains of Mammals first make their appearance, to that in which we again get indubitable evidence of their existence, a lapse of time incalculably vast has occurred. We trace it by the successive deposition from seas and estuaries, of enormous masses of rocks of various kinds, the graveyards of as various extinct forms of animal and vegetable life. The shelly limestone of Stonesfield, which contains the bones of the *Amphitheria* and *Phascolotheria*, lies upon Inferior Oolite. Upon it have been accumulated the strata of the Great Oolite, the Cornbrash and the Forest Marble; and upon these have been successively piled the Oxford group of Clay,* Calcareous Grit and Coral Rag, the Kimmeridge Clay and Portland Stone. In the extensive range of Wealden Rocks, deposited after the formation of the Portland Sands by the waters of an immense estuary, and rising to the height of eight hundred feet,† no true indications of warm-blooded animals have been hitherto discovered.‡ Four hundred feet deep of Gault and Greensand rest upon the Wealden, but reveal no trace of Cetacean or other form of Mammalian life.

* The fossil in the Woodwardian Museum, referred to at p. 520, gives the sole indication of a marine mammal at this period. Although the circumstances of its discovery are far from being satisfactory, I am unwilling to lose sight of this indication, because the cervical vertebræ, whilst they evince by their extreme compression and anchylosis, the cetacean characters, present well-marked specific distinctions from all known recent or fossil species.

† Lyell, 'Elements of Geology,' 8vo. 1838, p. 345; Fitton, 'Geology of Hastings,' p. 58; and, especially, Mantell, 'Geology of Sussex,' 4to. 1822. and subsequent Works of this original and successful explorer of the Wealden.

‡ See my paper in the 'Proceedings of the Geological Society,' Dec. 17th, 1845, "On the supposed Bones of Wading Birds from the Wealden."

Over these foundations of the present south-eastern part of our Island the ocean continued to roll, but under influences of heat and light favourable to the development of corals and microscopic shells, during a period of time which has permitted the successive accumulation of layers of these skeletons, in a more or less decomposed state, with probable additions from submarine calcareous and siliceous springs, to the height of one thousand feet. But although amongst the remains of higher organized animals that have become enveloped in the cretaceous deposits, there have been recognised Birds, Pterodactyles, and a land Lizard, probably washed down from some neighbouring shore, no trace of a Mammalian quadruped has yet been discovered in them.

The surface of the chalk, after it had become consolidated, was long exposed to the eroding action of waves and currents. Into deep indentations so formed have been rolled fragments of chalk and flint, with much sand. The perforations of marine animals on that surface have been filled with fine sand; and there are many other proofs of the lapse of a long interval of time between the completion of the chalk deposits of Britain and the commencement of the next or tertiary era. Of this era our present Island gives the first indication in traces of mighty rivers, which defiled the fair surface of the rising chalk by pouring over it the debris of the great continent which they drained,—a continent which has again sunk, and probably now lies beneath the Atlantic.

The masses of clay and sand that have been thus deposited upon the chalk are accumulated chiefly in two tracts, called the London and Hampshire Basins, which seem to have been two estuaries or mouths of the great river: the one extends from Cambridgeshire through Hertfordshire and Suffolk to the North Downs, the other from the South Downs, along the range of chalk hills, into Dorsetshire. Some parts of these deposits attain the height of more than one thousand feet, indicating the great depth of the ocean into which they were poured.*

At the time when these vast but gradual operations were taking place, an arm of sea extended from the north to the area called the Basin of Paris, which received the overflow of a chain of lakes extending thither from the highest part of the central mountain group of France.† An enormous mass of mixed or alternating marine and freshwater deposits was accumulated in this basin, coeval, if we may judge from the identity of the species of shells, with the outpouring of the London and plastic clays upon the English chalk. Each division of the French eocene deposits is characterised either by the exclusive possession or the predominance of particular fossils, and the entire series must have required a long lapse of ages for its accumulation.

Yet the sudden introduction, as it seems, of various forms of Mammalia, at this period of the earth's history, corroborates the inference, from more direct evidence, of the long interval of time that elapsed between the cessation of the British chalk formation, and the commencement of the tertiary deposits.

The proofs of the abundant Mammalian inhabitants of the eocene continent were first obtained by Cuvier from the fossilized remains in the deposits that fill the enormous Parisian excavation of the chalk. But the forms which that great Anatomist restored were all new and strange, —specifically, and for the most part generically, distinct

* Lyell, 'Principles of Geology,' vol. iv. ch. xx. and xxi.

† Omalius d'Halloy, cited by Lyell, l. c. p. 165.

from all known existing quadrupeds. By these restorations the Naturalist was first made acquainted with the aquatic cloven-hoofed animal which Cuvier has called Anoplothere, and with its light and graceful congeners, the Dichobunes and Xiphiodon, with the great Palæotheres, which may be likened to hornless Rhinoceroses, with the more tapiroid Lophiodon, with the large peccari-like pachyderm called Chœropotamus, and with about a score of other genera and species.

Long before any discovery had been made of remains of terrestrial Mammals in the contemporary London and plastic clays, the existence of neighbouring dry land had been inferred from the occurrence, in those deposits, of bones of crocodiles and turtles, and from the immense number of fossil seeds and fruits, resembling those of tropical trees, as pandani, cocoa-nuts, &c.

The remains of a few of the Mammals of the ancient palm-groves that bordered the mighty eocene river or estuary, have since been recovered from its sediments. One of these quadrupeds is a Lophiodon, another a nearly allied pachyderm (*Coryphodon*) larger than any existing tapir; a third (*Hyracotherium*) has the closest affinity to the Chœropotamus, but was not much larger than a hare. In a sandy deposit, probably near the margin of the estuary, and where Kingston in Sussex now stands, the remains of a smaller species of Hyracothere, about the size of a rabbit, have been found: and both here and in the eocene clay at Sheppey, and at Bracklesham, vertebræ of large serpents like the Boa Constrictor have been discovered. The combination of organic remains in these vast accumulations of the detritus of the eocene continent is, in fact, quite analogous to what may be expected to be found in the outpourings of the Ganges or the Amazon, when those sedimentary deposits are in their turn raised from the bed of the recipient ocean, and made dry land.

Scanty as are the eocene Mammalia hitherto discovered in the London clay, they are highly interesting from their identity or close affinity with some of the peculiar extinct genera of the Paris basin. In the fresh-water and marine beds, at the north side of the Isle of Wight, and at the opposite coast of Hampshire, there occur the remains of the same species of quadrupeds as have been found in the contemporaneous Parisian formations. One of the rarest and most remarkable of the Pachyderms, whose peculiar characters were obscurely indicated by Cuvier from scanty fossils yielded by the Montmartre gypsum, has had its claims to generic distinction established, and its nature and affinities fully illustrated, by more perfect specimens from the eocene limestone of the Isle of Wight: in no other part of Great Britain has any portion of this animal, the Chœropotamus, been found, except in the above limited locality, which alone corresponds with the formations of the Paris basin in mineral character, as well as in date of origin. This discovery becomes, therefore, peculiarly interesting and suggestive. For, were the common notion true, that all the fossil remains of quadrupeds not now existing in our island had been brought hither during a single catastrophe, and had been strewed with the detritus of a general deluge over its surface, what would have been the chance of finding the solitary bone of a Chœropotamus in the very spot, and in the very limited locality, where alone in all England the same kind of fresh-water deposits existed as those in which the unique upper jaw of the same extinct species had been found in France? With the Chœropotamus are associated in the Binstead and Seafield quarries of the Isle of Wight remains of Anoplotherium, Dichobune, Palæotherium, and Lophiodon, showing, with the fossils from the London clay, that the same peculiar generic forms of the class Mammalia prevailed during the eocene epoch in England as in France.

Almost the sole exception to the generic distinction of the Eocene Mammalia which occurred in the researches of Cuvier, was the famous Didelphys of Montmartre: and what made this discovery the more remarkable was the fact that all the known existing species of that marsupial genus are now confined to America, and the greater part to the southern division of that continent. An Opossum appears to have been associated with the peccari-like Hyracotherium in the eocene sand of Suffolk; where, likewise, some teeth of a Monkey, apparently a Macacus, have been found. It is not uninteresting to remark that the Peccari, the nearest existing ally to the old Hyracothere, is, like the Opossum, now peculiar to America; and that two species of Tapir, the nearest living allies to the Lophiodon, exist in South America. We gain little, however, from the comparison of the eocene with the existing Mammalia, in reference to their geographical distribution, except a strong indication that the relative distribution of land and sea, as well as the climate of English latitudes, were then widely different from what they are at the present day.

The marine deposits of the eocene epoch, in contrast with those of the preceding secondary periods, also bespeak the great advance of animal life, and show the remains of great Whales. Petrified cetaceous bones have been found *in situ* in the London clay at Harwich; and similarly petrified teeth and ear-bones, "cetotolites," have been washed out of the eocene clay into the Red-crag at Felixstow. These fossils, however, belong to species distinct from any known existing Cetacea, and which, probably, like some of the eocene quadrupeds, retained fully developed characters which are embryonic and transitory in existing cognate Mammals.

With the last layer of the eocene deposits we lose, in this island, every trace of the Mammalia of that remote period. The imagination strives in vain to form an idea commensurate with the evidence of the intervening operations which Continental Geology teaches to have gradually and successively taken place, of the length of time that elapsed before the foundations of England were again sufficiently settled to serve as the theatre of life to another race of warm-blooded quadrupeds. The miocene strata of the basins of the Danube and the Rhine, and the valley of the Bormida, attest the share which the sea took in the contribution of these deposits, between the end of the eocene period and the time when we again find Mammalian fossils in England. Lakes and rivers intercalated their sediments with those of the sea, as at Saucats, south of Bordeaux; whilst active volcanoes in Auvergne, Hungary, and Transylvania, were adding their share of solid matter to the rising continent.*

Our knowledge of the progression of Mammalian life in Europe during this period, is derived exclusively from continental fossils. These teach us that one or two of the generic forms most frequent in the older tertiary strata still lingered on the earth, but that the rest of the eocene Mammalia had been superseded by a new race, some of which present characters intermediate between those of eocene and those of pliocene genera. The Dinotherium and narrow-toothed Mastodon, for example, diminish the interval between the Lophiodon and the Elephant; the

* Lyell, loc. cit. ch. xv.

Anthracotherium and Hippohyus, that between Chœropotamus and Hippopotamus; the Acerotherium was a link connecting Palæotherium with Rhinoceros. With these and other forms, as Halitherium, a kind of Dugong with molar teeth like those of the Hippopotamus, there likewise appear a few genera that predominate in the pliocene strata, and which are still represented on the earth; though by species quite distinct from those that existed during either of the tertiary periods. Our own island yields but a dim and confused indication of the geological operations that took place between the eocene and pliocene periods, in the wreck of strata that constitute part of the so-called Crag-formations on its eastern coast. In the oldest, and probably miocene portion, called the "Red-crag," numerous remains of three or four extinct species of Cetaceous mammals occur; but these were probably washed out of the subjacent eocene beds. From the Red-crag there have, likewise, been obtained a few rolled fragments of teeth referable to a Bear, to a species of *Felis* of the size of a Leopard, to a Hog, and a Deer. In the Norwich, or fluvio-marine Crag, referred by Mr. Lyell to his oldest pliocene period, there are found teeth and tusks of a Mastodon of the same species as that which is associated with the Dinotherium in the miocene deposits at Eppelsheim; and no remains of Mastodon have been found in any other formation in this island. This rare British Fossil Mammal, occurring in a deposit which is very near, if not identical in point of time, with the continental formations containing more abundant and perfect remains of the same Mastodon, is a fact very analogous to that of the Chœropotamus and Anoplothere in our fresh-water eocene beds; and is equally illustrative of the relation of particular species to particular epochs.

When the eocene and other foundations of our present island had risen from the deep and become the seat of fresh-water lakes, receiving their tranquil deposits with the abundant shells of their testaceous colonies, and during the long progress of that slow and unequal elevation which converted chains of lakes into river-courses, an extensive and varied Mammalian Fauna, as distinct from the miocene as this from the eocene series, ranged the banks or swam the waters of those ancient lakes and rivers. Of these pliocene Mammals, we have abundant evidence in the bones and teeth of successive generations which have been accumulated in the undisturbed stratified lacustrine and fluviatile formations. The like evidence is given by the existence of similar remains in local drifts, composed of gravel, exclusively derived from rocks in the immediate vicinity of such drift, without a single intermixture of any far transported fragments. Equally conclusive and more readily appreciable proof, that the now extinct pliocene and pleistocene Mammalia actually lived and died in this country, has been brought to light from the dark recesses of the caves which served as lurking-places for the predaceous species, and as charnel-houses to their prey.

At the period indicated by those superficial stratified and unstratified deposits, the Mastodon had probably disappeared from England: but gigantic Elephants of nearly twice the bulk of the largest individuals that now exist in Ceylon and Africa, roamed here in herds, if we may judge from the abundance of their remains. Two-horned Rhinoceroses, of at least two species, forced their way through the ancient forests, or wallowed in the swamps. The lakes and rivers were tenanted by Hippopotamuses as bulky and with as formidable tusks as those of Africa.

Three kinds of wild Oxen, two of which were of colossal size and strength, and one of these maned and villous like the Bonassus, found subsistence in the plains. Deer, as gigantic in proportion to existing species, were the contemporaries of the old *Uri* and *Bisontes*, and may have disputed with them the pasturage of that ancient land: one of these extinct Deer is well known under the name of "Irish Elk," by the enormous expanse of its broad-palmed antlers; * another had horns more like those of the Wapiti, but surpassed that great Canadian Deer in bulk; a third extinct species more resembled the Indian Hippelaphus; and with these were associated the Red-deer, the Rein-deer, the Roe-buck, and the Goat. A Wild Horse, a Wild Ass or Quagga, and the Wild Boar, entered also into the series of British Pliocene hoofed Mammalia.

The Carnivora, organized to enjoy a life of rapine at the expense of the vegetable-feeders, to restrain their undue increase, and abridge the pangs of the maimed and sickly, were duly adjusted in numbers, size, and ferocity to the fell task assigned to them in the organic economy of the pre-Adamitic world. Besides a British Tiger of larger size, and with proportionally larger paws than that of Bengal, there existed a stranger Feline animal (*Machairodus*) of equal size, which, from the great length and sharpness of its sabre-shaped canines, was probably the most ferocious and destructive of its peculiarly carnivorous family. Of the smaller Felines we recognise the remains of a Leopard or large Lynx, and of a Wild Cat.

Troops of Hyænas, larger than the fierce Crocuta of South Africa, which they most resembled, crunched the bones of the carcases relinquished by the nobler beasts of prey; and, doubtless, often themselves waged the war of

* See cut in Title-page, and fig. 182.

destruction on the feebler quadrupeds. A savage Bear, surpassing in size the *Ursus ferox* of the Rocky Mountains, found its hiding-place, like the Hyæna, in many of the existing limestone caverns of England. With the *Ursus spelæus* was associated another Bear, more like the common European species, but larger than the present individuals of the *Ursus Arctos.* Wolves and Foxes, the Badger, the Otter, the Foumart, and the Stoat, complete the category of the known pliocene Carnivora of Britain.

Bats, Moles, and Shrews, were then, as now, the forms that preyed upon the insect world in this island. Good evidence of a fossil Hedgehog has not yet been obtained; but remains of an extinct Insectivore of equal size, and with closer affinities to the Mole-tribe, have been discovered in a pliocene formation in Norfolk. Two kinds of Beaver, Hares and Rabbits, Water-voles and Field-voles, Rats and Mice, richly represented the Rodent Order. The greater Beaver (*Trogontherium*) and the Tail-less Hare (*Lagomys*) were the only subgeneric forms, perhaps the only species, of the pliocene *Glires* that have not been recognised as existing in Britain within the historic period The newer tertiary seas were tenanted by Cetacea, either generically or specifically identical with those that are now taken or cast upon our shores.

In the subsequent pages of this work will be found the details of the various kinds of evidence which concur to prove that the Mammalia just enumerated actually lived, generation after generation, for a long succession of years, in the land that now constitutes Great Britain. It may be sufficient, here, to adduce one fact, derived from the peculiar economy of the Deer-tribe, which rebuts the notion that the fossil remains of extinct species have belonged to carcases of drowned animals drifted from a distance.

It is well known that the antlers of deer are shed and renewed annually; and a male may be reckoned to leave about eight pairs of antlers, besides its bones, to testify its former existence upon the earth: but, as the female has usually no antlers, our expectations might be limited to the discovery of four times as many pairs of antlers as skeletons in the superficial deposits of the countries in which such deer have lived and died. The actual proportion of the fossil antlers of the great extinct species of British pliocene Deer, which antlers are proved by the form of their base to have been shed by the living animals, to the fossil bones of the same species is somewhat greater than in the above calculation. Although, therefore, it may be contended that the swollen carcase of a drowned exotic Deer might be borne along a diluvial wave to a considerable distance, and its bones ultimately be deposited far from its native soil, it is not credible that all the solid shed antlers of such species of Deer could be carried by the same cause to the same distance; or that any of them could be rolled for a short distance with other heavy débris of a mighty torrent, without fracture and signs of friction. But the shed antlers of the large extinct species of Deer found in this island and in Ireland have commonly their points or branches entire as when they fell; and the fractured specimens are generally found in caves, and show marks of the teeth of the ossivorous Hyænas, by which they had been gnawed,—thus at the same time revealing the mode in which they were introduced into those caves, and proving the contemporaneous existence in this island of both kinds of Mammalia.*

The perfect condition, and the sharply defined processes,

* See the beautiful and conclusive reasoning of Dr. Buckland on this subject, in his 'Reliquiæ Diluvianæ,' pp. 19—24.

often in high relief, of many of the bones of Elephants, Rhinoceroses, and Hippopotamuses from our tranquil fresh-water deposits, concur, with the nature of their beds, to refute the hypothesis of their having been borne hither by a diluvial current from regions of the earth to which the same genera of quadrupeds are now limited. The very abundance of their fossil remains in our island, is incompatible with the notion of their forming its share of the carcases of one generation of tropical beasts drowned and dispersed by a single catastrophe of waters. This abundance indicates, on the contrary, that the deposits containing them formed the grave-yard, as it were, of many successive generations. But I may here remark, that, notwithstanding we are led to believe, from the extraordinary number of their remains, that Mammoths existed in Britain in herds, like their gregarious congeners in Asia and Africa, yet the multitude of co-existing individuals is not to be reckoned from the absolute quantity of their fossil remains in a given locality. As reasonably might we infer the former populousness of a deserted village from the quantity of human bones in its churchyard.

Having offered the foregoing remarks, chiefly for the Reader who may not be versed in Geology, in justification of the title of the present work, according to its full signification, that not merely the Fossils, but the Species reconstructed by their interpretation, were British,—I proceed to consider the question which will next naturally suggest itself, viz.: how the various members of that ancient Fauna came into this Island? The Geologist, cognizant of the great changes in the relative position of land and sea which continued to be in operation during the pliocene and post-pliocene periods, will probably reply, that Britain was not insulated from the Continent when it received its pliocene Mammalia; and the Zoologist finds this answer to accord with the known powers and habits of those Mammalia. It is true that the Elephant crosses rivers too deep for it to ford; but it swims heavily and slowly, the head and body quite immersed, and only the end of the trunk raised out of the water. The Hippopotamus has been observed to go a short way out to sea from the mouth of its native African river. "The Tiger is seen swimming about among the islands and creeks in the delta of the Ganges; and the Jaguar traverses with ease the largest streams in South America. The Bear, also, and the Bison cross the current of the Mississippi."* But these facts seem to me to form inadequate grounds for belief that those animals could cross a tidal current of sea, twenty miles in breadth. Still less can we suppose that the ponderous Rhinoceroses, the Hyænas, Wolves, Foxes, Badgers, Oxen, Horses, Hogs, and Goats; the smaller Deer, Hares, Rabbits, Pikas, or even the aquatic Rodents, could have reached this island from the Continent, if the present oceanic barrier had interposed. The idea of a separate creation of the same series of Mammalia which existed on the Continent, in and for a small contiguous island, will hardly be accepted. M. Desmarest deduced an argument in proof that France and England were once united, from the correspondence of their Wolves, Bears, and other species known to have existed in this island within the period of history: the conclusion becomes irresistible when the same correspondence is found to extend through the entire series of Proboscidian, Pachydermal, Equine, Bovine, Cervine, Carnivorous, and Rodent Mammalia, which characterized the two countries during the pliocene period of Geology. Thus the science of

* Lyell, Principles of Geology, vol. iii. p. 33.

Anatomy not only reveals the great fact of the former existence in our present island of the same extinct species of quadrupeds that co-existed on the Continent, but becomes in an unexpected degree auxiliary to geographical science ; it throws light upon the former physical configuration of Europe, and on the changes which it has since undergone, and shows that the most striking of those changes have taken place at a comparatively modern period in the history of this planet.

Amongst the purely geological phenomena which indicate the movements and disturbances of the southern and south-eastern parts of England during the pliocene period, may be cited the patches of London clay, with overlying lacustrine strata, which are met with on highly elevated mounds of chalk, indicating considerable up-heaval of those marine formations subsequent to their reception of pliocene fresh-water deposits. Some of the deposits which, from the abundance of Mammoth fossils in them, have received from Dr. Mantell the name of "Elephant-bed,"* have been spread out confusedly, either by successive waves, or by ice-floes carried along by ocean currents. Mr. Lyell, generalizing the various particular phenomena indicative of these changes, says:—"First, the south-eastern part of England had acquired its actual configuration when the ancient chalk-cliff was formed, a beach of sand and shingle having been thrown up at the base of the cliff. Afterwards the whole coast, or at least that part of it where the Elephant-bed now extends, subsided to the depth of fifty or sixty feet, and during the period of submergence, successive layers of white calcareous rubble were accumulated so as to cover the ancient beach. Subsequently, the coast was again raised, so that the ancient

* Geology of the South East of England, 8vo., 1833, p. 31.

shore was elevated to a level somewhat higher than its original position." *

In this interpretation of the phenomena of the supra-cretaceous deposits of Sussex, Mr. Dixon, of Worthing,† who has concentrated the observations of many years, upon the geology of that county, fully coincides, and bears testimony to the comparatively modern character of certain remarkable changes which have taken place on our southern coast.

To a series of successive elevations and depressions, like those elucidated by the observations of the Geologists above cited, may be attributed the final establishment of the British Channel. And, in referring to that event as comparatively recent, the term must not be judged of in relation to so small a fraction of the world's time as has been marked down in the records of the present infancy of the human race: we shall better appreciate it, perhaps, by recalling the ideas of perpetuity which we attach to our ocean barrier, when, gazing on its waves, we sum up the known changes which they have produced on the coast line within the period of history or tradition.

Indications of Geological changes during the pliocene period are not limited in England to the southern parts of the island. Mr. Lyell, in his elucidation of the 'Boulder formation of Eastern Norfolk,'‡ says:—"The fluvio-marine contents of the Norwich Crag imply the former existence of an estuary on the present site of parts of Norfolk and Suffolk, including the eastern coast of Norfolk. Into this estuary or bay, one or many rivers entered; and in the strata then formed were imbedded

* Op. cit. vol. vi. cit. p, 261.

† On the cretaceous and tertiary formations of Sussex, 4to.

‡ 'Philosophical Magazine,' vol. xvi. May 1840, p. 373.

the remains of animals and shells of the land, river, and sea. Certain parts of this area seem at length to have been changed from sea into low marshy land, either because the sea was filled up with sediment, or because its bottom was up-heaved, or by the influence of both these causes."

The present position of the fresh-water white marls in Lancashire, in the Isle of Man, and in Ireland, in which marls the remains of the *Megaceros* are so common, attest the great changes which have taken place in the geographical condition of those lands since the period when that now-extinct Deer left its remains in those newer pliocene lacustrine deposits.*

The extraordinary phenomena of the great northern drift show that, whilst the eastern portion of England, and so much of the western part as Mr. Murchison has called *Siluria*,† were dry land, and inhabited by the pliocene Mammalia, the eastern part of Lancashire, nearly all Cheshire, the north of Shropshire, and a large part of Staffordshire, Worcestershire, and Gloucestershire, were under the sea.

The indications of such changes, mighty in comparison with any of which human history takes cognizance, prepare us to view with less surprise the corresponding changes which have taken place in our Mammalian Fauna; but we are still ignorant of the cause of the extirpation of so large a proportion of it as has become extinct. It is an important fact, however, that a part and not the whole of the terrestrial species have thus perished,‡ whence it may be concluded that the cause of their destruction has not been a violent and universal catastrophe from which none could escape. There is no small analogy, indeed,

* Professor Ed. Forbes, cited at p. 467. † 'Silurian System,' 4to. p. 523.

‡ This fact was established by several of the determinations in my 'Reports on the British Fossil Mammalia,' communicated to the Meetings of the British Association in 1842 and 1843.

between the course of the extirpation of the Pliocene Mammals, and that which history shows to have reduced the numbers of the wild animals of continents and islands in connection with the progress of man's dominion. The largest, the most ferocious, and the least useful of the pliocene species have perished; but the Horse, the Ass, the Hog, probably the smaller Wild Ox, the Goat, the Red-deer, and Roe, and many of the diminutive quadrupeds, remain. The present negative evidence supports the belief that the Human species had not been called into existence when the Mammoth, the tichorhine and leptorhine Rhinoceroses, and the great northern Hippopotamus became extinct. Cuvier drew the same conclusion as to the Quadrumanous Order from the same grounds; but the recent discovery of a true fossil portion of a Monkey's skeleton, (figs. 1, 2, and 3, p. xlvi,) in the same lacustrine deposits which abound in the remains of extinct Pachyderms, with similar discoveries noticed in the first section of the present Work, should teach caution in the application of conclusions from merely negative facts. It is probable that the Horse and Ass are descendants of a species of pliocene antiquity in Europe. There is no anatomical character by which the present Wild Boar can be distinguished specifically from that which was contemporary with the Mammoth. All the species of European pliocene *Bovidæ* came down to the Historical period, and the Aurochs and Musk-Ox still exist; but the one owes its preservation to special Imperial protection, and the other has been driven, like the Rein-deer, to high northern latitudes.* There is evidence that the great *Bos*

* The observations of Mr. Murchison, in his great work on the Geology of Russia, 4to., 1845, pp. 471, 492 to 507, bearing upon the question of the specific identity of the existing with the fossil Aurochs, are highly interesting, and support the conclusions to which I had arrived from anatomical comparisons.

primigenius, and the small *Bos longifrons*, which date, by fossils, from the time of the Mammoth, continued to exist in this island after it became inhabited by Man.* The small shorthorned pliocene Ox is most probably still preserved in the mountain varieties of our domestic cattle. The great Urus seems never to have been tamed, but to have been finally extirpated in Scotland. Of the Cervine tribe, the Red-deer and the Roebuck still exist in the mountainous districts of the north, but, like the Aurochs in Lithuania, by grace of special protective laws. The Rein-deer has, relatively to Britain, become extinct, nor will our present climate permit its naturalization. The *Megaceros*, the still larger *Strongyloceros*, and the remarkable *Cervus Bucklandi*, have absolutely perished. With the diminution of the great Herbivora, which would naturally follow the limitation of their range of pasturage, when England became an island, that of the Carnivora dependent on them for food, would inevitably follow. But the sabre-toothed

* Both the Urus (*Bos primigenius*) and the *Bison priscus* appear to have been contemporary with Man in the North of continental Europe. Their skeletons have been found, with that of the large variety of Rein-deer which existed in Germany in the time of Tacitus, in a bog in Scania by Professor Nillson, and are preserved in the Museum at Lund. My friend Mr. Murchison writes to me:—"This Urus is most remarkable in exhibiting a wound of the apophysis of the second dorsal vertebra, apparently inflicted by a javelin of one of the aborigines, the hole left by which (offering its larger orifice towards the head of the Ox, and the smaller orifice towards its rump,) was exactly fitted by Nillson with one of the heads of the ancient stone javelins collected and described by that excellent Naturalist, in his Work, entitled, "Skandinaviska Norden's Ur-Invoandre, Lund, 1843." This instrument fractured the bone and penetrated to the apophysis of the third dorsal vertebra, which is also injured. The fractured portions are so well cemented that Nillson thinks the animal probably lived two or three years after. The wound must have been inflicted over the horns, and the javelin must have been hurled with prodigious force."

I am much disposed to assent to this interpretation of the wound of the great extinct Ox. It is hard to conceive how such a wound could have been inflicted by the horn of another Urus; but, in interpreting these evidences of primeval hostility, the combative instincts and pointed weapons of the Ox and Deer-tribe, are always to be taken into the account.

Machairodus, the great Spelæan Tiger, Hyæna, and Bear, together with the gigantic pliocene Pachyderms, became extinct here and elsewhere, as it would seem, before the creation of Man,—which would indicate that the extirpating cause, if it were extrinsic to their own constitution, had been due to changes of the configuration and climate of the great continent over which they ranged. We can only associate with the insular condition of Britain the subsequent progress of extirpation, through the agency of Man, by which the smaller kind of Bear and the Wolf have ceased to exist with us. Whilst the Fox, the Badger, the Otter, the Polecat, the Wild Cat, and the Stoat, owe their prolonged existence, as British species, to their comparatively less noxious character and insignificant size.

With regard to the Rodentia, the great Trogonthere seems to have become extinct in England and the Europæo-Asiatic continent before the historical period, whilst the smaller pliocene Beaver continued to exist with us like the Wolf, until hunted down by man: it still survives in a few of the great continental rivers.* Of the little Lagomys of our ossiferous caves no living example remains in either England or Europe: the species, indeed, may be extinct: its genus is now limited to central and southern Asia. I am unable to detect any specific distinction in the fossil bones of the pliocene species of *Lepus* and *Arvicola* from those of the Hares, Rabbits, and Voles that still exist in this island. Native species are still obviously departing, whilst varieties of the domesticated animals are coming in.

We learn, then, from history, that part of the reduction

* The Beaver of North America, (*Castor fiber,*) is a distinct species from the *Castor Europæus.*

of a former rich series of British Mammalia to its present scanty proportion, has been caused by human agency; and we may reasonably conjecture that the rest of the great change has been the consequence of a series of gradual and consecutive dyings-out of species; since certain conditions of the pliocene and post-pliocene Mammalia are irreconcilable with the hypothesis that they all simultaneously perished by a sudden and violent catastrophe, like that which Cuvier deduced from the phenomenon of the frozen Mammoth.* Evidence will be given in the present work in proof, that the Elephants and Rhinoceroses of pliocene Britain, were adapted to live in a northern or temperate climate; and since the Hippopotamus, their contemporary and associate, was a different species from the present African one, it might also have been able to exist beneath a less sultry sky than that of Africa.

Thus, in the endeavour to trace the origin of our existing Mammalia, I have been led by the researches detailed in the present work, to view them as descendants of a fraction of a peculiar and extensive Mammalian Fauna which overspread Europe and Asia at a period geologically recent, yet incalculably remote and long anterior to any evidence or record of the human race. It would appear, indeed, from the comparisons which the present state of Palæontology permits to be instituted between the recent and extinct Mammalian Faunæ of other great natural divisions of the dry land, that these divisions also severally possessed a series of Mammalia, as distinct and peculiar in each, during the pliocene period, as at the present day.†

When such a comparison is restricted to the Fauna of a limited locality, especially an insular one like Great

* See the interpretation of that striking fact in pp. 261, 270.

† See 'Report of the British Association,' 8vo. 1844, p. 237.

Britain, the discrepancy between the pliocene extinct and the existing groups of Mammalia appears to be extreme. But if we regard Great Britain in connection with the rest of Europe, and if we extend our view of the geographical distribution of extinct Mammals beyond the limits of technical geography,—and it needs but a glance at the map to detect the artificial character of the line which divides Europe from Asia,—we shall then find a close and interesting correspondence between the extinct Europæo-Asiatic Mammalian Fauna of the pliocene period, and that of the present day. The very fact of the pliocene Fossil Mammalia of England being almost as rich in generic and specific forms as those of Europe, leads, as already stated, to the inference that the intersecting branch of the ocean which now divides this island from the continent did not then exist as a barrier to the migration of the Mastodons, Mammoths, Rhinoceroses, Hippopotamuses, Bisons, Oxen, Horses, Tigers, Hyænas, Bears, &c., which have left such abundant traces of their former existence in the superficial deposits and caves of Great Britain.* Now, it is a most interesting fact, that, in the Europæo-Asiatic expanse of dry land, species continue to exist of nearly all those genera which are represented by pliocene and post-pliocene Mammalian fossils of the same natural continent and of the immediately adjacent island of Great Britain. The Bear has its haunts in both Europe and Asia; the Beaver of the Rhone and Danube represents the great Trogontherium; the Lagomys and the Tiger exist on both sides of the Himalayan mountain chain; a Hyæna ranges through

* Mr. Lyell infers the former existence of an isthmus between Dover and Calais on other grounds. See his Memoir on the relative ages of the "Crag" of Norfolk and Suffolk. Mag. Nat. Hist. 1839, p. 326.

Syria and Hindostan; the Bactrian Camel typifies the huge *Merycotherium* of the Siberian drift; the Elephant and Rhinoceros are still represented in Asia, though now confined to the south of the Himalayas. The true Macacques are peculiar to Asia, and, though most abundant in the southern parts of the continent and the Indian Archipelago, also exist in Japan; a closely allied subgenus (*Inuus*,) is naturalised on the rock of Gibraltar at the present day. A fossil species of Macacus was associated with the Elephant and Rhinoceros in England during the period of the deposition of the newer pliocene fresh-water beds.* The more extraordinary extinct forms of Mammalia called *Elasmotherium* and *Sivatherium*, have their nearest existing pachydermal and ruminant analogues in the same continent to which those fossils are peculiar. Cuvier places the Elasmothere between the Horse and Rhinoceros: the existing four-horned Antelopes, like their gigantic extinct analogues, the Sivathere and Bramathere, are peculiar to India.

The Mediterranean and Red Seas constitute a less artificial boundary between Africa and the Europæo-Asiatic continent, than that which, on our maps, divides Europe from Asia; yet those narrow seas form a slight demarcation as compared with the vast oceans which divide the old from the new worlds of the geographer, or these from the Australian continents. The continuity of Africa with Asia is still, indeed, preserved by a narrow isthmus, near to which, within the historical period, the Hippopotamus descended, venturing down the Nile almost to its mouth. May it not be regarded, then, as part of the same general concordance of geographical distribution, that

* See 'Comptes Rendus de l'Académie des Sciences,' Paris, Sept. 1845, p. 573, and fig. 1, 2, 3, p. xliv.

the genus *Hippopotamus*, extinct in England, in Europe, and in Asia,* should continue to be represented in Africa and in none of the remoter continents of the earth?—Africa also having its Hyæna, its Elephant, its Rhinoceroses, and its great feline Carnivores. The discovery of extinct species of *Camelopardalis* in both Europe and Asia, of which genus the sole existing representative is now, like the Hippopotamus, confined to Africa, adds to the propriety of regarding the three continuous continental divisions of the Old World as forming, in respect to the geographical distribution of pliocene, post-pliocene and recent Mammalian genera, one great natural province. The only large Edentate animal (*Pangolin gigantesque*, Cuvier, *Macrotherium*, Lartet) hitherto found in the tertiary deposits of Europe, but in those of an earlier period (older pliocene or miocene) than the deposits to whose Mammalian Fossils the present comparison more immediately refers, manifests its nearest affinities to the genus *Manis*, which is exclusively Asiatic and African.

Extending our comparison between the existing and the latest of the extinct series of Mammalia to the continent of South America, it may first be remarked, that with the exception of some of the carnivorous and Cervine species, no representatives of the above-cited Mammalian genera of the Old World of the geographer have yet been found in South America. Buffon † long since enunciated a similar generalization with regard to the existing species and genera of Mammalia; it is almost

* Marsden, in his 'History of Sumatra,' mentions a species of Hippopotamus as still existing in the Sunda Isles; but this has much need of confirmation: the fossil sub-genus of Hippopotamus (*Hexaprotodon* of Cautley and Falconer) gives a new stimulus, however, to the inquiry after the Hippopotamus or Succatyro of the Indian Archipelago.

† Cited by Lyell in the 'Principles of Geology,' 1837, vol. iii. p. 27.

equally true in respect of the fossil. Not a relic of an Elephant, a Rhinoceros, a Hippopotamus, a Bison, a Hyæna,* or a Lagomys, has yet been detected in the caves or the more recent tertiary deposits of South America. On the contrary, most of the Fossil Mammalia from those formations are as distinct from the Europæo-Asiatic forms, as they are closely allied to the peculiarly South American existing genera of Mammalia.

The genera *Equus*, *Tapirus*, and the still more ubiquitous *Mastodon*, form the chief, if not sole exceptions. The representation of *Equus*, during the pliocene period, by distinct species in Asia (*E. primigenius*) and in South America (*E. curvidens*), is analogous to the geographical distribution of the species of *Tapirus* at the present day. Fossil Tapirs have been found both in Europe and in South America.

Pangolins still exist in Asia and in Africa, and, as we have seen, a gigantic extinct species has been found in the middle tertiary beds of Europe, but not a trace of a scaly Anteater, recent or extinct, has been discovered in South America, where the Edentate order is so richly represented by other generic and specific forms.

South America alone is now inhabited by species of Sloth, of Armadillo, of Cavy, Aguti, Ctenomys, and Platyrrhine Monkey; but no fossil remains of a quadruped referable to any of these genera have yet been discovered

† Dr. Lund ('Danish Transactions,' Œrsted, Kiöbenh, 1842, p. 16,) discovered the remains of an extinct Carnivore in a Brazilian cavern, which he at first announced as a species of *Hyæna*, but he has since recognised very distinctive dental characters, and refers it to a new genus, which he calls *Smilodon*. From the figures which he has given of the canine and incisor teeth, it seems to belong to the same genus (*Machairodus*) as the so-called *Ursus cultridens* of Europe, and this is certainly the case with portions of the skull, lower jaw, and teeth, since discovered in the Pampas of Buenos Ayres, and now in the British Museum.

in Europe, Asia, or Africa. The types of *Bradypus* and *Dasypus* were, however, richly represented by diversified and gigantic specific forms in South America, during the geological period immediately preceding the present; and fossil remains of extinct species of *Cavia*, *Cœlogenys*, *Ctenomys*, and *Cebus*, have hitherto been detected exclusively in the continent where these genera still as exclusively exist. *Auchenia* more remotely typifies *Macrauchenia*. The murine fossils in the rich collection of remains from Brazilian caverns, lately received at the British Museum, all belong to the genus *Hesperomys*, the aboriginal living representative of the *Muridæ* in South America; not a single fossil is referable to a true Old World *Mus*, though numbers of the common Rat and Mouse have been imported into South America since its discovery by Europeans. With regard to the Sloths and Armadillos, they now seem, after the rich harvest of bulky Glyptodons, Mylodons, Pachytheriums, and the more gigantic Megatherioid quadrupeds, to be the last remnants of a Mammalian Fauna, which once almost equalled in the size and number of its species that of the Europæo-Asiatic expanse, and was as peculiarly characteristic of the remote continent in which almost all its representatives have been entombed.

In North America the most abundant Mammalian fossils of the corresponding recent geological epoch belong to a species of *Mastodon* (*M. giganteus*) peculiar to that continent. Since, however, North America borders closely upon Asia at its northern basis, and is connected by its opposite apex with South America, it perfectly accords with the analogies of the geographical relations of the last-extirpated series of Mammals of the Old World that the Asiatic Mammoth and the South American Megatherium should have migrated from opposite extremes, and have met in the temperate latitudes of North America, where, however, their remains are much more scanty than in their own proper provinces.

Australia in like manner, yields evidence of an analogous correspondence between its last extinct and its present aboriginal Mammalian Fauna, which is the more interesting on account of the very peculiar organization of most of the native quadrupeds of that division of the globe. That the Marsupialia form one great natural group, is now generally admitted by zoologists; the representatives in that group of many of the orders of the more extensive placental sub-class of the Mammalia of the larger continents have also been recognised in the existing genera and species:—the Dasyures, for example, play the parts of the *Carnivora*, the Bandicoots of the *Insectivora*, the Phalangers of the *Quadrumana*, the Wombat of the *Rodentia*, and the Kangaroos, in a remoter degree, that of the *Ruminantia*. The first collection of Mammalian Fossils from the ossiferous caves of Australia brought to light the former existence on that continent of larger species of the same peculiar marsupial genera:—some, as the *Thylacine*, and the Dasyurine sub-genus represented by the *Das. ursinus*, are now extinct on the Australian continent, but one species of each still exists on the adjacent island of Tasmania; the rest were extinct Wombats, Phalangers, Potoroos and Kangaroos, some of the latter being of gigantic stature. Subsequently, and after a brief interval, we obtain a knowledge of the former existence in Australia of a type of the marsupial group, exemplified by the genera *Diprotodon* and *Nototherium*,* which represented the Pachyderms of

* See "Catalogue of Fossils in the Museum of the College of Surgeons," 4to., 1845, pp. 291, 336, pls. vi, x.

the larger continents, and which seems now to have disappeared from the face of the earth.

The genus Mastodon forms an exception to that continental localization, not only of existing, but of pliocene extinct genera of Mammalia above briefly dwelt upon. The solitary character, however, of this exception serves rather to establish the rule: at least, I know of no other extinct genus of Mammal which was so cosmopolitan as the Mastodon: it was represented by species, for the most part very closely allied, if actually distinct, in Europe, in Asia, in North and South America, and in Australia: it is the only aboriginal genus of quadruped in Australia which was represented by other species in other parts of the world.*

The most remarkable local existing Fauna, in regard to terrestrial vertebrated animals, is that of the islands of New Zealand, with which geologists have been made familiar by Mr. Lyell's indication of its close analogy with the state of animal life during the period of the Wealden formation.† The only terrestrial Mammalian quadruped hitherto discovered in New Zealand, whose recent introduction into that island is at all doubtful, is a small Rat. The unequivocally indigenous representatives of the warm-blooded vertebrata are Birds, of which the *Apteryx* is the most peculiar. It is the smallest known species of the Struthious or wingless order, has the feeblest rudiments of the anterior members, and not any of its bones are permeated by air-cells. This bird forms the most striking and characteristic type of the proper or primitive Fauna of New Zealand.

* See 'Report on Australian Fossil Mammalia,' in the 'Transactions of the British Association,' 1844, p. 239.

† 'Elements of Geology,' 8vo, 1838, p. 366, and 'Principles of Geology,' 1837, vol. i. p. 204.

The organic remains of the most recent deposits of the North Island, which are most probably contemporary with the post-pliocene formations of Australia and Europe, are referable to an apparently extinct genus of Struthious birds, having the nearest affinities to the Apteryx. The remains of this genus (*Dinornis*) appear to be very abundant, notwithstanding the stupendous stature of some of the species.* It is reported that a large *Dinornis* still exists in the South Island of New Zealand; and some of the species may have been living in the North Island, when the human aborigines first set foot there. But the bones which have reached me from that Island, although retaining much of their animal matter, are more or less impregnated with ferruginous salts, and may have lain in an argillaceous soil for as long a period as some of the latest extinct Mammals of Australia, South America and Europe. Not a trace of a fossil quadruped has been found in New Zealand; but our present knowledge of the living and the last-exterminated Faunæ of the warm-blooded animals of that small but far distant and isolated portion of earth, shows that the same close analogy existed between them, as has been exemplified in the corresponding Faunæ of larger natural divisions of the dry land on the present surface of this planet.

Additional facts, and the means of extending our comparisons, by the collection of the fossils of distant lands, are most desirable in order to precisely define the laws of the geographical distribution of the Mammalia of the older

* I estimate the *Dinornis ingens* to have stood nine feet, and the *Din. giganteus* ten feet, in height. See Zoological Transactions, vol. iii. part 3; in which, also, the peculiar and suggestive geographical distribution of other existing and extinct Struthious birds is discussed, p. 268, *et seq.*

and newer pliocene periods; and to speak of the sum of the present observations under the term "law," may, perhaps, be deemed premature. But the generalizations first enunciated in my Report to the British Association in 1844, seemed to be sufficiently extensive and unexceptionable to render them of importance in a scientific consideration of the present distribution of the highest organized and last-created class of animals; and to show that, with extinct as with existing Mammalia, particular forms were assigned to particular provinces, and, what is still more interesting and suggestive, that *the same forms were restricted to the same provinces at the pliocene periods, as they are at the present day.**

In carrying back the retrospective comparison of recent and extinct Mammals to those of the eocene and oolitic strata, in relation to their local distribution, we obtain indications of extensive changes in the relative position of sea and land during those epochs, in the degree of incongruity between the generic forms of the Mammalia which then existed in Europe, and any that actually exist on the great natural continent of which Europe now forms part. It would appear, indeed, from our present knowledge, that the further we penetrate into time for the recovery of extinct Mammalia, the further we must go into space to find their existing analogues. To match the eocene Palæotheres and Lophiodons, we must bring Tapirs from Sumatra or South America, and we must travel to the antipodes for Myrmecobians and Dasyures, the nearest living

* Humboldt, in citing the *Mylodon*, *Dinornis*, and *Diprotodon*, briefly repeats my generalizations from those discoveries, and says: "Es herrscht in Südamerika und in den Australländern eine grosse Aehnlichkeit zwischen den dort lebenden und den untergegangenen Thieren." "In South America and the Australian lands there prevails a great resemblance between the existing and the extinct animals."—*Kosmos*, 8vo. 1845, p. 303.

analogues to the Amphitheres and Phascolotheres of our oolitic strata.

If ever the first types of the primary groups of the class Mammalia radiated from a common centre, it must have been at a period incalculably remote, and there is small hope of our ever being able to determine its site, by reason of the enormous alternations of land and sea that have come to pass since the class was first introduced into our planet. We find, however, that, from the period when the great masses of dry land assumed the general form and position that they now present, the same peculiar forms of Mammalia characterized their respective Faunæ: and the evidence of the distribution of the recent and extinct pliocene Mammalia favours the conclusion that New Zealand, Australia, South America, and the Old World of the geographers had been as many distinct centres of creation.

By the same evidence we are compelled to admit, that the difficulties which beset the Linnæan view of the actual diffusion of organized beings * are insurmountable. According to the hypothesis that all existing land animals radiated from a common Asiatic centre within the historical period, we must be prepared to believe that the nocturnal Apteryx, which is neither organized for flying nor swimming, migrated across wide seas, and found its sole resting-place in the Island of New Zealand, where alone the remains of similar wingless birds have been found fossil: —that the Wombats, Dasyures, and Kangaroos as exclusively travelled to Australia, where only have been found, in pliocene strata and bone caves, the remains of extinct

* See Linnæus' preface to the 'Museum Regis Adolphi Frederici,' 1754: and the excellent remarks in Dr. Pritchard's 'Physical History of Man,' vol. i. 1826, pp. 16, 81.

and gigantic species of the same genera or families of Marsupialia :—and that the modern Sloths, Armadillos, and Anteaters, chose the route to South America, where only, and in the warmer parts of North America, are to be found the fossil remains of extinct species of those very peculiar edentate genera. It is not less striking and suggestive, though at first sight less subversive of the recent-dispersion theory, to find the Macacus, Elephant, Rhinoceros, Hippopotamus, Hyæna, Beaver, Pika, Hare and Rabbit, Vole and Mole still restricted to that great natural division of dry land, the old world of geography, to which the fossil remains of the same genera or species appear to be peculiar. These generalizations, and the special facts which are treated of in the following pages, must be interpreted agreeably with right reason, and not warped to suit with preconceived views.

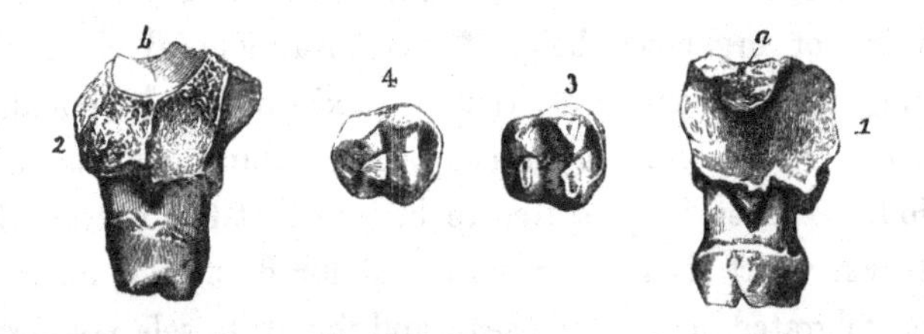

Portion of right upper maxillary bone with the penultimate true molar of a fossil Monkey (*Macacus pliocenus*) ; from the newer pliocene brick-earth at Grays, Essex. 1, front view ; *a*, base of malar process. 2, back view ; *b*, smooth surface of the antrum maxillare. 3, grinding surface of the fossil tooth. 4, grinding surface of the corresponding tooth of a recent Monkey (*Macacus sinicus*).

CONSPECTUS OF BRITISH FOSSIL MAMMALIA, ACCORDING

OOLITE.	EOCENE.	MIOCENE.	PLIOCENE.
Stonesfield Slate.	*Clays and Marls.*	*Red Crag.*	*Fluvio-marine Crag.*
Amphitherium Prevostii.	Macacus eocænus.	Ursus.†	Mastodon angustidens.
,, Broderipii.	Didelphys? Colchesteri.	Meles.†	Elephas primigenius.‡
Phascolotherium Buckland.	Coryphodon eocænus.	Felis par-	Rhinoceros tichorhinus.‡
	Lophiodon.	doides.†	Equus fossilis.‡
	Lophiodon minimus.	Sus.†	Cervus elaphus.‡
	Palæotherium magnum.	Cervus.†	Arvicola.‡
	,, medium.		Lutra.‡
	,, crassum.		
	,, minus.		
	Chœropotamus Cuvieri.		
	Hyracotherium leporinum.		
	,, Cuniculus.		
	Anoplotherium commune.		
	Dichobune cervinum.		
	Balænodon affinis.*		
	,, definita.*		
	,, emarginata.*		
	,, gibbosa.*		
	,, physaloïdes.		

* These are described in the text as species of the existing genus *Balæna*, but reasons are there
Whales as the fossil teeth found with them: most of them occur in the Miocene crag, but there is
† The nature of the stratum renders the actual age of these fossils doubtful.
§ From Durdham Down Cave, near Bristol, on the authority of E. T. Higgins, Esq., of Clifton.

TO THEIR GEOLOGICAL POSITION.

NEWER PLIOCENE.		ALLUVIUM.
Drift and Fresh-water Deposits.	*Caves.*	*Fen and Turbary.*
Macacus pliocænus.	Vespertilio Noctula.	Sorex remifer.
Sorex.	Rhinolophus Ferrum-equinum.	Talpa europæa.
Talpa europæa.	Ursus priscus.	Ursus Arctos.
Palæospalax.	,, spelæus.	Meles taxus.
Ursus splæus.	Meles taxus.	Putorius vulgaris.
Canis Lupus.	Putorius vulgaris.	Lutra vulgaris.
Hyæna spelæa.	,, ermineus.	Canis Lupus.
Felis spelæa.	Lutra vulgaris.§	Felis Catus.
,, catus.	Canis Lupus.	Arvicola amphibia.
Trogontherium.	,, Vulpes.	Arvicola agrestis.
Castor europæus.	Hyæna spelæa.	Castor Europæus.
Arvicola.	Felis spelæa.	Lepus cuniculus.
Elephas primigenius.	,, Catus.	,, timidus.
Rhinoceros tichorhinus.	Machairodus latidens.	Equus Caballus,
,, leptorhinus.	Mus musculus.	Equus Asinus.
Equus fossilis.	Arvicola amphibia.	Sus scrofa.
Asinus fossilis.	,, agrestis.	Cervus Elaphus.
Hippopotamus major.	,, pratensis.	,, Capreolus.
Sus Scrofa.	Lepus timidus.	Capra Hircus.
Megaceros Hibernicus.	,, cuniculus.	Bos longifron
Cervus elaphus.	Lagomys spelæus.	Phocæna crassidens.
,, Tarandus.	Elephas primigenius.	Balænoptera Boops.
,, Capreolus.	Rhinoceros tichorhinus.	Balæna mysticetus.
Capra Hircus.	Equus fossilis (Caballus?)	
Bison priscus.	,, plicidens.	
Bos primigenius.	Asinus fossilis.	
,, longifrons.	Hippopotamus major.	
Phocæna crassidens.	Sus Scrofa.	
Monodon monoceros.	Megaceros Hibernicus.	
Physeter macrocephalus.	Strongyloceros spelæus.	
Balænoptera Boops.	Cervus Elaphus.	
Balæna mysticetus.	,, Tarandus.	
	,, Capreolus.	
	,, Bucklandi.	
	Bison priscus.	
	,, minor.	
	Bos primigenius.	

assigned which make it probable that they belonged to the same kinds of
little doubt that they were washed out of the underlying eocene clay.

‡ Probably derived from overlying blue clay.

ERRATA.

Page 29, in *fig.* 15 omit the letter *e*.
„ 30, for Cut 15, read *fig.* 16.
„ 46, „ Cut 12, „ *fig.* 15.
„ 140, „ *fig.* 60, „ *fig.* 62.
„ 168, „ *Limnæas*, „ *Limnæa*.
„ 536, in foot-note, for Ziphius read φαλαινη, and *dele* fossil.
„ 537, in *fig.* 227, a dotted line should connect letter *d* with the dark central spot.

BRITISH FOSSIL MAMMALIA.

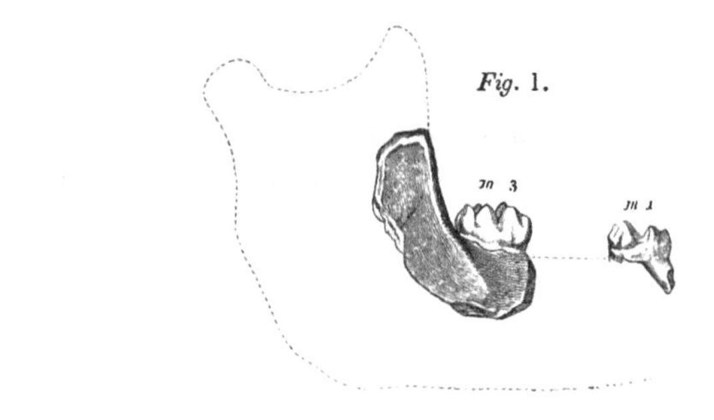

MACACUS EOCÆNUS.

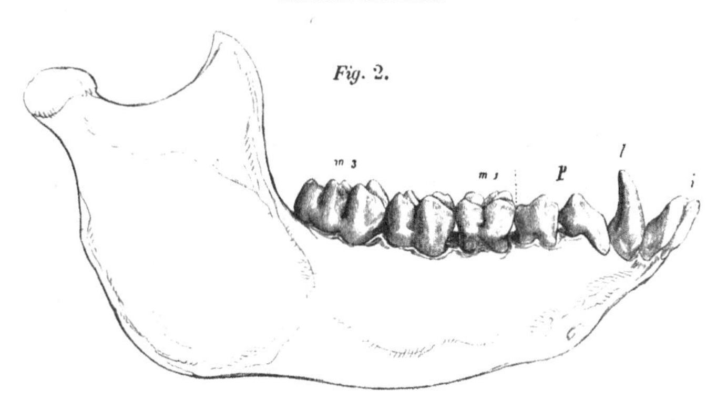

MACACUS RHESUS.

QUADRUMANA. (Apes, Monkeys.)

Cuvier, the great founder of that department of the Science of Organic Remains which relates to the interpretation of the fossil Bones and Teeth of the Vertebrated animals, had met with no evidence of any species more highly organised than a Bear or a Bat, in the fossiliferous strata which formed the theatre of animal life anterior to the record of the Human Race. Not a bone, not a tooth of an Ape, Monkey, or Lemur, had ever presented themselves to his notice during the long period of his researches; * whence it came to be generally believed that the QUADRUMANA, or those Mammals which most nearly resemble Man in their organization, were scarcely, if at all, anterior to the Human Species in the order of Creation. Mr. Lyell, however, in 1830,† had remarked, that the evidence of the total absence of the Anthropomorphous tribes was inconclusive. He rightly stated that the bones of quadrupeds met with in tertiary deposits, were chiefly those which frequent marshes, rivers, or the borders of lakes; as the Elephant, Rhinoceros, Hippopotamus, Tapir, the Ox, &c., while the species which live in trees were extremely rare; that we had, as yet, no data for determining how great a number of the one kind we ought to find, before we had a right to expect a single individual of the other. And this distinguished Geologist concluded by the remarkable anticipatory observation that, "if we are led to infer from the presence of Crocodiles and Turtles in the London Clay, and from the Cocoa Nuts and Spices found in the Isle of Sheppy, that at the period when our older tertiary strata were formed, the climate was hot enough for the Quadrumanous tribe; we, nevertheless, could not hope to discover any of their skeletons until we had made considerable progress in ascertaining what were the contemporaneous Pachydermata,"—not one of which at the period when the foregoing passage was

* "Aucun os, aucune dent de Singe ni de Maki se sont jamais présentés à moi dans mes longues recherches." Cuvier, Discours sur les Révolutions du Globe, p. 159.

† Principles of Geology. First edition, 1830, vol. i. p. 152.

penned had been discovered in any of the marine strata of the Eocene epoch in England.

I have been so fortunate, in my researches on the Fossil Mammalia of Great Britain, as to determine not only the remains of extinct Pachydermal animals (*Lophiodon* and *Hyracotherium*) in the Eocene beds called the London Clay, but, likewise, of a Quadrumane, or Monkey, in a sandy stratum of the same formation, the epoch of which had been shown by Mr. Lyell, from the evidence of other organic remains, to have had a temperature sufficiently high for arboreal Mammalia of the four-handed order.

The fossils manifesting the quadrumanous characters were discovered, in 1839, by Mr. William Colchester, in a bed of whitish sand beneath a stratum of tenacious blue clay, situated by the side of the river Deben, about a mile from Woodbridge, in the parish of Kingston, commonly called Kyson, in Suffolk.*

The first of these fossils submitted to my inspection, (*fig.* 1, *m*, 3,) was the fragment of the right side of the lower jaw, including the anterior part of the base of the coronoid process, and the last molar tooth entire in its socket. This tooth is, fortunately, a very characteristic one; and after a comparison of it with the corresponding tubercular tooth in the lower jaw of the Coati (*Nasua*), Racoon (*Procyon*), Ratel, Opossum, Phalanger, and other small unguiculate quadrupeds of a mixed or partially carnivorous diet, I proceeded to an examination of the Quadrumana, and found in that order the desired correspondence.

The extreme rarity of the fossil remains of such highly organised animals in any part of the world, and the previous total absence of any in a land so far from the Equator

* In August 1839, see Magazine of Natural History for September, 1839, p. 446. These rare fossils are now in the possession of Mr. Colchester.

as England, prevented my examination, in the first instance, of the skeletons of the recent Quadrumana; and it was not until I had tried all the more probable analogues of the fossil fragment in the lower forms of the Mammalia, that I began to test it by the side of the jaws of the Apes and Monkeys.

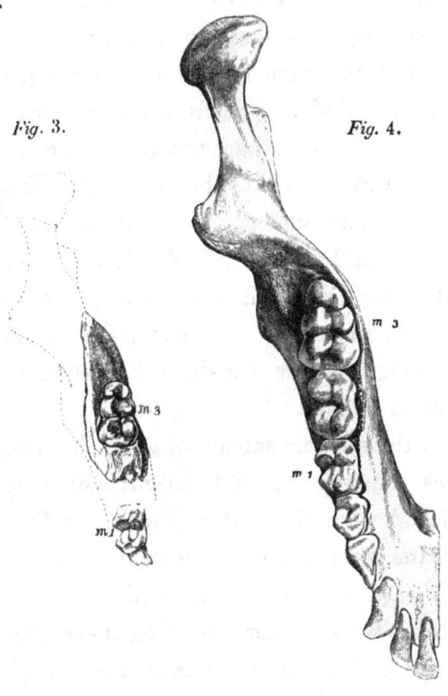

The grinding surface of the fossil tooth (*fig.* 3, *m*, 3,) supports five tubercles, the four anterior ones being arranged in two transverse pairs, the fifth forming a posterior heel, or talon. This conformation of the crown of the last molar in the lower jaw characterises two families of Catarrhine, or old world Monkeys, *viz.*, the *Semnopithecidæ*, including the genera *Colobus* and *Semnopithecus*, and the *Macacidæ*, including the genera *Macacus*, *Cynocephalus*, and *Papio*.

The next step was to ascertain whether any special marks of resemblance would yield a further insight into the affinities of the fossil, and justify its reference to any of the genera of either family. A difference in the shape of the hinder tubercle of the tooth, was first noticed in the recent Quadrumana. In the *Semnopithecidæ* it was large, but simple; in most of the *Macacidæ* it was partially subdivided into two cusps, the outer one being the largest. As this character was well marked in the fossil, it seemed decisive of its closer affinity to the *Macacidæ;* and, as the smallest species in this family belong to the typical genus, I referred the fossil to the *Macacus*, and now propose to designate the extinct species represented by it "*Macacus eocænus*," the Eocene * Monkey or Macacque. The portion of the fossil jaw is narrower from side to side, or more compressed, than in any of the existing Macacques, and the internal wall of the socket of the tooth, in the fossil, is flatter and thinner. The ridge on the outer side of the alveolus, which forms the commence ment of the anterior margin of the coronoid process, begins closer to the tooth.

These characters establish the specific distinction of the extinct Macacque to which the fossil fragment of the jaw belonged, and afford additional proof, if such were wanting, that it could not have been accidentally introduced, in recent times, into the stratum out of which it was disinterred.†

* *Eocene*, a term invented by Mr. Lyell, from the Greek words *ηως*, aurora, or the dawn, and *καινος*, recent, expressive of the lowest division of the tertiary strata, in which the extremely small proportion of fossil remains referrible to species yet living, indicates the first commencement, or dawn, of the existing state of the animal creation.

† A newspaper critic, when this discovery was first announced, suggested that the supposed fossil might be nothing more than the remains of some monkey belonging to a travelling menagerie, which had died, and been cast out in the progress through Suffolk.

Another specimen of a fossil tooth of the *Macacus eocænus* had been previously discovered, 1838, in the same stratum and locality as the fossil above described. It was submitted to my inspection by Mr. Lyell, who has communicated the result of my comparisons in the "Annals of Natural History," for November 1839, proving it, likewise, to be the molar of a Monkey of the genus *Macacus*, thus constituting at once the first terrestrial mammal which had been found in the London Clay, and the first Quadrumanous animal hitherto discovered in any country in tertiary strata so old as the Eocene period.

The specimen in question consists of the crown and one fang of the first true molar tooth, and is marked *m*, 1, in the cuts figs. 1 and 3. The series of teeth in the recent lower jaw, figs. 2 and 4, figured for comparison, is divided into two incisors, marked *i*, one laniary, or canine, marked *l*, two premolars, or false molars, (called bicuspides in human anatomy,) *p*, and three molars or true molars, of which the analogues of the fossil teeth are marked respectively *m*, 1, and *m*, 3. The crown of the false molar of the fossil *Macacus*, (*m*, 1, figs. 1 and 3,) presented four tubercles, arranged in two transverse pairs, the anterior pair being the highest; there was, also, a very small ridge across the anterior, and another across the posterior part of the crown. The latter is placed between, and connects together the two posterior tubercles. The fangs were two in number, strong, and divergent: the tooth had belonged to an animal that had passed its maturity, the tubercles having been worn at their summits, and the posterior concavity having been smoothly deepened by attrition.

It differed from the corresponding tooth in the existing Macacques, in having the ridge along the base of the forepart of the crown, and by being relatively narrower from side to side, which is also the case with the hindmost fossil grinder, as is illustrated in the cuts. As, moreover, the two fossil molars bore the same proportions to one another as the corresponding teeth from the same jaw of the recent Macacque bear to each other, it was reasonable to conclude that the two fossils appertained to the same extinct species of *Macacus.*

The evidence on which the fossil Monkey in the Eocene strata of England has been determined, is of the same kind as that which has brought to light the former existence of another and apparently higher species of Quadrumane, in the South of France, and is equally conclusive with that by which Quadrumanous fossils have also been recognised in India, and in South America.

In all the instances, however, of the discovery of Anthropomorphous fossils in foreign countries, the amount of the evidence yielded by the fossils has been greater than that which has hitherto been obtained from the tertiary strata of Britain. Lieutenants Baker and Durand, who first announced the fact of a fossil quadrumane in 1836,* supported their highly important statement by the description and figures of an almost entire right superior maxillary bone, containing the five molar teeth and part of the canine, and demonstrating the anterior aspect of the orbits, which is so marked a peculiarity of the Quadrumana. This rare and valuable fossil was obtained from the tertiary strata of mixed calcareous sandstone and clay, in the Sub-Himálayan hills near the Sutlej.

In the year following, Captain Cautley and Dr. Falconer discovered in the same formation of the Sub-Himálayan district, a considerable portion of the lower jaw, with all

* Journal of the Asiatic Society of Bengal, for November 1836, p. 739, pl. XLVII.

the molars of the right side, and a part of the dental series of the left side, together with the two middle incisors, and the right canine. Fragments of two other lower jaws and an entire astragalus were subsequently discovered by these gentlemen. All these remains were entirely fossilized, and impregnated with the hydrate of iron, and they satisfactorily confirmed the conclusions of Lieutenants Baker and Durand, that a large species of *Semnopithecus* had coexisted with the *Sivatherium* and the Hippopotamus, and had, like these and other strange quadrupeds of the tertiary period in India, become extinct.

The fossil Quadrumane of the fresh-water tertiary strata of the South of France, was determined by M. Lartet,* upon the conclusive evidence of an almost complete lower jaw with all the teeth in situ. This fossil, which was originally referred to the Gibbons (*Hylobates*), which immediately follow the Orangs in the Quadrumanous series, is more correctly regarded by M. de Blainville, on account of the conformation of the crown of the last molar tooth, which is much more like that of the Eocene Macacque or Semnopithèque, than that of a Gibbon, as the representative of an extinct genus intermediate between *Hylobates* and *Semnopithecus*.

As if it were intended that the antiquity of the Quadrumanous order should be put beyond all doubt, the independent testimony of Dr. Lund, a Danish naturalist resident in Brazil, was added to those of the observers in the East Indies and South of France. Very shortly after the announcement of the fossil Quadrumana in those countries, Dr. Lund, unacquainted with their discoveries, thus addressed the Academy of Sciences at Copenhagen, on the subject of his own palæontological researches:—

* Comptes rendus de l'Académie des Sciences, January and April 1837; and De Blainville, Ostéographie, Primates fossiles, p. 53.

"I am at length enabled to solve the important question as to the existence of the highest order of Mammalia (*Quadrumana*,) in those ancient times to which these fossils belong; a question which has, as yet, been unanswered, or to which most philosophers have replied in the negative. It is certain that this order was then in existence; and the first animal of the class recovered is of gigantic size; a character belonging to the organization of the period. It considerably exceeds the largest individuals of the Orang Outang or Chimpanzee yet seen; from which, also, as well as from the Gibbons, or long-armed apes (*Hylobates*), it is generically distinct. As it also differs from the existing Monkeys of this continent (South America), I would place it for the present in a genus of its own, for which I propose the name of *Protopithecus*."

In letters communicated to the Academy of Sciences, Dr. Lund states that the large fossil Brazilian Monkey belongs to the Platyrrhine or New World group of Quadrumana, all the species of which have three premolars on each side of the upper and lower jaws, and that it surpassed any known *Cebus* or *Mycetes* in size, since it must have been four feet in height.

These dimensions, however, do not exceed those of the full grown Chimpanzees and Orangs; but it is interesting to find that the fossil Semnopithecus of India, and the fossil Protopithecus, or Capuchin Monkey of Brazil, are, like the associated lower organized extinct Mammalia, of gigantic size, as compared with the nearest existing analogues of the same localities. It is not less interesting to find that the representatives of the Quadrumanous order in latitudes, the climate of which is now unfit for the existence of apes and monkeys in a state of nature, were of smaller size than their own nearest analogues, which seems to indicate that although the climate was warmer than at present, it was not of so strictly tropical a character as to favour the full development of the Quadrumanous type.

The formation at Sansan near Auch, in which M. Lartet discovered the quadrumanous fossils allied to the Gibbon, is regarded by Mr. Lyell as probably of the Miocene, or middle tertiary period. In the same formation were found remains of *Mastodon*, *Dinotherium*, and many other extinct quadrupeds.

With respect to the deposit at Kyson, in which the remains of the *Macacus eocænus* were discovered, it consists of layers of white and yellow sand, which had been pierced to the depth of twelve feet without reaching the bottom. Above the sand is a bed of brown clay, which has been laid open to the depth of twelve feet. Both the clay and sand are dug for making bricks. Mr. Lyell says, "as the clay at Kyson is covered by red crag at a short distance from the pits, and as I had seen clay of the same colour beneath the crag in the neighbouring cliffs of Bawdsey, and also at Felixstow and Harwich, containing Septaria, and, as at Harwich, the imbedded shells, fruits, and bones of Turtle, are such as characterise the London Clay, I entertained no doubt that the Kyson formation belonged to the Eocene period." My subsequent discovery of the *Hyracotherium*, an extinct genus of Pachyderms, whose fossil remains have hitherto been met with only in the London Clay, and of the vertebræ of the great extinct British Boa-constrictor (*Palæophis*),* equally characteristic of that formation, in the same bed at Kyson from which the fossil Macacque was obtained, places its geological antiquity beyond question.

* Annals of Natural History, vol. iv. p. 189.

Fig. 5.

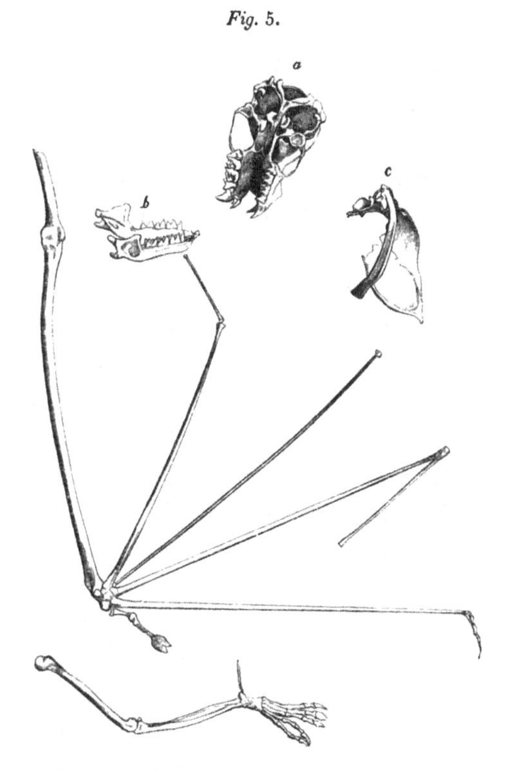

VESPERTILIO NOCTULA.

CHEIROPTERA.—Bats.

If the fossil remains of the small Mammals which live in trees are rare, still rarer, one might have supposed, would be those of the much smaller species which are organized for flight, and whose bones are necessarily light and fragile. The skill of Cuvier, however, long since exposed a considerable portion of the skeleton of a Bat, allied to the Serotine, which was petrified and imbedded in a block of the eocene gypsum at Montmartre. A few fossil teeth of a very small insectivorous Mammal, somewhat resembling those of a Bat, have been found in the eocene sand at Kyson.* But the most numerous and authentic remains of the small species of the Cheiropterous order have been met with in England in the limestone caverns containing the fossil bones of extinct Bears, Hyænas, &c. In these situations, however, as likewise in the cave of Köstritz in Germany, the Bat's bones occur mixed with those of existing as well as of extinct animals, and may, therefore, have been introduced at a recent period.

The chemical condition of such small and delicate remains cannot be relied upon as evidence of their antiquity, since they are altered by surrounding agencies, and especially by contact with calcareous stalactite, more rapidly than are the bones of larger quadrupeds. We must pause, therefore, before we adopt the conclusion at which Dr. Schmerling has too hastily arrived, that the skulls and other bones of Bats, which have lost a greater or less proportion of their animal matter, are coeval with those of the large extinct spelæan or cave-haunting quadrupeds in the same absorbent state, which are associated with them.

With regard to the more satisfactory test of the comparison of Cheiropterous remains with the skeletons of existing species, I have failed to detect in the more complete skulls and skeletons from cave localities any character by which they could be distinctly referred to unknown species of Bats, or to such as do not now exist in England: and after much pains bestowed on the less complete and more abundant fragmentary and detached parts of the enduring framework of the Cheiroptera, I have been seldom able,—partly, indeed, from the still imperfect state of the Osteology of this Order,—to arrive at any sound specific determinations.

* Annals of Natural History, 1839, p. 194.

One of the most complete examples of the skeleton of a Bat, from a crevice of a bone-cave in the Mendips, although partially fossilized, is here figured rather with a view to aid the collector of Mammalian remains in the recognition of the Cheiropterous characters, than as an example of a species coeval with the great Bear and Mammoth of the same cavern.

The short and expanded cranium (*fig.* 5, *a*), with the wide inferior apertures caused by the loss of the large and naturally loose bony vesicles of the ear-drum,—the short and broad upper jaw, with the characteristic wide and deep anterior notch, occupied in ordinary Mammalia by the intermaxillary bones,—and the teeth, bristling with sharp points, all yield unequivocal characters of the insectivorous Bat.

The large and broad scapula, the long and strong clavicle (*fig.* 5, *c*), bespeak the muscular forces, and the resistance required for the use of the arm in the vigorous actions of flight: the bones of the fore-arm and hand, and those of the hinder extremity, equally illustrate that remarkable organization, the final purposes of which have been so well explained by the author of the History of the existing Mammalia of Britain.

"The sternum, the ribs, and the bones composing the shoulder," says Professor Bell,* "are all developed for the attachment of powerful muscles, adapted to the rapid and continued movements of the anterior extremity, which, although consisting essentially of the same parts as that of Man, has its different bones so modified in form and extent as to afford the most admirable and complete support to an extensive expansion of the skin, which thus forms a perfect and efficient pair of wings. This modification principally

* Bell's British Quadrupeds, p. 3.

consists in the extraordinary development of the fingers, which are greatly elongated for the purpose; and upon which the skin is stretched like the silk on the rods of an umbrella." This fossil, which forms the chief and central figure in cut 5, includes the distal end of the humerus, or arm-bone, the entire radius, or chief bone of the fore-arm, the little bones of the carpus, or wrist, the small thumb with its broad flattened phalanx for the prehensile claw, and the long and slender metacarpal bones, and a few of the phalanges of the fingers, which Professor Bell has so aptly compared to umbrella-rods.

"The hinder-toes," continues the same author, "are short, of nearly equal length, and are chiefly used as suspending organs, the Bats hanging by them, from the trees or walls on which they rest, with the head downwards." This character is likewise displayed in the well-preserved hinder limb of the skeleton figured, together with another peculiarity, viz., a slender rod of bone extending from the heel to sustain the inter-femoral web.

Any one of these characters singly would suffice to determine the ordinal relations of the bony relics presenting them. To obtain a deeper insight into the affinities of the fossil, much closer and more minute comparisons must be instituted. In the specimen under consideration, the two pairs of incisors in the upper jaw, and the three pairs indicated by the sockets in the lower jaw, *b*, where they are combined with two premolar teeth on each side, prove it to belong to the true *Vespertiliones*, and distinguish it from the *Nycterides*, which have but one premolar in each ramus of the lower jaw. In the *Noctiliones* there is only one pair of incisors in the lower jaw: in the *Molossines*, the *Megaderms*, and the *Rhinolophines*, there is only one pair of incisors in the upper jaw: the *Taphians* have no upper incisors at all.

The fossil having been thus brought to a particular section of the unfoliated or simple-nosed Bats, its affinity to some particular genus or species of this family remained to be considered. The *Barbastelle*, the *Pipistrelle*, and the *Noctule*, offer three modifications of the anterior upper premolar; * it is rudimental, hardly discernible in the first, of large size and more outwardly situated in the second, of intermediate size but not visible from the outside of the jaw, in the third species. The fossil comes nearest the Noctule in this character. The canines and large molar teeth afford no grounds of discrimination amongst these genera.

The skull, by the somewhat greater length of the cranium and its strong sagittal crest, confirms the indication given by the teeth and the heel-spine of the affinity of the fossil or pseudo-fossil to the true *Vespertiliones*, and herein, more especially to the Great Bat of Pennant (*Vespertilio noctula*), the first of the British existing species described by Professor Bell.

RHINOLOPHUS FERRUM-EQUINUM.

From amongst the more fragmentary fossils of Cave *Cheiroptera*, I select a ramus, or half lower jaw (fig. 6) with the coronoid process broken off, but with the series of teeth perfect, since these manifest characters which indicate not only a species of Bat distinct from the preceding, but, likewise, one that belongs to a different section of the order. There are two false molars in this lower jaw, as in the *Vespertilio Noctula*, but of different

Fig. 6.

* The "molares spurii," or "false molars," "bicuspides," in Human Anatomy; they are situated before the true molars, between these and the canine.

proportions, the first being much smaller, the second somewhat larger.

In regard to these teeth, the jaw in question resembles that of the *Molossi*, especially *Mol. Daubentoni*, but it differs from all the species of that genus that I have seen in the more produced angle of the jaw. In this character, as well as in the number, shape, and size of the teeth, it agrees closely with the *Rhinolophi*, especially the species called "Greater Horse-shoe Bat." It is too large for any of our native species of *Vespertilio*, save the *Noctule*, to which the proportions of the premolar forbid a reference: but it corresponds in the size as well as shape of the bone, and in the modifications of the teeth, with the *Rhinolophus Ferrum-equinum*.

Unequivocal remains of this species of Bat, from the Bone-cave called Kent's Hole near Torquay, Devon, are contained in the British Museum: some of the specimens appear to be in the same absorbent condition, as the bones of the *Hyæna*, *Rhinoceros*, &c., from the same cave; others are evidently more recent. It is worthy of remark that the Greater Horse-shoe Bat is most commonly met with in the Devonshire caves at the present day, and is the only species known to frequent Kent's Hole.*

In every other example of remains of bats from bone-caves, where the condition of the specimen has permitted a direct or approximate identification, it has been with some existing British species; and the general result of this part of my palæontological researches—the most tedious, but yielding the least important results—is, that no remains of Bats have hitherto been found, however situated in caverns, or altered in chemical constitution, which establish the former existence of any species not now known to exist,

* See Bell's British Quadrupeds, p. 71.

and which does not, in most instances, frequent the same caverns.

FOSSIL CHEIROPTEROUS (?) INSECTIVORE.

a *Fig.* 7. *b*

Twice nat. size.

As we pass to lower and older geological formations, our comparisons lead to different conclusions. Such remains as may with any probability be referred to the Cheiropterous Order, cannot be satisfactorily identified with known existing species, unquestionably not with any that are indigenous. In regard to the small molars, already referred to, (p. 12,) which were associated with the *Macacus* and *Hyracotherium*, in the Eocene sand at Kyson in Suffolk, one of these (a penultimate or antepenultimate grinder, *fig.* 7, *a*) has the crown composed of four triangular prisms, placed in two transverse rows, with an angle turned outwards, and a side or flat surface inwards, the summits being sharp-pointed. The exterior prisms are the largest; the crown swells out abruptly above the fangs, defending them, as it were, by an overhanging ridge. There is a small transverse eminence, or talon, at the anterior part of the crown; and a very small tubercle is placed between the bases of the two external prisms.

The second molar (*fig.* 7, *b*) differs from the preceding in having the two posterior prisms suppressed, and replaced by a flattened triangular surface. The anterior prisms are present, and their apices project far beyond the level of the posterior surface. There is a small ridge at the anterior part of the tooth.

These teeth agree very nearly with the antepenultimate and last molars of the larger insectivorous bats:

they differ chiefly in the presence of the small tubercle at the basal interspace of the exterior prisms; a difference which M. de Blainville regards as ground for doubting the legitimacy of their approximation to the Cheiropterous order at all.* Since, however, an anatomist so familiar by his recent researches with all the modifications of the teeth of the Mammalia had been unable to refer the fossil molars in question to any of the terrestrial or aquatic genera of *Insectivora*, but had given the figures of these molars a place in the plate illustrating the ancient *Vespertiliones* in his great work, the "Ostéographie," I deduced from that fact, when preparing my Report on Fossil Mammalia for the British Association, additional confidence in my original determination.

An extinct genus, new to science, of a Mole-like Insectivore, has lately come under my notice, in which the grinding teeth present the above described peculiar character of the minute tubercle at the basal interspace of the two exterior prismatic cusps. They are not, in other respects, identical, and additional fossils from the Kyson sand will be required to establish even the generic identity of the present small teeth from that formation, with the Palæospalax of the lacustrine beds at Ostend.

* Ostéographie des Cheiroptères, p. 93, pl. xv. fig. ix.

Fig. 8.

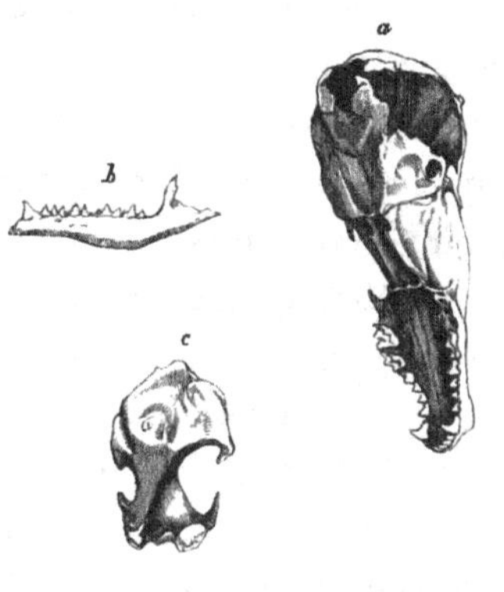

FOSSIL.

Fig. 9.

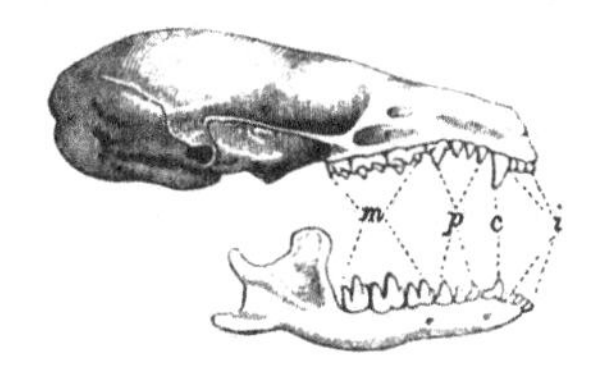

RECENT.

THE COMMON MOLE.

Talpa vulgaris. Brisson.

Since the period when Cuvier first detected parts of the fossil skeleton of a Shrew (*Sorex*) in the osseous breccias of Sardinia, the remains of Moles and Hedgehogs have likewise been described in works on Fossil Mammalia, especially in those of Schmerling, Schlotheim, the Abbé Croizet, and M. de Blainville.

Remains of the skeletons of the three principal genera of Insectivora have been brought under my notice at different times from caverns, and the more recent geological formations in various parts of England, but, with one exception, they have not offered any specific difference from the Common Mole, Hedgehog, and Shrews, that exist at the present day in this country.

With respect to the genus *Talpa*, those remains which are mentioned by Dr. Buckland* as occurring with the bones of various birds, water-rats (*Arvicolæ*), in a bed of brown earth, at the bottom of the cave at Paviland, belong to the common existing species, and their presence in that almost inaccessible spot, is explained by Dr. Buckland on the supposition of their having been introduced by hawks and other birds of prey. It is most probable that the almost entire skull, and other portions of the skeleton described and figured by Dr. Schmerling,† and by him identified with the existing Mole, belonged to individuals whose introduction into the Belgian caverns is to be referred to a similar agency. And the remains of moles found in the soil covering the floor of the cavern at Köstritz, may belong to an equally recent period.

The nearly entire skull, lower jaw, and humerus, figured in cut 8, have a better claim to be regarded as fossils, although, in fact, not differing from the recent species.

The skull, *a*, from a raised beach near Plymouth, appears to have belonged to the same epoch as the fossil *Mustela* subsequently to be described.

In its size and general form, in the characteristic flattening and elongation of the cranium, in the slenderness of the zygomatic arches, the extremities of which were still preserved in the fossil, and in the dentition of the upper jaw, the correspondence with the recent *Talpa communis* is

* *Reliquiæ Diluvianæ*, 4to., 1823, p. 93.

† Recherches sur les Ossemens fossiles de Cavernes de Liège, 1833, p. 80, pl. v.

complete, as the figures of the skull of this species (figs. 9 & 10) demonstrate.

These figures may afford acceptable aid to the collectors of fossil bones, who have not the recent skeleton at hand for comparison. The dentition of the fossil, as in the recent mole, consists of eleven teeth on each side of both upper and lower jaws. The first three in the upper jaw are small, simple, and implanted by a single fang: the fourth resembles a canine tooth by the size and shape of its crown, but it has two fangs, like the three succeeding premolars of the upper jaw; the last three teeth are implanted by three fangs, and their large and complicated crowns and their mode of succeeding the deciduous teeth, prove them unquestionably to be true molars. In the lower jaw the first four teeth are small, simple, and with single fangs; the fifth corresponds in shape and development of the crown with the canine-shaped tooth above, but it has also two fangs, and moreover passes behind that tooth when the mouth is closed, which is contrary to the relative position of the true canine teeth in Carnivora; all the remaining teeth of the lower jaw are implanted by two fangs each, the last three being evidently true molars. The letters in cut 9 indicate the classification of the Mole's teeth, according to the views adopted by Professor Bell;* the letter *i* indicating the incisors; *c*, the canines; *p*, the premolars, and *m*, the true molars.

Fig. 10.

Professor de Blainville regards the upper canine as an incisor; I much regret that I have not hitherto had an opportunity of examining a Mole young enough to shew

* British Quadrupeds, p. 85.

the exact limits of the intermaxillary bone: until this be done the true character of the double-fanged canine-shaped upper tooth cannot be decided. The importance of an exact acquaintance with the dentition of our small Insectivora was forced upon my attention some time since by the fossil, figured at *b*, cut 8, a drawing of which was transmitted to me by Professor Sedgwick, with the following note:—"At the same spot (in the brown diluvial clay on the coast of Norfolk, near a village called Bacton,) was found a pretty perfect skeleton of a reptile, of which I send you a drawing; but its legs, pelvis, and sternal bones, have been put together in a monstrous fashion. The little jaw in the corner of the plate was drawn on the supposition of its belonging to the reptile; but I have seen it, and it seems to be the jaw of no reptile, but of a small Insectivorous Mammal."—Extract of letter, dated Norwich, Feb. 12, 1842.

The accuracy of the Professor's opinion was soon established, by the comparison of the drawing with the dentition of the Mole; the fossil in question presented the double-fanged canine-shaped tooth, followed by three small premolars and three true molars; corresponding precisely in number and proportions with those teeth in the lower jaw of the Mole. The teeth in Reptilia are not usually implanted in sockets, and when they are, it is always by a single fang. The value of this character will be more strikingly manifested when we come to the consideration of more problematical fossils than that of the supposed Bacton reptile.

Having communicated the result of the comparison, with a request to have the supposed sternal and pelvic bones of the reptile transmitted to me, these proved to be, as I had suspected, characteristic parts of the skeleton of a Mole: the anomalously developed humeri having been mistaken for the broad and flat bones of the pectoral and pelvic arches in the Saurian reptiles. To any one unacquainted with the extraordinary and exceptional development of the humerus, or arm-bone, of a mole, the real nature of the bone is little likely to be divined: from its shape it should be ranked rather with the flat than the long bones of the skeleton. Its prodigiously developed tuberosities and condyles relate to the mass and force of the muscles which are required to work the spade-shaped paw in the act of excavating the soil. One of the fossil humeri of the Bacton skeleton is figured at *c*, cut 8, and I here subjoin the representation of the whole bony framework of the fossorial anterior extremities of the mole: *s* indicates the

Fig. 11.

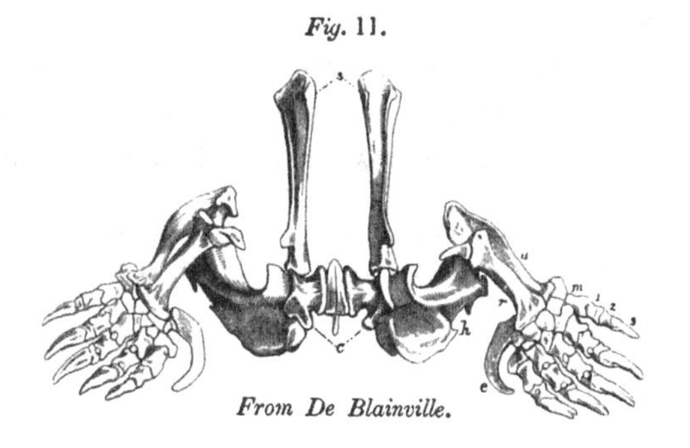

From De Blainville.

scapulæ, or blade-bones; *c*, the clavicles, or collar-bones; *h*, the humeri, or arm-bones; *u*, the ulna, and *r*, the radius, both bones of the fore-arm; *m*, the outermost of the five metacarpal bones, between which and the bones of the fore-arm the small bones of the wrist, or carpal bones, are situated, of which a most extraordinary sabre-shaped one, *e*, is peculiar to the Mole, and strengthens that margin of the broad palm which first digs into the earth like the spade's edge: the short and strong phalanges of the fingers are indicated by the numerals 1, 2, 3.

If the reader will compare this figure with the skeleton of the bat's hand in cut 5, he will see the two extremes, as to length and breadth, in the development of the bones of the anterior member in the Mammalian class; yet the analogy of their respective organizations is perfect, and carried out to the least of the component ossicles: the same parts being adapted by different proportions to their very different functions. The unity of plan bespeaks the One Great Cause, as the Supreme Wisdom is testified by the perfect fitness of the instruments for their specific end; nor is the combination of typical conformity, with exact adaptation to the destined function, less manifest in the hand which guides the pen, than in that which moves the Bat through the air and the Mole through the earth.

The most complete fossil skeleton of the Mole is that, of which the parts are above described, now in the Norfolk and Norwich Museum: it was discovered by Mr. Green.*

* The specimen, with the bones collocated according to the first notion of their nature, forms the frontispiece to Mr. Green's 'Geology of Bacton.'

Fig. 12.

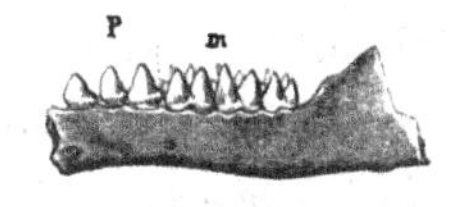

Nat. size.

PALÆOSPALAX MAGNUS.

At Ostend, near Bacton, on the coast of Norfolk, there is a lacustrine deposit of dark clay and greenish sand, with the ruins of an ancient forest, indicated by overthrown charred trunks, compressed branches, and leaves of trees; this forms a very rich mine of organic remains. The stupendous Mammoth, two or three species of Deer, and the graceful Roebuck, have left, in their abundant and well-preserved bones and teeth, the evidences of the extinct population of that forest. Its streams were tenanted by gigantic Beavers, and were also frequented by a water-mole, which as much surpassed any known existing species in size, as the *Trogontherium* did the *Castor* of Canada or continental Europe. This extinct Insectivore, for which I propose the name of *Palæospalax*,* is clearly referable to the *Talpidæ*, or Mole tribe, by the most important part of its dental system, but was as large as a Hedgehog.†

This interesting addition to the extinct British *Insectivora*, which is the only example of a form in that order,

* Greek, *palaios*, ancient, *spalax*, mole.

† It is probably referred to by Mr. Green, in his 'Geology of Bacton,' 8vo., 1842, p. 12, "Rodentia,—bones, jaws, and teeth, of four species, probably arvicola, shrew, hedgehog, and mole." At least, I have seen no true remains of the Hedgehog in the collections of Bacton fossils in the British Museum, or in that of Norwich.

no longer represented in this island by living species, is established by a single fossil in the British Museum, consisting of a portion of the left branch of the lower jaw, (figs. 12 and 13,) containing the three true molars, *m*, and three premolar teeth, *p*; it was discovered by the Rev. Mr. Green of Bacton, in the lacustrine formation above described at Ostend.

Fig. 13.

p

m

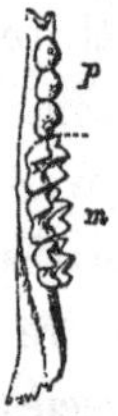

The size of the fossil, and the obvious insectivorous character presented by the sharp cusps with which the crowns of the molar teeth are bristled, might naturally lead, in the first instance, to its comparison with the common Hedgehog; from this the fossil is distinguished by its relatively larger and more complicated last molar, and by the smaller and more simple fourth molar in advance, which unequivocally represents, in the fossil, the last of the series of false molars, whilst in the Hedgehog, the corresponding tooth has the same quadricuspid crown as the antepenultimate true molar. The form of the jaw is, also, different in the Hedgehog, the lower contour, beneath the true molars, being more convex. From the genera of exotic Hedgehogs, called *Centetes*, *Ericulus*, and *Echinops*, the fossil is still more distinct, by the smaller number, and larger relative size, the square crown, and quinque-cuspid structure of the true molar teeth: from *Gymnurus* it differs in the smaller relative size of the premolars; and by the same character it is sufficiently, though less markedly, distinguished from *Glisorex tana*. The teeth of the fossil make a nearer approach to those of *Tupaia javanica*, but differ in the closer approximation of the three premolars, and in the small size of the middle one. The closest resemblance to the forms and proportions of the six teeth preserved in the fossil is found in the family of *Talpidæ*, in which I include the Water-moles, or Desmans (*Mygale*); the fossil differs from the common Mole (*Talpa*), and resembles the *Mygale pyrænaica* in the size of the first true molar, which nearly equals the second, and in the larger size of the three premolars; it precisely resembles the common Mole in the position of the two outlets of the dental canal which are preserved in the fossil. The fossil differs, however, from both the typical Moles and the Desmans, not only in its larger size, but in some slight modifications of the crown of the true molars; there is a minute but sufficiently obvious tubercle at the bottom of the outer fissure, between the two principal cusps of each molar, of which there is no trace in recent *Insectivora*. The specific name is founded on the leading character of this extinct Mole-like Insectivore, viz., its large size.

For the knowledge of this specimen I am indebted to Mr. Waterhouse, the able assistant in the Fossil Department of the British Museum.

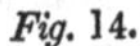

Fig. 14.

Twice nat. size.

Genus. SOREX.

The bones of Shrews, mixed with those of Field-mice, are sometimes found aggregated in extraordinary numbers in hedge-bottoms, beneath the foundation of walls or other parts of the soil. I examined, with Dr. Buckland, a remarkable accumulation of this kind in a mound, indicating the remains of an old Roman encampment, near Cirencester. Dr. E. D. Clarke transmitted to Sir Everard Home a quantity of similar remains, as "bones of a species of *Sorex*, found regularly deposited in the soil in Cambridgeshire." These specimens, which are preserved in the Museum of the Royal College of Surgeons, consist almost exclusively of remains of a small species of *Arvicola*. None of them can be regarded as true fossils.

The remains of Shrew-mice, which have been found in the bone-cave called Kent's Hole, near Torquay, and in the raised beaches near Plymouth, have offered no indication of species distinct from those now existing in Great Britain. The best preserved specimen which I have seen is identical with the *Sorex vulgaris*.*

The remains of Shrews from the lacustrine formations of Bacton and Ostend, Norfolk, appear to be referable to the *Sorex fodiens*, (cut 14, fig. 1, fossil, fig. 2, recent, magnified,) and to the *Sorex remifer* (fig. 3); the dentition of the jaws figured is not, however, in so complete a state as to allow of an unequivocal determination.

* *Sorex araneus*, Bell, British Quadrupeds.

INSECTIVORA. *AMPHITHERIIDÆ.*

Fig. 15.

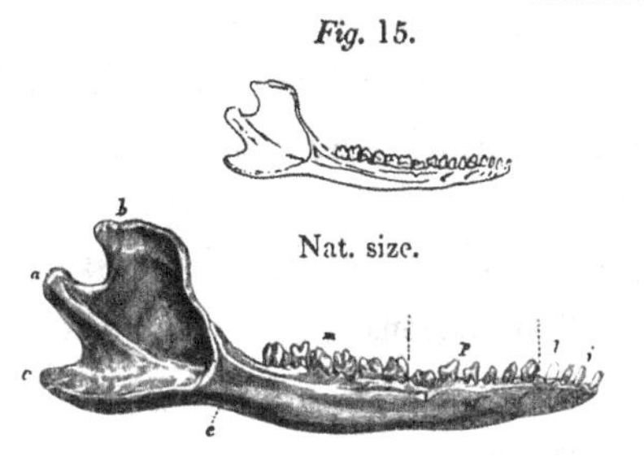
Nat. size.

AMPHITHERIUM PREVOSTII.

Genus. AMPHITHERIUM.

If the genera of *Insectivora* now represented by living species have hitherto yielded very few additions to the catalogue of British fossils—but one new species of mole, and no lost shrew, or hedgehog, having been well authenticated from any of our recent tertiary formations—the Order has assumed a more than common importance in the eyes of the Geologist, by the strange and unexpected forms of small quadrupeds referable thereto, which have been detected in strata, far more ancient than any heretofore known to have concealed relics of animals so highly organized as the Mammalia.

The insect-eating quadrupeds may be the rarest, but they unquestionably include the most ancient of Mammalian fossils; for, if the pedimanous *Cheirotheria* have failed to endure the test of later scrutiny, the most rigid criticism has but tended to rivet more firmly the links which attach the *Amphitheria* and *Phascolotheria* to the Mammalian series.

The rare and interesting fossils on which those genera have been founded, which have been the subjects of such close and repeated examination, which have exercised the discriminative and analogical powers of so many philosophic naturalists, and have excited such warm discussion, are the well-known small under-jaws from the oolitic calcareous slate at Stonesfield, near Oxford, first indicated as evidence of the Mammalian class by Dr. Buckland, in his celebrated memoir on the *Megalosaurus*, published in 1823, in the "Transactions of the Geological Society of London,"* and there referred, on the authority of Cuvier, to the genus *Didelphys*.

In regard to the value of that authority in this particular instance, M. Prevost has informed us that Baron Cuvier examined the specimen, (cut 15,) at that time unique, during a visit which he paid to the University of Oxford, in 1818, and that a cursory inspection led that learned anatomist to say, that it had some resemblance or affinity to the jaw of a *Didelphys*.† Cuvier, himself, has added to the last volume of the second edition of the Ossemens fossiles, 4to., 1825, the following note: "M. Prevost, who is at present travelling in England, has just sent me a drawing of one of these jaws; it confirms me in the idea which my first inspection gave me of it. It is that of a small Carnivore, (Carnassier,) the jaws of which bear much resemblance to those of the Opossums; but it has ten teeth in a row, a number which no known Carnivore displays. At all events, if this animal be really from the schist of Stonesfield, it is a most remarkable exception to an otherwise very general rule, that the strata of that high antiquity do not contain the remains of Mammals."

The statement did, in fact, soon excite close and sceptical

* Vol. i. Second Series, p. 399.

† Cette pièce unique était conservée dans la collection de l'université d'Oxford, lorsque M. Cuvier la vit en 1818. Une inspection rapide fit dire à ce savant anatomiste qu'elle avait des rapports avec la mâchoire de quelque Didelphe." Prevost in "Annales des Sciences," iv. 1825, p. 396.

inquiry, first in regard to the geologica relations of the alleged oolitic stratum, and next, as to the true zoological affinities of the fossils.

The first exception to a generalization that has assumed the character of a law is always admitted with difficulty, and, by a rigid systematist, with reluctance. The geological arguments by which M. Prevost endeavoured to invalidate the conclusions of Dr. Buckland, as to the relative position of the Stonesfield slate, were soon and satisfactorily rebutted by Dr. Fitton; the antiquity of the oolitic masses could not be diminished to correspond with the presumed exclusive Mammalian epoch,—the mountain refused to move to Mahomet, and the question as to the real age of the rock containing the alleged marsupial fossils has not since been agitated. The attempts to do away with the anomalous exception, by interpreting the characters of the fossil jaws as indications of an extinct species of reptile, or other cold-blooded oviparous animal, have been more frequent and persevering; and they assumed the appearance of so systematic a refutation of the Cuvierian view, in the memoirs communicated by M. de Blainville to the French Academy, in the year 1838, that a close and thorough reexamination and comparison of the fossils in question seemed to be imperatively called for, in order that the validity of the doubts cast upon their Mammalian nature might be fully and rigorously tested.

By a very singular coincidence the fossil 'bones of contention,' from the Stonesfield slate, are all of them portions of the lower jaw; whether belonging to individuals of different species, or of different genera, or even, as appears by examination of new specimens acquired since the publication of Professor de Blainville's and my own memoirs of 1838, of different orders of Mammalia.

The first fossil was referred originally to the genus *Didelphys*, from the resemblance of the grinders to those of the opossums; but we have seen that Cuvier expressly stated that they exceeded in number the molar series in that or any other known carnivorous genus of Mammalia.

M. Agassiz,* originally regarding this fossil as insufficient to determine the nature of the animal to which it belonged, subsequently proposed,† nevertheless, a generic name, *Amphigonus*, for that animal, expressive of its supposed ambiguous nature.

M. de Blainville,‡ likewise, though participating in the incertitude or doubt which M. Agassiz had cast upon the original determination of the Stonesfield fossil, felt as little hesitation in suggesting a name for the new genus which it seemed to indicate, whatever might subsequently prove to be its characters or affinities; and it is remarkable that the Greek compound "*Amphitherium*," should imply by its terminal element a relation to the class Mammalia, which the memoir, read to the French Academy by its inventor, was especially designed to disprove; as the following summary with which the author concludes his Memoir sufficiently manifests:

"Meanwhile, in the present state of our information, it appears to me that we are authorized in drawing the following conclusions —

"1st. The two solitary fragments found at Stonesfield, and referred to the genus *Didelphys* of the class Mammalia, have none of the characters of animals of this class, and certainly ought not to be arranged among them.

* Neue Jahrbuch Mineral. and Geolog. von Leonhard und Bronn, 1835, iii. p. 185.

† German Translation of Dr. Buckland's Bridgewater Treatise.

‡ "Doutes sur le prétendu Didelphe fossile de Stonesfield." Comptes rendus de l'Acad. de Sciences, Aug. 20, 1838.

"2nd. Neither can they be referred to an insectivorous *Monodelph* allied to the *Tupaia* or *Centetes*.

"3rd. If we deem ourselves justified in regarding them as of the class Mammalia, the molar portion of their dental system brings them nearer to the family of the Seals than to any other.

"4th. But it is infinitely more probable, from analogy with what we know of the *Basilosaurus* found in America, in a formation likewise secondary, that they ought to be referred to a genus of the sub-order of Saurians.

"5th. That in any case they must be distinguished by a different generic name, for which purpose we propose that of *Amphitherium*, as indicating their ambiguous nature.

"Lastly; the existence of the remains of Mammalia anterior to the formation of tertiary strata is not at all proved by the Stonesfield fossils on which we have now treated, although we are far from asserting that Mammalia were not in existence during the secondary period."

Dr. Buckland, shortly after the publication of M. de Blainville's doubts, visited Paris, taking with him the original specimen seen by Cuvier, and a second specimen, also from Stonesfield, more perfect as regards the jaw-bone, but less perfect in reference to the teeth: and he submitted both these specimens, in the absence of M. de Blainville, to M. M. Valenciennes and Laurillard. The results of their comparisons were communicated by M. Valenciennes to the Academy of Sciences,* in September 1838. The second specimen was referred to the species (*Didelphys Bucklandi* Brod.) which had been described and figured by my friend Mr. Broderip in the Zoological Journal;† but in this latter determination I cannot agree with M. Valenciennes, who has,

* Comptes rendus de l'Acad. des Sciences, Sept., 1838, p. 572.

† Vol. iii. p. 408, pl. xi.

indeed, himself afforded sufficient grounds for such dissent by stating, that "he had convinced himself that the second jaw must have had ten molar teeth, as in the first specimen;" the *Did. Bucklandi* having had only seven, or at most eight, molars.

In regard to the question of the general affinities of these fossils, M. Valenciennes arrived at the conclusion that the jaw, described and figured by M. Prevost and Dr. Buckland, not only belonged to a mammalian but likewise to a marsupial animal, and accordingly proposed for it a third generic name, indicative of these presumed affinities, viz., *Thylacotherium*.

The arguments of M. Valenciennes were opposed, in a second detailed memoir by M. de Blainville,* who concluded by stating, "that he felt himself compelled to pause, at least until fresh evidence was produced, in the conviction that the portions of the fossil jaws found at Stonesfield, certainly did not belong to a marsupial—probably not to a mammalian genus, either insectivorous or amphibious—that, on the contrary, it was most likely the animal had been oviparous, and, in regard to the opinion, founded on the analogy of the *Basilosaurus*, a large fossil reptile of America, the teeth of which display the peculiarity of possessing a double root, that it might have been an animal of the Saurian order:"—and "that had not M. Agassiz decidedly given his opinion against the fossils in question belonging to fishes, he would rather have been led to suppose that they might have been the remains of an animal of that class."

"In conclusion," adds M. de Blainville with naiveté, "I ought also to announce to the Academy, that the scientific

* "Nouveaux Doutes sur le prétendu Didelphe de Stonesfield; Comptes rendus," October 6th, 1838, p. 727.

conductor of the English Journal called the 'Athenæum,' has already laid before his readers the point under discussion, having no doubt but that there will soon be discovered, in the Stonesfield quarries, some fragment that will be sufficiently demonstrative; and, in the mean time, he himself proposes, to avoid, he says, being accused of partiality towards either of the three already proposed,—the name *Botheratiotherium* for the supposed *Didelphys* of the Oolite; so that Science is already embarrassed with four or five denominations for an animal, of which our knowledge is most imperfect; since, by one party it is referred to the *Mammalia*, by another to the insectivorous *Monodelphs*, or to the *Amphibia;* and by a third to the *Didelphs* allied to the opossums, or to a genus representing the seals, in the sub-class of *Marsupialia;* whilst others make a *Saurian*, or even a *Fish* of it; which, it may be remarked *en passant*, appears much more in accordance with the age and the geological character of the formation which contains the fossils in question, as well as with the organized bodies with which they are associated."

This was an unlooked-for result of the journey to Paris, undertaken by Dr. Buckland for the purpose of affording the Comparative Anatomists of that celebrated school of Natural History and Palæontology the opportunity of studying, not only the original fossil examined by Cuvier, but the second and more perfect jaw from the same ancient Oolitic stratum.

The final judgment of M. de Blainville met with approbation and support from the stricter systematists, since it harmonized with their preconceived opinions on the progressive appearance of organized forms on this planet. It seemed to afford a striking example of the alleged inefficacy of the Cuvierian principle of interpretation of organic remains, and gave to its promulgator occasion to reflect on those persons "who are little versed in the study of organic structures, and who place too implicit a reliance on, perhaps, rather a presumptuous assertion, that by the aid of a single bone, or of a simple articular surface of a bone, the skeleton of an animal can be reconstructed, and consequently its class, order, family, genus, and even species determined. Such persons," M. de Blainville states, "may, very probably, think it strange that four or five half-jaws, more or less furnished with teeth, should be insufficient to indicate promptly and with certainty to what class the animal to which they belonged should be referred."

Such thoughts were, in fact, so strongly entertained by the discoverer of the *Megalosaurus*, than whom no one could have better grounds for reliance on the Cuvierian axiom, that he brought the two specimens to London, and favoured me by leaving them in my hands for a close re-examination and comparison. With a view to obtain as many incontestable facts as possible, on which to base the arguments that might establish the desired demonstration of the nature and affinities of the supposed enigmatical fossils, I soon after visited York, and examined the specimen in the Museum of the Philosophical Society of that City, and finally devoted a close scrutiny to the most perfect of the Stonesfield jaws, which had been presented to the British Museum by Mr. Broderip. The results of these observations, with figures of the four specimens most carefully executed by the late Mr. Charles Curtis, were published in the Transactions of the Geological Society.*

The accuracy of the descriptions can be tested by reference to the original specimens; the soundness of the conclusions must be left to the judgment of the unbiassed

* 2nd Series, vol. vi. pp. 47—65, pl. 5, 6.

and experienced Physiologist and Comparative Anatomist.

The following dissentient opinions have, however, subsequently been recorded, and ought here to be noticed. Mr. Ogilby, the learned Secretary of the Zoological Society, in a paper read before the Geological Society, at the conclusion of my second memoir, after calling attention to the relative extent and position of the incisive and canine teeth in the fossil jaws, as objections, "because, among all Mammals, the incisors occupy the front of the jaw, and stand at right angles to the line of the molars," stated "After a due consideration of the whole of the evidence, that the fossils present so many important and distinctive characters in common with Mammals on the one hand, and cold-blooded animals on the other, that he does not think Naturalists are justified, at present, in pronouncing definitively to which Class the fossils really belong."*

Thus the question, in the opinion of the Naturalist just cited appeared not to have been advanced beyond the doubts of M. de Blainville, which had led to the examinations and conclusions dissented from. Professor Grant has recorded a more decided opinion on the mooted question, in his "General View of the Characters and the Distribution of extinct animals," published in Thompson's British Annual for 1839. "The jaws and teeth of *Amphitherium*, mistaken," he says, "by some for a mammiferous *Didelphis*, occur in the Stonesfield Oolite; they are distinctly associated with *Trigoniæ* and other marine shells, and are imagined to have been detached and mutilated by drifting.

* Proceedings of the Geological Society, Dec. 1838, Mr. Ogilby has lately informed me that his opinion, with regard to the non-mammalian nature of the Stonesfield fossils, was expressed in the above abstract more strongly than he intended, but I am not aware that he afterwards corrected or published any modification of his views.

These jaws have the coarse fibrous structure, and dark glistening surface from the abundant proportion of animal matter common in fossil cold-blooded Vertebrata, and their composite structure is obvious, from the distinct deep fissure extending along their base between the dental and opercular pieces; the articulated pieces of these compound jaws more or less resemble the coronoid, condyloid, and angular, processes of carnivorous Mammalia, as they do also in most osseous fishes, but most distinctly in Reptiles, where the detached elements of the jaw are more numerous. The teeth are often black, glistening and bituminous from their abundance of animal matter and carbon, as in most fossil fishes; their crowns are compressed, free, multicuspid, and their cervix much contracted and long, as in the Amboyna lizard, the iguana, iguanodon, many fishes, &c., and their surface is minutely furrowed with close vertical grooves near their cervix, as in most Saurian reptiles and Sauroid fishes. The fangs are deeply implanted in the jaw, as in all the *Acanthuri*, &c., and they are bifid, as in many Squali, and in the closely allied *Basilosaurus.* At least eleven similar multicuspid molars are seen in a fragment of one side of the lower jaw, as commonly observed in fishes and reptiles, but never in mammiferous quadrupeds.

"I have examined four of these jaws in England which have been referred to *didelphis;* and the jaw at Paris, from the same locality as the others, is acknowledged by all to belong to a reptile, as demonstrated by Blainville. The supposed incisores, instead of being small, symmetrical, approximated, and parallel, as in insectivorous and carnivorous mammalia, are long, conical, irregular, widely separated from each other at their base, almost as long and large as the supposed canine, and diverging, as the front teeth of *anarrhichas* and many other fishes. There is a hinge-like, dentated appearance, consisting of three similar and equally distant grooves, as between the rami of the lower jaw of most osseous fishes, distinctly perceptible in *Amphitherium Bucklandi*, at the broken surface of the symphysis, and the lower jaw is arched, as in most fishes and Saurian reptiles. The coronoid, or complementary piece, is deeply concave interiorly, and its anterior suture is seen extending to the third posterior molar tooth, as in the *Iguana.* The condyloid, or articular surface, passes obliquely into the imbedding hard rock, and may be concave, as in reptiles and fishes, but is not exposed. This animal has received its name from the mixed and ambiguous character of its relics, and the foot-marks of *Chirotherium*, left on the new red sandstone, have been referred to a similar *didelphis* existing at that early period. The great jaws, teeth, and vertebræ of *Basilosaurus*, approaching closely in its characters to *Amphitherium*, were found in the oolite of the New World."

The high importance of the question touching the antiquity of Mammalian organization calls for a due notice of the foregoing statements relative to the most interesting fossils which have yet been discovered, and the more imperatively in this place, since they are peculiar to Great Britain, and, despite the numerous objections, are here admitted into the series of its fossil Mammalia.

First, then, as to the alleged facts respecting these fossils of the Stonesfield Oolite, repeated scrutiny enables me to state, that, instead of presenting 'the coarse fibrous structure' common in fossil cold-blooded Vertebrata, they have the peculiarly fine, compact structure which the jaws of insectivorous and marsupial Mammalia manifest. The alleged 'distinct, deep fissure, extending along their base, between the dental and opercular pieces,' is no fissure at all; but, in the two specimens of *Amphitherium Prevostii* of the Oxford Museum, and in the larger specimen of *Amphitherium* at York, which exhibit the inner side of the ramus of the jaw, is a distinct groove with an entire surface, answering to that which exists in the corresponding part of the jaw of the marsupial *Myrmecobius* and the Wombat.

The *Myrmecobius* is an insectivorous Mammal, and also marsupial, and it does not possess approximated and parallel incisores, but widely separated and diverging ones; they are, indeed, symmetrical with those of the opposite ramus, as in other Mammalia, but, as no one has yet seen an entire jaw of an *Amphitherium* or *Phascolotherium*, it is hard to understand the meaning of the assertion, that their incisores are not symmetrical: they are undoubtedly small; but, if they are almost as large as the supposed canine, such likewise are their porportions in *Myrmecobius*, and many *Insectivora*. The lower jaw of *Amphitherium Bucklandi*, (*Phascolotherium*, mihi,) is not more arched than in the recent *Dasyurus*, whose jaw is placed beside the fossil in the British Museum; and it will be plainly seen that the condyloid, or articular surface, instead of 'passing obliquely into the hard imbedding rock,' stands boldly out, and demonstrates a form diametrically opposite to that concave one which characterizes the jaws of reptiles and fishes. Thus much I have thought it necessary to state as to matters of fact, respecting which, however, the specimens speak plainly for themselves.

Not any of the jaws hitherto discovered present eleven similar multicuspid molares: even in the fragment of one side of the lower jaw figured by M. Prevost, which contains ten of the series of twelve molares, which is the true number in the genus *Amphitherium*, the fractured state of the posterior teeth still permits the recognition of a different conformation of the crown in them, as compared with the four remaining anterior molares. The difference in the configuration of the perfect crowns of the molar teeth, in the maxillary ramus of the *Amphitherium* at York, is such as to render both easy and certain the distinction of the *molares spurii* from the *molares veri*, which is commonly observed in Mammalian quadrupeds, but never in fishes and reptiles. The term 'multicuspid' cannot properly be applied to the anterior or false molars of the *Amphitherium*, which have but one principal cusp, and a minute tubercle, or talon, at one or both sides of its base.

Of the value of the argument drawn from the colour of the fossils, any one conversant with the varieties of shade, from brown to deep black, which Mammalian fossil teeth present, may judge; and, on this point, Mr. Ogilby has remarked, "the composition of the teeth cannot be advanced successfully against the mammiferous nature of the fossils, because animal matter preponderates over mineral in the teeth of the great majority of the Insectivorous *Cheiroptera*, as well as in those of the *Myrmecobius* and other small Marsupials."

If it were true that the crown of the teeth of the *Amphitherium* was supported by 'a long and much contracted cervix,' before the fangs were formed, these might be said to be bifid; but the original specimen of *Amphitherium* in the Oxford Museum demonstrates the independent origin of two fangs from the base of the crown, and the same fact is as plainly shewn in the *Phascolotherium Bucklandi* in the British Museum, where the origins of the double fangs are plainly visible above the sockets.

The cervix of the teeth is extremely short; in fact the

fangs diverge so immediately from the base of the crown, that this presents scarcely any contracted prolongation to which the term 'cervix' can be properly applied; the contrast between the teeth of the *Amphitherium*, and those of most Saurian Reptiles, is very striking in this respect. The enamelled surface of the teeth of the *Amphitherium* near the cervix is smooth and polished, and entirely devoid of any close vertical grooves.

The bifid osseous base which supports the true dental tissues of the teeth of *Squali* cannot be adduced as a corresponding structure to the two-fanged lower molars of the *Amphitherium* except by a forced and overstrained analogy; the real bearing of the two-fanged structure of the teeth of the *Amphitherium* upon the question of its affinities, is kept out of sight by such a comparison; for it is the implantation of the teeth in deep double sockets of a bony jaw by the double fangs which demonstrates the mammalian character of the animal: * the bifid osseous base of the teeth of sharks is attached, as is well known, by ligaments, to a cartilaginous jaw.

I was well aware, when replying to the objections of M. de Blainville, that portions of the jaws of a gigantic fossil Vertebrate animal, shewing teeth implanted by two fangs, had been discovered in the Alabama tertiary deposits, associated with *Corbulæ*, *Modioli*, sharks teeth, &c., and that these fossils had been referred by Dr. Harlan to a genus of Saurian Reptiles which he had called *Basilosaurus:* but the very fact of the implantation of the teeth by double fangs—the first alleged example of such a structure in the Reptilian Class—led me to receive the ascription of

* "The teeth, composed of dense ivory with crowns covered with a thick coat of enamel, are every where distinct from the substance of the jaw, but have two fangs deeply imbedded in it." Geol. Proceedings, Dec. 1838, p. 17.

such a structure to a Saurian reptile with a degree of scepticism, which the configuration of the vertebræ and other bones figured in Dr. Harlan's Memoir tended to increase. An unbiassed Anatomist, after a critical perusal of that memoir, would have been justified in maintaining a cautious hesitation in applying the conclusions of Dr. Harlan, as to the Saurian nature of his gigantic fossil animal with two-fanged teeth, to depreciate the value of the mammalian evidence yielded by the Stonesfield fossils.

That Author's Memoirs, in the Transactions of the American Philosophical Society, and in his "Medical and Physical Researches," cannot, however, be made responsible for the statements that the *Basilosaurus* is closely allied to the *Squali*, or that it is found in the Oolite of the New World; for Dr. Harlan, in his second and more extended Memoir, and in his Communication to the Geological Society, expressly leaves the geological question open, and contents himself with the statement — "In the matrix of the vertebra from the Washeta river was a fossil *Corbula*, common to the Alabama tertiary deposits." And the only character by which the so called *Basilosaurus* approaches to *Amphitherium*, is the implantation of the molar teeth by two fangs, which they exhibit in common with most Mammalia.

That the *Basilosarus* is, in fact, a mammiferous animal, I had the satisfaction of demonstrating,* in January 1839, by a close examination of the bones and teeth described by Dr. Harlan, on which occasion I proposed for it the name of *Zeuglodon*. All the subsequent discoveries of the remains of that gigantic species, — and an almost entire skeleton has been recently brought to light,†—have added

* Geological Transactions, 2nd Series, vol. vi. p. 69.

† Silliman's American Journal, vol. xliv. p. 411.

fresh proof of its cetacean character. All the weight, therefore, which the *Basilosaurus* was supposed to add to the Saurian hypothesis of the Stonesfield jaws, must now be transferred to the scale of the Mammalian view.

And having now answered the statements and arguments which have been put forth by those whom the Memoirs of M. Valenciennes and myself failed to convince, I shall proceed to describe successively all the specimens of the remains of the small insectivorous animals, from the Stonesfield Oolitic slate, that have hitherto come under my observation.

Fig. 16.

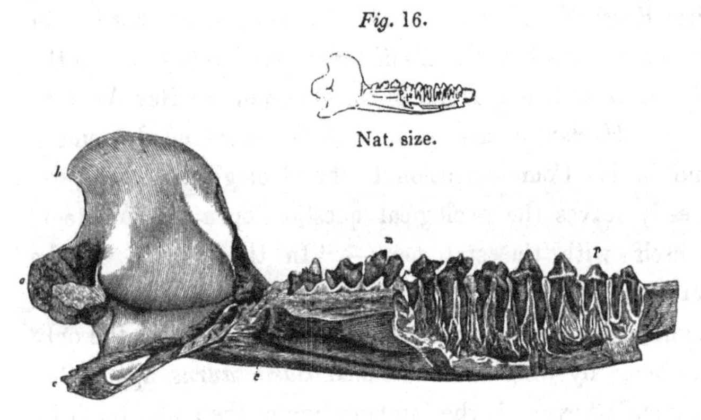

Nat. size.

AMPHITHERIUM PREVOSTII. No. 1.

The above wood-cut, (No. 16,) represents the original specimen of the remains of the *Amphitherium Prevostii*, examined by Baron Cuvier in the year 1818,* first noticed by Dr. Buckland in 1823, and figured by M. Prevost in 1825. The cut is carefully copied from the engraving in the Geological Transactions; the natural size of the fossil is given in outline, and it is enlarged four diameters in the finished figure below.

The fossil partly exhibits, partly represents by impression

* Ossemens fossiles, 4to. vol. v. pt. ii. Ed. 1824, p. 349.

in the matrix, the left half or ramus of the lower jaw, wanting the anterior or symphysial extremity, which is broken off. A thin layer of the original bone adheres to that part of the impression which indicates the articular condyle of the jaw *a*; the impression alone, which is well defined, gives the size and shape of the broad elevated and slightly recurved coronoid process (*b*), the base of which extends from the condyle to near the posterior commencement of the molar series of teeth. There is a slight remnant of the original angle of the jaw at *c*, which is continued backwards, in the form of a process, to nearly the vertical parallel of the condyle. The part of the jaw containing the three hindmost grinders is nearly entire, only the outer wall of the rest of the ramus is left imbedded in the Oolite, and fortunately retains seven of the molars, with their roots entire, and undisturbed in their sockets.

The undulations of the impression of the coronoid process shew that its anterior margin projected externally as a smooth convex ridge, and that between this ridge and the condyle the outer surface was slightly concave. That part of the angular process, which was naturally extended inwards, or towards the observer, is broken away, so that the degree of the inward inflection is left undetermined. The canal for the dental artery and nerve is exposed at the posterior fractured margin of the jaw, filled with the whitish Oolitic matrix. Below this aperture begins a smooth moderately wide and deep groove, which is continued forwards, gradually contracting to a point, at the lower magin of the jaw opposite the interspace between the true and false molar teeth. This groove has been described as a suture, or line of union, between two separate parts or elements of a composite jaw: such sutures, or harmoniæ, in the composite jaws of reptiles and fishes, are simple linear fissures, penetrating the substance of the bone; but the bottom of the groove in the jaw of the *Amphitherium* is quite entire, and the decreasing breadth of the groove indicates its origin from the pressure of a nerve or vessel.

A similar but relatively wider and shorter groove impresses the same part of the lower jaw in the insectivorous marsupial quadruped called *Myrmecobius:* and I have observed a narrower impression extending forwards, from the posterior entry of the dental canal, upon the inner surface of the ramus of the jaw in the Wombat (*Phascolomys*), another Mammal of the Marsupial order.

The following is the exact condition of the teeth, in this much-referred to specimen of *Amphitherium:* there are ten molars *in situ*, the seven anterior ones imbedded by two long and slender fangs in deep and distinct sockets. The molars gradually increase in size from the foremost to the sixth, in the present specimen: the rest are equal, except the last, which is somewhat smaller. The nearly perfect specimen of the jaw of the *Amphitherium* in the collection of Dr. Buckland, lately discovered and figured at the head of the present chapter (Cut 12), has demonstrated the accuracy of my deductions from the less complete specimen described in the Geological Transactions,* viz. that the *Amphitherium* had sixteen teeth in each ramus. Of the ten teeth, contained in the fossil under consideration, the first four, counting backwards, correspond with the third, fourth, fifth, and sixth premolars, and the remaining six to the true molars; the crowns of the fourth, fifth, and sixth premolars, enumerating them according to the true dental formula of the genus, are entire, and shew them to be simple, compressed, consisting of a single or principal conical cusp, with a minute tubercle or

* Loc. cit. p. 56.

talon at the hinder part of its base, and a more minute and hardly recognizable one in front; the base of the crown is slightly tumid, and from it are continued, without the intervention of a cervix, the two long slender almost parallel or slightly diverging fangs. The remains of the vertically split crown of the third premolar indicate the same form as that of the *fourth.* Traces of the double alveolus of the second premolar are preserved at the broken anterior end of the fossil. The fractured crown of the *first* true molar shews more distinct anterior and posterior cusps, at the base of the large middle cusp. The breadth of the base of the crown is displayed by the fracture of the *third* true molar, and refutes the notion of their being compressed like the premolars. The fourth true molar gives a view of the anterior, and of the large middle external cusp, with part of the posterior external cusp. In the fifth molar, the middle external cusp is nearly entire to its sharp apex: part of the anterior cusp and the base of the internal posterior cusp are preserved; the thicker and more complicated crowns of the *molares veri*, as compared with the *molares spurii*, are unequivocally demonstrated in all the last three molars.

The fangs descend half-way or more towards the lower margin of the ramus; their chief constituent, (dentine,) is clearly contrasted, by its texture and deeper colour, with the surrounding bone, from which they are plainly separated by a thin layer of a distinct coloured substance, infiltrated, apparently from the matrix, into the sockets of the teeth, like that in the vascular canals of the jaw. The minute cylindrical remains of the pulp-cavity are discernible in many of the exposed fangs.

In one of the genera of Seals, (*Stenorhynchus*, F. Cuv.,) all the molar teeth are compressed, and tricuspid or multicuspid, according to the species. M. M. Agassiz and De Blainville have supposed that the Stonesfield fossils presented a form of tooth resembling most those of such seals amongst Mammalia; but the teeth of all the Seal-tribe offer a well-marked peculiarity in their thick and ventricose fangs, to which character those of the *Amphitherium* offer no approximation, but, on the contrary, have long and slender fangs, as in the small marsupial and placental *Insectivora:* besides, no species of Seal presents the backward prolongation of the angle of the jaw demonstrated by the fossil *Amphitheria.*

The term 'Amphibia,' in the concluding summary of M. de Blainville's second memoir, has reference not to the cold-blooded Amphibia of Linnæus and the German naturalists, but to the above-cited and last-expressed opinion of M. Agassiz, who, admitting the Stonesfield fossils to be certainly those of mammals, rejects them from the marsupial and insectivorous orders, observing that "each separate tooth resembles the greater part of those of seals, near which group (amphibious Carnivora) the animal to which the jaws belonged should form a distinct genus. In fact," adds M. Agassiz, "the aspect of these fossil fragments is so peculiar, that it draws our attention towards aquatic animals rather than away from them."

But, in addition to the anatomical objections above adduced, it may be urged, that, though an extinct mammiferous animal, not larger than the water-shrew, should have been of aquatic habits, it does not follow that, therefore, it was piscivorous, and endowed with the instincts and organization of a Seal; in the absence of any evidence of the locomotive extremities, the affinity of the diminutive Mammalia of the Stonesfield epoch to the *Phocidæ*, could, at best, but be matter of conjecture, and this conjecture a close examination of the dental and maxillary characters entirely disproves.

The only inference which can be legitimately drawn from the remarkable fossil above described is that which the great Cuvier has left on record, viz., that it belonged to a small ferine Mammal * with a jaw much resembling that of an Opossum, but differing from all known ferine genera in the great number of the molar teeth, of which it had, at least, ten in a row. All that is now contended for in respect of the present fossil is, that it offers to the Comparative Anatomist sufficient evidence of the accuracy of Cuvier's conclusion.

I next proceed to consider the additional proof which the subsequently discovered fossils from the same locality have afforded.

Fig. 17.

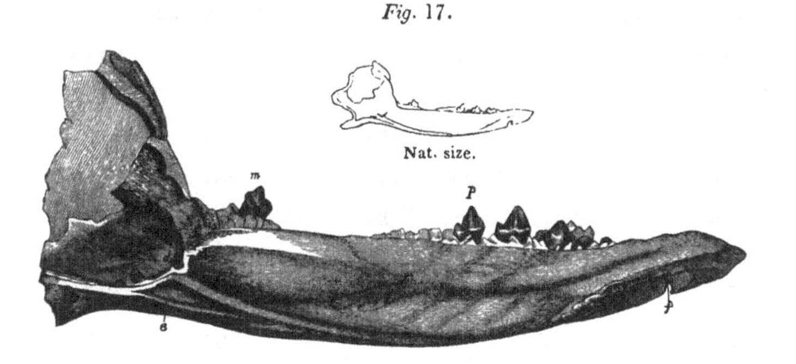

AMPHITHERIUM PREVOSTII, No. 2.

The above cut (No. 17) is copied from the figures of the second specimen of *Amphitherium Prevostii*, described in my Memoir in the Geological Transactions †: the outline gives the natural size, which corresponds precisely with that of the foregoing specimen (fig. 16): in the finished

* I use the word 'ferine' as equivalent to the French 'carnassier,' the term by which Cuvier signifies collectively the Cheiroptera, Insectivora, Carnivora, and Marsupialia.

† Loc. cit. Pl. 5, fig. 1.

figure, four times enlarged, the condyle and angle of the jaw have been left out for want of room in the page, but their shape is accurately given in the outline above.

In this specimen the whole of the exposed surface of the left ramus of the lower jaw, with the exception of the coronoid, articular, and angular processes, is entire; the smooth surface near the anterior extremity of the jaw is in bold relief, and slopes away at nearly a right angle from the rougher articular surface of the elongated symphysis. It may be supposed that this symphysial surface, which at once determines the side of the jaw, might be obscured in the plaster cast studied by M. de Blainville, who has contended, in opposition to the opinion of M. Valenciennes, that the outside of the jaw was here displayed, but there is no possibility of mistaking it in the fossil itself; it is long and narrow, and is continued forwards in the same line with the gently convex inferior margin of the jaw, which thus tapers gradually to a pointed anterior extremity, precisely as in the jaws of the *Didelphys* as well as in other *Insectivora*, both of the marsupial and placental series. Its lower margin presents a small but pretty deep notch, (*f*,) which possesses every appearance of a natural structure, and a corresponding but shallower notch, is present in the same part of the jaw of the *Myrmecobius*. In the relative length of the symphysis, as in its form and position, the jaw of the *Amphitherium* corresponds with that of the *Didelphys*, *Myrmecobius*, and *Gymnurus*. A greater proportion of the convex articular condyle is preserved in this than in the foregoing specimen, and it projects backward to a greater extent. The precise contour of the coronoid process is not so neatly defined in this as in the first specimen of *Amphitherium*, but sufficient remains to show that it had the same height and width.

The exposed surface of the coronoid process is slightly convex. The surface of the ascending ramus of the jaw is entire above the angle, whence we may conclude that, if the process from the latter part had been continued directly backwards, it would also have been entire; but the extremity of the angular process is broken off, proving it to have originally inclined inwards, or towards the observer: as, however, the greater part of the angle is entire, it could not have been inflected to the same extent as in the *Didelphys*, *Dasyurus*, or the *Phascolotherium* next to be described. A groove is extended from the lower end of the articular condyle forward to the orifice of the canal for the dental artery, where it divides; the upper branch terminates in the dental orifice, the lower and larger division is continued forward near the lower margin of the jaw, gradually contracting and disappearing half way towards the symphysis: this smooth vascular groove has as little resemblance to an articular fissure as in the former specimen. There is a broader and shorter groove in the corresponding part of the jaw of the *Myrmecobius;* and a narrower groove in that of the Wombat. The alveolar wall of the posterior grinders makes a convex projection, characteristic of the inner surface of the ramus of the lower jaw. The posterior grinder in the present jaw is fortunately more complete than in the first example, and shows a small, middle, internal cusp, with part of a large external cusp, both projecting from the crown of the tooth in nearly the same transverse line. The enamel covering the internal cusp, which is vertically fractured, is beautifully distinct from the ivory, and considerably thicker in proportion to the size of the tooth than is the enamel or its analogue in the teeth of any species of reptile, recent or fossil. The six molars anterior to the one in place, are broken off close to the sockets; both the fifth and fourth false molars are entire; the anterior cusp presents the same superior size as in the first specimen. The thick external enamel, and the silky, iridescent lustre of the compact ivory, are beautifully shown in these teeth. The third and second grinders are more fractured than in the first specimen, but sufficient remains to show that they possess the same form and relative size; but the most interesting evidence, as regards the teeth, which the present jaw affords, is the existence of the sockets of not less than seven teeth, anterior to those above described. Of these sockets the four anterior ones are small and simple, like those of the mole, and are more equal in their size and interspaces than in the *Didelphys:* the fifth socket contained a small premolar with double fangs; the next is a similar socket, and then come four other premolars in place with more or less perfect crowns: between the last of these premolars and the last molar the empty alveoli agree in number with, and occupy the same extent as, the the first five true molars in the jaw, cut 16. This fossil afforded evidence, therefore, that the dental formula of the *Amphitherium* included thirty-two teeth in the lower jaw; sixteen on each side.

Thus the *Amphitherium* differs more considerably than the evidence in Cuvier's possession showed, from the genus *Didelphys* in the number of its teeth. Indeed at the time when the great Palæontologist wrote respecting it, believing it to have had ten molars, no mammiferous ferine quadruped was known to possess a greater number of these teeth than the Cape Mole or Chrysochlore, which has nine molars on each side of the upper jaw, and eight molars on each side of the lower jaw. The Chrysochlore, however, is not the only species in which the molars exceed the number usually found in the unguiculate Mammalia. The marsupial genus, *Myrmecobius*, (*fig.* 18,) has nine molars on each side of the lower jaw, besides one small canine and three conical incisors.

Fig. 18.

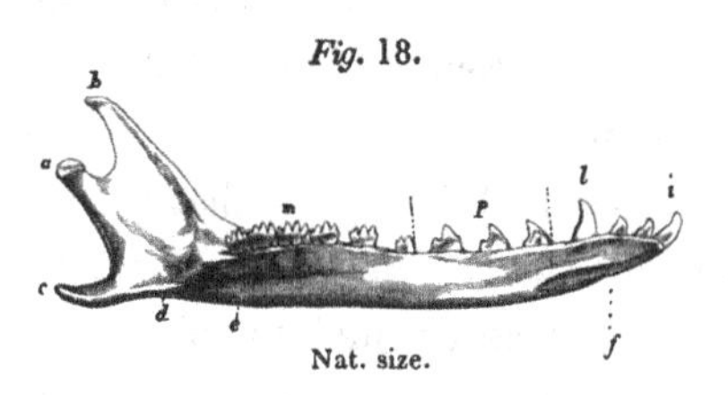

Nat. size.

MYRMECOBIUS FASCIATUS.

The teeth of *Amphitherium*, moreover, differ from those of *Didelphys* not only in number but also in size, being relatively smaller. The teeth of *Myrmecobius*, besides their approximation in number to those of *Amphitherium*, resemble them in their small relative size more than do those of *Didelphys*, but they are still smaller than in *Amphitherium*, which in this respect, as well as in the structure of the teeth, appears to hold an intermediate position between *Didelphys* and *Myrmecobius*. The incisors (*i*) of the *Myrmecobius* are conical, separated at their base, diverging, the anterior one almost as long as the canine (*l*) ; the first three molars (*p*) have compressed, conical, bicuspid, or tricuspid, crowns, the middle cusp being the largest, and they have each two fangs ; they belong to the series of premolars : the remaining six teeth are multicuspid and true molars, as in *Amphitherium*.

The discovery of an existing quadruped in the marsupial series, presenting so many resemblances in the number, size, shape, and proportions of the teeth, to the *Amphitherium*, so far as the dental characters of that genus are elucidated by the two specimens above described, adds to the probability of the marsupial nature of the fossil: the symphysial emargination (*f*), and the groove (*e*), are characters common to the jaws of both genera, to which, therefore, due weight must be assigned. And, if the *Myrmecobius* differs from the *Amphitherium* in the higher position of the articular condyle (*a*), and the narrower coronoid process (*b*), it is, in other genera of the Marsupial Order, as *Thylacinus* and *Dasyurus*, that the closest agreement with *Amphitherium* is in this respect to be found. The term *Didelphys* was originally applied to the Stonesfield Mammalian genus under consideration, in its wide Linnæan sense, which is almost equivalent to the ordinal term *Marsupialia.* Three generic names, in the proper or restricted sense had been proposed in earnest, and a fourth in jest, for the ancient Insectivore, before its affinities were agreed upon, or its true dental formula known. In my Memoir of 1838, I ventured to observe, in reference to the new name proposed by M. Valenciennes, that it would have been more prudent to have chosen a less descriptive one than *Thylacotherium*, since the affinities of the fossil Insectivore to the marsupial order were indicated only with a certain degree of probability, and required further evidence before the desired demonstration could be attained. But the determination of the particular *order* of mammals to which the fossils in question belonged, was a matter of very inferior importance to the discovery of the *class* of vertebrate animals in which the species they represented ought to rank. In reference to this point the evidence afforded by the two jaws above described decisively proves, in my opinion, that they belong to a true, warm-blooded, mammiferous species, referrible also to the higher or unguiculate division of the class Mammalia, and to an insectivorous genus; with a probability of the marsupial character of such genus.

The probability entertained in 1838, and supported by the degree of resemblance between *Amphitherium* and *Myrmecobius* in the number and form of the molar series of teeth, has since been diminished by the discovery of the right ramus of a lower jaw, presenting its external surface to the observer, and the most complete of all the extant specimens of the *Amphitherium:* it is figured, of the natural size in outline, and twice the natural size in the finished cut, at the head of the present section (fig. 15).

This jaw, which is in the choice collection of Professor Buckland, contains the whole series of twelve molar teeth, the last six (*m*) being quinque-cuspidate; the six anterior ones (*p*) unicuspidate, with one or two small basal accessory cusps; it, also, displays the socket of one small canine (*l*), and three small incisors (*i*), *in situ;* altogether amounting to sixteen teeth on each side of the lower jaw, as indicated by the sockets of the second specimen above described. The convex condyle, the broad and high coronoid process, the projecting angle, the varied kinds and double-rooted implantation of the teeth, all unequivocally displayed in this fossil, establish the conclusions deduced from the foregoing specimens, of the existence of a small insectivorous mammal during the oolitic epoch.

Here, likewise, was a specimen adapted to afford the much desired test of the form of the angular process of the lower jaw. The inward inflection of this process had been long ago pointed out by Cuvier as a character of the genus *Didelphys*, and I have established its generality in the entire marsupial series, and pointed out its characteristic modifications in the different genera.*

Dr. Buckland had transmitted this beautiful specimen to

* Geological Trans. 2nd Series, vol. vi. p. 50. Art. Marsupialia, Todd's Cyclopædia.

me soon after it came into his possession, and on being made acquainted with my wish, he kindly permitted me to take the requisite means to determine the shape of the angular process. The external surface, the only part exposed, was quite entire; and with a fine graving-tool, I cleared away sufficient of the matrix to show the extent to which it was imbedded. Although the inferior margin of the process is inflected so as to render the outer surface convex, the degree of inflection is less than in any of the known Marsupialia, and does not exceed that of the Mole or Hedgehog. This slightly inflected margin is broken away, in both the half-jaws that have their inner surface exposed, and if that indication of an inflected angle has been insisted upon too strongly as a marsupial character, we may be warned thereby to avoid the opposite extreme of concluding too absolutely, that a Mammal, with such peculiar dental characters as those of the *Amphitherium*, may not have combined the more essential points of the Marsupial organization, with the lowest development of that peculiar character of the existing species, which is afforded by the angle of the jaw.*

The main fact in the present inquiry,—the antiquity of the Mammalian type of organization,—is, if possible, more unequivocably established by the present than by the preceding fossils. The whole outer surface of the ramus of the jaw is beautifully entire: not a trace of the alleged distinct, deep fissures which, in Lizards and other cold-blooded Ovipara, separate the coronoid or complementary and other elements of the jaw can here be discerned. The broad and simple coronoid process shows the wide concavity and the

* The ossa marsupialia even, may be absent, without the loss of the essential organic Characters of the Marsupial Order, as I have lately ascertained by dissection of the great carnivorous Opossum (*Thylacinus Harrisii*) of Van Dieman's Land.

anterior marginal ridge, which are indicated by the impressed matrix in the before described jaws of the *Amphitherium:* the entire and prominent convex condyle is now seen to rise higher above the level of the molar teeth than was indicated by its incomplete remains in the former specimens: and the outer surface of the present instructive fossil demonstrates several small anterior foramina or outlets of the dental canal; one beneath the third premolar, and others nearer the end of the jaw, as in that of the *Myrmecobius.* Such foramina are still more mumerous, and extend further back in the jaws of most Saurians.

INSECTIVORA. *AMPHITHERIIDÆ.*

Fig. 19.

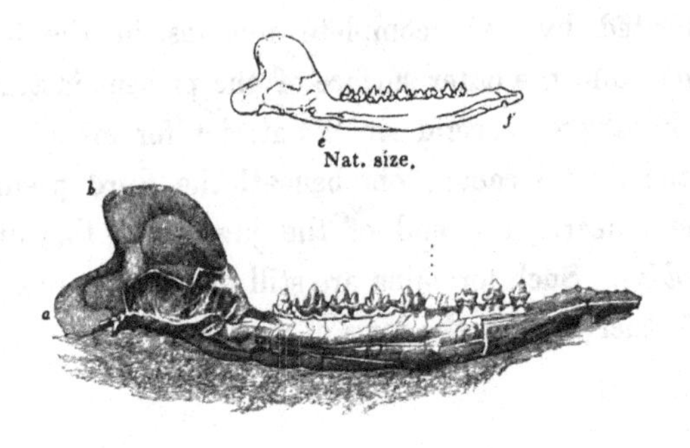

AMPHITHERIUM BRODERIPII.

Amphitherium Broderipii. Owen.
Thylacotherium ,, ,, Geol. Trans. 2nd Series, vol. vi. pl. 6. fig. 1.

The fossil figured above, of the natural size in outline, and magnified in the finished cut, was discovered, like the preceding jaws, in the Oolitic slate of Stonesfield, and was presented by the Rev. H. Sykes to the Philosophical Institution at York, in whose Museum it is now preserved.

In this, as in the first two specimens of the *Thylacotherium Prevostii*, the left ramus of the lower jaw offers its inner surface to the observer: it presents at its anterior part the sockets of three incisors and one canine, of small and nearly equal size, each having a simple fang; then follow the empty sockets of three small premolars, each with two fangs; to these succeed the three larger premolars, in place, each having two fangs protruded to a certain extent from their sockets, and fixed by the adherent matrix in that position, which proves that they were not anchylosed to the osseous substance: for these teeth, no doubt, became loosened and slightly displaced after decomposition of the soft parts; and the anterior teeth, which are missing, were probably lost from the same cause, before the jaw was finally encased in the oolite. There is a small anterior as well as posterior tubercle at the base of the large middle cusp or cone, in each of the three premolars which are in place: the middle cusp of the posterior one is fractured: there is a slight ridge along the inner side of its base in that tooth, indicating the transition to the true molar series, the commencement of which is indicated by the dotted line. The first true molar is wanting; the next four present the inner surface of their crowns in a perfect and uninjured state: the large middle cusp has a smaller one at the anterior and posterior part of its base; this is traversed by a strong ridge along the inner side, which supports three small cusps; one of these rises at the middle of the base of the large external cusp, and the other two form the anterior and posterior extremities of the crown of the tooth. This form of grinder resembles that of the *Phascolotherium* except in the presence of the middle internal cusp, more than that of the molars of the true *Didelphys*. The sharp points of these multicuspid teeth are well adapted for crushing the cases of coleopterous insects, and correspond essentially, though with a generic modification of form, with the teeth of the existing Insectivora, as Bats and Shrews. "The existence of the wing-covers of Insects in the secondary series, in the Oolitic slate of Stonesfield," Dr. Buckland states, "has been long known; these are all Coleopterous, and in the opinion of Mr. Curtis, many of them approach most nearly to the *Buprestis*, a genus now most abundant in warm latitudes."* In the present example of the jaw of the small co-existing Insectivores, the con-

* Bridgewater Treatise, vol. i. p. 411.

dyloid (*a*) and coronoid (*b*) processes have both left their impressions on the matrix: the angle of the jaw is fractured: there is the same shallow, wide and smooth groove (*e*) near the lower margin of the jaw, and the same notch (*f*) in the symphysis, as in the *Amphitherium Prevostii*, and the *Myrmecobius*. The chief value of the specimen in the Museum at York, arises out of the very perfect state of the crowns of the molar teeth, the peculiar form of which, giving one of the characters of the extinct genus, could not be satisfactorily determined from the specimens before described. That the fossil in question belongs to the genus *Amphitherium* is proved by the number and nature of the teeth which it contained; but its difference of size, as compared with the jaw of *Amphitherium Prevostii*, is greater than has been observed in mature individuals of the same species of Placental or Marsupial Insectivores. I have, therefore, indicated the species which the present fossil represents by the name of *Amphitherium Broderipii*, in honour of the Naturalist and Geologist, to whom we are indebted for the first accurate description and figure of a Stonesfield Mammalian fossil.*

For the opportunity of describing and figuring the half-jaw of the *Amphitherium Broderipii*, in the Geological Transactions, I was indebted to Professor Phillips, to whom I again beg to record my obligations for the facilities afforded me in studying this additional evidence of the oldest Mammalian inhabitants of this Planet.

* See the "Observations on the Jaw of a Fossil Mammiferous Animal, found in the Stonesfield Slate. By W. J. Broderip, Esq., Sec. G.S. F.L.S., &c." Zoological Journal, 1828, vol. iii. p. 408, pl. xi.

MARSUPIALIA. *ENTOMOPHAGA.*

Fig. 20.

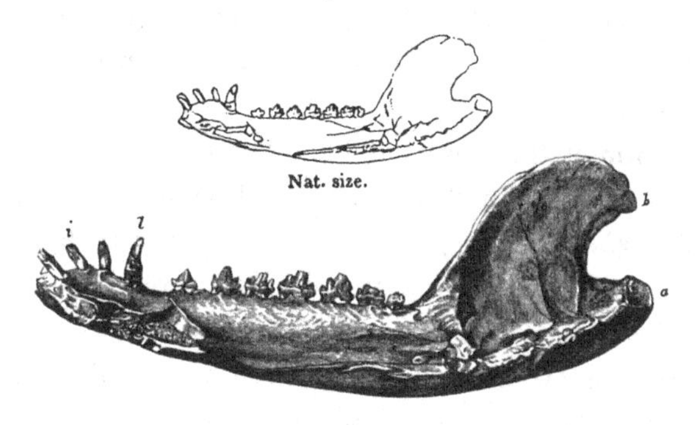

PHASCOLOTHERIUM BUCKLANDI.

Genus PHASCOLOTHERIUM.

Phascolotherium Bucklandi.	OWEN, Geol. Trans. 2nd Series, vol. vi. p. 58, pl. 6.
Didelphis „	BRODERIP, Zool. Journ. vol. iii. p. 408.

ALTHOUGH the evidence of the very slight degree of inflection of the angular process in the lower jaw of the *Amphitherium Prevostii*, (fig. 15,) turns the scale in favour of its affinities to the placental Insectivora, yet the range of variety, to which the character of the inflected angle of the jaw is subject in the different existing genera of *Marsupialia*, warns us against laying undue stress upon its feeble development in the extinct genus of the Oolitic epoch, and incites us to look with redoubled interest at whatever other indications of a Marsupial character may be present in the fossil remains of other genera and species of Mammalia that have been detected in the Stonesfield slate.

In that remarkably perfect and interesting specimen, presented to the British Museum by Mr. Broderip, its original describer, and which is figured at the head of the present section, the Marsupial characters are more strongly manifested in the general form of the jaw, and in the extent and position of its inflected angle; while the agreement with the genus *Didelphys* in the number of the premolar and molar teeth is complete.

The form of the crowns of these teeth corresponds, however, so closely with that which has been described in the *Amphitherium Broderipii*, as to argue strongly for their close natural affinity; and accordingly whatever additional approximation to the Marsupial Order is made by the *Phascolotherium*, may be held to support the Marsupial nature of the *Thylacotherium* although the proof be yet absent.

Respecting the jaw of the *Phascolotherium*, — "Some years have elapsed," writes my friend Mr. Broderip, in 1828, "since an ancient stone-mason, living at Heddington, who used to collect for me, made his appearance in my rooms at Oxford, with two specimens of the lower jaws of mammiferous animals, imbedded in Stonesfield slate, fresh from the quarry. At the same time he brought several other very fine Stonesfield fossils, the result of the same trip. One of the jaws was purchased by my friend Professor Buckland, who exclaimed against my retaining both, and the other I lent to him some time ago. Dr. Buckland's specimen, which wants incisor and canine teeth, has been examined by M. Cuvier, and is figured by M. Prevost as an illustration to his "Observations sur les Schistes calcaires Oolitiques de Stonesfield en Angleterre," &c.,* the other was lost, after the Professor had returned it; and the loss was, most unjustly as I must now acknowledge, attributed to him. To my no small gratification, this specimen has

* Ann. des Sciences, Nat. Avr. 1825.

just been found and forms the subject of the following sketch.—

"The ten teeth, represented in the figure accompanying M. Prevost's memoir,* are evidently grinders, and somewhat resemble the molar teeth of my specimen, which has, however, only seven grinders; and, when it was lent to Dr. Buckland, they were the only teeth apparent. He, however, caused the stone to be carefully scraped away, and there appear, in addition, a canine tooth" (cut 20, *l*) "and three incisors" (*ib. i*). "There is room also for a fourth: the end of the jaw is fractured, and there are traces of what may be the alveolus of a fourth incisor. With this addition, the specimen would give the exact number of teeth in the half of a lower jaw of a *Didelphis*, *viz.*, four incisors, one canine, seven grinders. The fossil, which is in high preservation, is imbedded in a slab of Stonesfield slate, together with *Trigoniæ* and other marine exuviæ; the whole mass exhibiting the Oolitic structure in the most satisfactory manner."

"My specimen consists of the right half of a lower jaw, the inside of which is presented to view. To say nothing of the difference of form in the jaw-bone, M. Prevost's figure gives us the representation of a portion of a lower jaw with ten grinders therein: my fossil has only seven, and appears to have been part of an animal generically different. The teeth are distinctly separated, and those who are best qualified to judge, are of opinion that the jaw did not belong to a young individual. The well defined ridges and decided features of the bone denote a full-grown animal: the sharpness of the teeth makes it probable that it was not an aged one."—"As the history of this animal,"

* "Pl. 18. fig. 1, 2." See ante, fig. 16.

concludes Mr. Broderip, "rests only upon the portion of its lower jaw, figured in the plate accompanying the present memoir, (for the specimen figured by M. Prevost appears to have belonged to a different animal,) it would be presumptuous in me to pronounce on its generic identity with *Didelphis* Cuv. But, until some more able anatomist shall correct the generic name, I may be permitted, for the sake of convenience and perspicuity, to name it *Didelphis Bucklandi*."

The statements and arguments of those Anatomists who first applied their skill to the reconsideration of the Stonesfield jaws, and who have not only rejected the reference of the present fossil to the genus *Didelphys* but to the Class *Mammalia*, have already been discussed, and I shall now cite those observations which, while they favour its claims to be admitted, not only into the Mammalian Class, but into the Marsupial Order, at the same time establish its generic distinction, and necessitate the imposition of a new generic name.

The condyle of the jaw of the *Phascolotherium* here described, (fig. 20, *a*,) instead of being vertically split, as in the specimens of *Amphitherium*, is fortunately entire, and stands out in bold relief from the Oolitic matrix; it presents exactly the same form and degree of convexity as in the genera *Didelphys* and *Dasyurus*. In its relative position to the series of molar teeth, with which it is on a level, it corresponds with *Dasyurus* more nearly than with *Didelphys*: in the *Dasyurus ursinus*, in fact, as well as in the allied Marsupial genus *Thylacinus*, the condyle has precisely the same relative position to the molar series; so that this particular structure in the jaw of the *Phascolotherium* affords no argument against its admission into the Marsupial series.

The general form and proportions of the coronoid process (*fig.* 20, *b*,) resemble those in the zoophagous Marsupials; but in the depth and form of the entering notch, between this process and the condyle, it corresponds most closely with the *Thylacinus*.

It is, indeed, a most interesting fact, that this rare and solitary genus, represented by a single species (the Hyæna of the Tasmanian colonists), whose term of existence seems fast waning to its close, should afford the only example of a form and backward extension of the coronoid process, and a corresponding deep emargination above the condyle, which would else exclusively characterize the ancient *Phascolotherium*.

The base of the inwardly-bent angle of the lower jaw progressively increases in *Didelphys*, *Dasyurus*, and *Thylacinus;* and judging from the fractured surface of the corresponding part in the fossil, it also resembles most nearly, in this respect, the *Thylacinus*.

The condyle of the jaw is nearer the plane of the inferior margin of the ramus in the Thylacine than in the Dasyures or Opossums; and, consequently, when the inflected angle is broken off, the curve of the line continued from the condyle along the lower margin of the jaw in the Thylacine is least; in this particular again the Phascolothere resembles the Thylacine. In the position of the dental foramen, the Phascolothere, like the Amphithere, differs from all the zoophagous Marsupials already cited, and also from the placental *Feræ;* but in the *Potoroo* (*Hypsiprymnus*), a marsupial Herbivore, the orifice of the dental canal is situated, as in the Stonesfield Marsupials, very near the vertical line, dropped from the last molar tooth.

A portion of the inner wall of the jaw, near its anterior margin, in the Phascolothere, has been broken off, so that the form of the symphysis cannot be precisely determined;

but, in the gentle curve by which the lower margin of the jaw is continued along the line of the symphysis to the anterior extremity of the jaw, the *Phascolotherium* resembles *Didelphys* more than *Dasyurus* or *Thylacinus.*

It is interesting to find that this analogy is associated with a correspondence in the condition of the teeth at the anterior part of the jaw. In examining the fossil we can scarcely refuse our assent to Mr. Broderip's opinion, that there were originally four incisors in each ramus of the jaw of *Phascolotherium*, as in *Didelphys.* Of the three incisors which are actually present in the fossil, only the internal and posterior surfaces are displayed, and not the whole breadth of the tooth; so that in the enlarged figure of the jaw detached from its matrix, the incisors appear both narrower and further apart than they really are. The incisors in the *Thylacinus* are of a prismatic form; and the surface, corresponding to that which is exposed on the fossil, forms one of the angles, from which the tooth increases in breadth to its anterior part, which forms one of the three facets.

Allowing for this circumstance, which must be borne in mind in an endeavour to arrive at the true affinities of the Phascolothere, the incisors in that fossil are evidently separated by wider intervals than in *Thylacinus*, *Dasyurus*, or *Didelphys;* and the Phascolothere resembles, in this respect, as in the smaller proportions of its canine, the genus *Myrmecobius.*

In the proportions of the grinders to each other, especially the small size of the hindmost molar, the Phascolothere resembles the *Myrmecobius* more than it does the Opossum, the Dasyure, or the Thylacine; but in the form of the crown it resembles the Thylacine more closely than any other genus of Marsupials. In the number of molar teeth the Phascolothere differs both from the Amphithere and the *Myrmecobius*, and resembles the Opossum and Thylacine, having three false and four true molars, or seven grinders altogether, in each maxillary ramus. The distinction between the false and true molars is however much less strongly marked, both in the Phascolothere and Thylacine, than in the Opossum. The difference between the false and true molars in the Opossum is chiefly indicated by the addition, in the true molars, of a pointed tubercle on the inner side of the middle large tubercle, and in the same transverse line with it; but in the Phascolothere, as in the Thylacine, there is no corresponding tubercle on the inner side of the large, middle, pointed cusp; its place is occupied in the *Phascolotherium* by a ridge, which extends along the inner side of the base of the crown of the true molars, and, projecting a little beyond both the anterior and posterior smaller cusps, gives the quinquecuspid appearance to the crown of the tooth, as represented by Dr. Buckland in his magnified view of the antepenultimate grinder of the *Phascolotherium*, given in the 2nd Plate of the illustrations of the Bridgewater Treatise. In the Thylacine the internal ridge is not continued across the base of the large middle cusp, but it extends along and beyond each of the lateral cusps, so as to give the tooth a similar quinquecuspid form to that which characterizes the true molars of the Phascolothere. Connecting the close resemblance which the molar teeth of the Phascolothere bear to those of the Thylacine with the similarities which have been already shown to exist in the several characteristic features of the ascending ramus of the jaw, I am of opinion that the marsupial extinct genus, indicated by the Stonesfield fossil here described, was nearly allied to *Thylacinus*, and that its position in the marsupial series is between *Thylacinus* and *Didelphys*.

There are two linear impressions on the inner side of the horizontal ramus of the jaw of the Phascolothere which have been mistaken for indications of *harmoniæ*, or toothless sutures, analogous to those which join together the component pieces of the compound jaws of reptiles and fishes. One of these is a faint, shallow, linear impression, continued from between the antepenultimate and penultimate molars, obliquely downwards and backwards, to the foramen for the dental artery. I conceive it to be due to an accidental crack; and if the portions of the bone which it separates were to be compared to the contiguous margins of the opercular and dentary pieces of a reptile's jaw, it would be seen that the only suture, which in Reptiles is continued from any part of the level of the dental series between these pieces, passes in a totally different direction; it is the suture which bounds the anterior part of the opercular piece, and which, in all reptiles, runs obliquely downwards and forwards, instead of downwards and backwards. The second impression in the jaw of the *Phascolotherium* is much more strongly marked than the preceding; it is a linear groove continued from the anterior extremity of the fractured base of the inflected angle obliquely downwards to the broken surface of the anterior part of the jaw. Whether this line be due to a vascular impression, or an accidental fracture, I do not offer an opinion; but this may be confidently affirmed, that there is not any suture in the compound jaw of a reptile which occupies a corresponding situation.

And lastly, with reference to the philosophy of pronouncing judgement on the saurian nature of the Stonesfield fossils from the appearances of sutures in the jaws themselves, I would offer one remark, the justness of which will be obvious alike to those who are and those who are not conversant with the details of Comparative Anatomy. The cumulative evidence of the true nature of the Stonesfield fossils, afforded by the shape of the condyle, coronoid process, angle of the jaw, different kinds of teeth, with the shape of their crowns, double fangs, and implantation in sockets, reposes on structures which cannot be due to accident, while those which favour the evidence of the compound structure of the jaw may arise from accidental circumstances.

The close approximation of the *Phascolotherium* to marsupial genera now confined to New South Wales and Van Dieman's Land, leads us to reflect upon the interesting correspondence between other organic remains of the British oolite, and other existing forms now confined to the Australian continent and adjoining seas. Here, for example, swims the *Cestracion*, which has given the key to the nature of the 'palates' from our oolite, now recognized as the teeth of congeneric gigantic forms of cartilaginous fishes.

Mr. Broderip, in his memoir above quoted, observes, "that it may not be uninteresting to note, that a recent species of *Trigonia* has very lately been discovered on the coast of Australia, that land of marsupial animals. Our specimen lies imbedded with a number of fossil shells of that genus."

Not only *Trigoniæ*, but living *Terebratulæ* exist, and the latter abundantly, in the Australian seas, yielding food to the *Cestracion*, as their extinct analogues doubtless did to the allied cartilaginous fishes, called *Acrodi*, *Psammodi*, &c. Araucariæ and cycadeous plants, likewise, flourish on the Australian continent, where marsupial quadrupeds abound, and thus appear to complete a picture of an ancient condition of the earth's surface, which has been superseded in our hemisphere by other strata and a higher type of Mammalian organization.

The subjoined cut, from Dr. Fitton's paper on the strata from whence the fossil jaws of the *Amphitheria* and *Phascolotheria* were obtained, represents the section of the superincumbent beds, taken through a shaft sunk to a pit or gallery in which the "slate," (as it is called,) is worked.

On the side opposite the right hand is marked the depth of the shaft to the horizontal gallery where the slate is dug, which contains the Mammalian fossils; on the opposite side, the strata are numbered in succession as follows:—

Fig. 21.

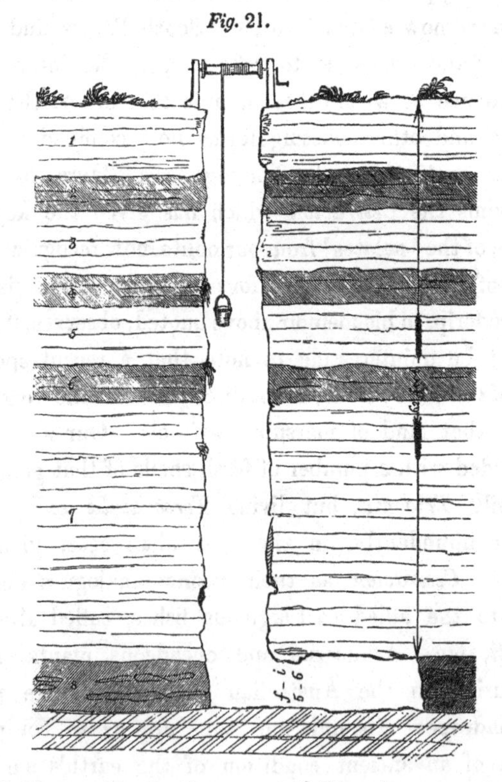

1. Rubbly limestone (Cornbrash).
2. Clay, with Terebratulites.
3. Limestone rock.
4. Blue clay.
5. Oolitic rock.
6. Blue clay.
7. Rag, or oolitic limestone.
8. Sandy bed, containing the "Stonesfield slate."

MARSUPIALIA ? *DIDELPHIDÆ ?*

Fig. 22.

Outside, nat. size.

Upper view.

Inside, nat. size.

DIDELPHYS ? COLCHESTERI, *Owen*.

Didelphys Colchesteri.	CHARLESWORTH, Mag. Nat. Hist., 1839, p. 450, fig. 60.
Opossum ?	LYELL, Annals of Nat. Hist. vol. iv. p. 190.

THE fossil above figured in three views, though small and not yielding decisive evidence as to its generic relations, forms one of the interesting unequivocal evidences of the Mammalian class, which the careful researches of Mr. Colchester, of Ipswich, have brought to light from the Eocene sand underlying the red-crag at Kyson, near Woodbridge, Suffolk.

It consists of a small portion of the right branch of a lower jaw, with one double-rooted premolar tooth and the sockets of two others, and it has been referred by Mr. Charlesworth,* who first described it, to the Opossum genus (*Didelphys*). He states, "The tooth, in its symmetrical form, united with the indication of an anterior, as well as posterior heel or talon, does not agree with any species of *Didelph* with which I have as yet been able to compare it, but I think no doubt can be entertained of the generic or family affinities indicated by the characters which it exhibits. Judging from the empty alveoli on either side, the tooth appears to be the one immediately succeeding the true molars: its posterior tubercle is

* Loc. cit.

strongly developed, and divided longitudinally by a prominent ridge, the continuation of which forms the posterior edge of the body of the tooth. At the base of the anterior root of the tooth, the opening of a foramen is seen, on the outer surface of the bone."

In subsequently studying this fossil, I have not been so satisfied as to its unequivocal indication of the genus or family of the small zoophagous Mammal of which it formed part. There is no tooth so little characteristic, or upon which a determination of the genus could be less safely founded, than one of the spurious molars of the smaller carnivorous and omnivorous *Feræ* and *Marsupialia.* A large, laterally compressed, sharp-pointed middle cone, or cusp, with a small posterior and sometimes also a small anterior talon, more or less distinctly developed, is the form common to these teeth in many of the genera of the above orders. It is on this account, and because the tooth of the fossil in question differs in the shape of the middle, and in the size of the accessory cusps, from that of any known species of *Didelphys*, that I regard its reference to that genus as premature, and the affinities of the species to which it belongs as wanting further evidence, before they can be determined beyond the reach of doubt. Besides the presence of the anterior tubercle, or talon, and the larger and more complicated posterior tubercle, the middle compressed cone is more equilateral and symmetrical than in the corresponding tooth of the Opossum.

The crown of the premolars of the placental *Feræ*, which present the same general form as the fossil, are thicker from side to side, in proportion to their breadth; the premolars of the *Dasyurus*, *Thylacinus*, and *Phascogale*, differ in like manner from the fossil. It is in the marsupial genera *Didelphys* and *Perameles* that the false molars present the same laterally compressed shape as in the fossil. In addition to the perfect tooth, the fossil includes the empty sockets of two other teeth, and the relative position of these sockets places the *Perameles* out of the pale of comparison. On the hypothesis that the present fossil represents a species of *Didelphys*, the tooth *in situ* unquestionably corresponds with the second, or middle false molar, right side, lower jaw. This is proved by the size and position of the anterior alveolus.

Had the tooth *in situ* been the one immediately preceding the true molars, the socket anterior to it should have been at least of equal size, and in juxta-position to the one containing the tooth. The anterior socket, however, is little more than half the size of the one in which the tooth is lodged; it is, also, separated from that socket by an interspace, equal to that which separates the first from the second false molar in the *Didelphys Virginiana*. This is well shown in the inside view. In the placental Mammalia, in which the first small false molar is similarly separated by a diastema from the second, the first false molar has only a single fang. In the present fossil, the empty socket of the first false molar proves that the tooth had two fangs, as in the marsupial *Feræ* and *Insectivora*. There is nothing in the size or form of the socket, posterior to the implanted tooth of the fossil, to forbid the supposition that it contained a false molar, resembling the one in place; had it been the socket of a true molar, then the fossil could not have belonged to *Didelphys*, or to any other known marsupial genus, because no known marsupial animal, which presents the posterior false molar of a similar form and in like juxta-position with the true molars, as the tooth in the present fossil, (on the supposition that it immediately preceded the true molars,) has the next false molar so small as it must have been in the fossil on that supposition.

Upon the whole, the conclusion that the present Eocene tertiary fossil is marsupial is the most probable one, but the evidence is insufficient to demonstrate that fact, much less the family or genus.

A record of the slightest indication of a marsupial animal, and especially of an Opossum, or true *Didelphys*, in a tertiary deposit of the Eocene period in this country, becomes valuable, if only as an incentive and aid to further researches and discoveries, which might place beyond doubt so interesting an additional concordance between the Mammalian fossils of that epoch in England and in Continental Europe.

The circumstances attending the discovery by Cuvier of the fossil remains of a small species of *Didelphys* in the gypsum of Paris, furnish so striking an exemplification of the power of the principle which guided that great anatomist in the interpretation of fossil bones and in the reconstruction of extinct animals, that a brief notice of them may not be unacceptable, as they are not entirely foreign to the present section of the History of British Fossil Mammalia.

The remains in question included a considerable proportion of the skeleton of a small quadruped, partially buried in two portions of a split block of gypsum.

The impression of the lower jaw indicated, by the elevation of the coronoid process above the condyle and by the backward prolongation of the angle of the jaw, that it belonged to a carnassial or ferine animal; but the elevation of the condyle above the level of the line of teeth excluded it from the true Carnivora, as Dogs, Cats, Bears, Weasels. &c., and brought it within the range of the

Chiroptera, and the small placental or marsupial Insectivora. In this category, the breadth of the coronoid process—a character equally developed in our small oolitic Mammalia—and the inward bending of the angle of the jaw, left no doubt in the mind of Cuvier of the affinity of the fossil to the small marsupial Insectivora, among which it offered the closest resemblance in the shape of the teeth to the Opossums.

But one of the strictest instances of the generalizations which Buffon had enunciated respecting the geographical distribution of animals, was the limitation of the true Opossums, (*Didelphys*, Cuv.,) to the American Continents, and the triumphs of the comparative anatomist, by the fulfilment of predictions founded on fragmentary beginnings, had not at that time occurred so frequently, as not to render it desirable to dispel any lurking scepticism in the minds of the scientific contemporaries of Cuvier by the demonstration of the marsupial bones themselves. These, in the genus *Didelphys*, are two slender, moderately long, flat bones, extending forwards from the fore part of the pelvis; and this part of the skeleton of the little animal in question was buried in the block of gypsum.

Cuvier had successively appreciated and demonstrated the characters which enabled him to pronounce as to the class, the subclass, the order, and, as he believed, the particular family to which the small Eocene quadruped had belonged; but the best proof of the accuracy of his determination was hidden in the stone. He thereupon called together a few friends, capable of appreciating the trial: he laid before them the recent skeleton of a small Opossum, and, predicting the result of his operations, commenced the removal of the matrix of the Montmartre fossil with due precaution, by means of a fine graving tool, and soon brought into view the anterior brim of the pelvis, with the two supernumerary or marsupial bones (*fig.* 23, *a*, *a*,) in their natural position, and having the precise form and proportions of those of the Opossums.*

He thus demonstrated that there had been entombed in the gypsum of Paris an animal whose genus at the present day is exclusively proper to America.

Well might Cuvier exclaim, when the delicate but clear outlines of the parts he sought became manifest, "Que ces linéamens sont précieux !" and as richly did he merit, as he must have enjoyed, the satisfaction of thus stamping, in the presence of his colleagues, his principle of palæontological research with the impress of its prophetic power.

* The reader is referred to the original memoir in the "Annales du Muséum," t. v. p. 277, for the more minute comparisons, establishing the generic relations and specific differences of the extinct *Didelphys* of the Parisian Eocene rocks.

Fig. 23.

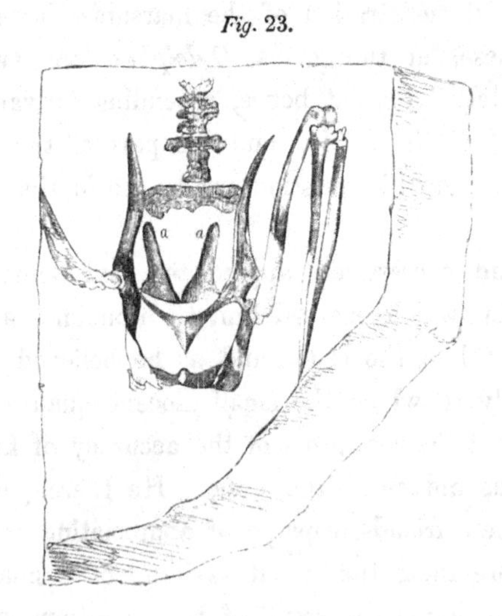

DIDELPHYS GYPSORUM.

Fig. 24.

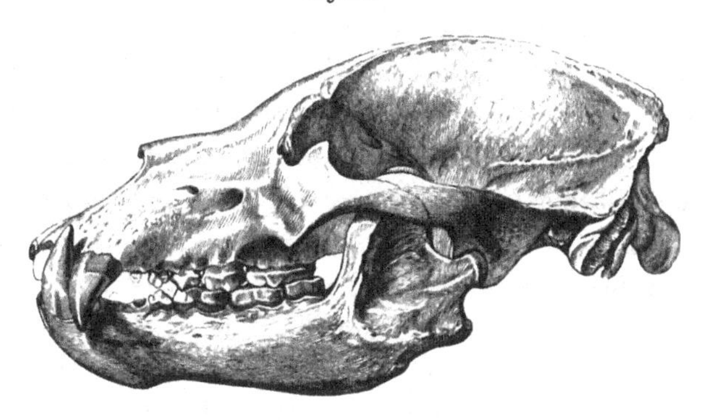

URSUS ARCTOS, ¼ Nat. size, Fens, Cambridgeshire.

Genus, URSUS.

Ursus Arctos.	OWEN, Report on Brit. Association, 1842.
„	MORRIS, Brit. Fossils, 8vo, 1843, p. 214.

As the order *Carnivora* includes the most noxious and dangerous quadrupeds, and those which most oppose themselves to the profitable domestication of the useful herbivorous species, it has suffered the greatest diminution through the hostility of man, wherever the arts of civilization, and especially those of agriculture, have made progress.

At the present day the three families of the Carnivorous Order, which include the largest and most formidable of the beasts of prey, have been so reduced, that they are severally represented in Great Britain in a wild state, by a single species of diminutive size. Of the *Canidæ*, or Dog-tribe, the Fox alone retains its primitive freedom and predaceous habits of life: the Wild-cat still lingers in remote mountain thickets, as the type of the *Felidæ*; and the harmless Badger is the sole representative, in our present indigenous Zoology, of the *Ursidæ*, or Bear-tribe.

History, however, points to a period when this island was infested by Bears and Wolves; but the superficial drift deposits, turbaries, limestone caverns, and the more recent tertiary strata have yielded evidence, not only of the remains of the common Bear and Wolf, but of other more strange and formidable beasts of prey, which appear to have perished anterior to the records of the Human race.

The Brown Bear (*Ursus Arctos*) infested the mountainous parts of Scotland, according to Pennant, so late as the year 1057, and the most recent formations in England contain remains which can scarcely be regarded as fossil, and which, if not perfectly identical with, indicate only a variety of the same species which is still common in many parts of the European Continent. Of these remains, the most perfect is the entire skull, figured at *fig.* 24, of a bear, discovered in Manea Fen, Cambridgeshire, five feet below the surface: it is preserved in the Woodwardian Museum, at Cambridge, and forms one of the very numerous and valuable additions to that collection made by Professor Sedgwick, to whom I am indebted for the opportunity of describing and figuring the specimen. I have, likewise, to acknowledge the liberal transmission by Sir P. de M. Grey Egerton, of a considerable part of the upper jaw and an entire under jaw of the same species of bear from the same locality, which have aided me in the comparisons instituted between these remains and the known existing and extinct species of *Ursus*.

In size the Bear of the Fen was very little inferior to the great extinct Cave-bear (*Ursus spelæus*), but it may be readily supposed that the Brown Bear and Black Bear of Europe have degenerated from the stature to which their progenitors, enjoying a wider range and more varied and nobler prey, attained.

On a closer comparison, especially of the dental system, differences appear which are not explicable on the known influence of external circumstances operating during a lengthened period of time.

The upper jaw of the Fen Bear differs from a similar sized one of the great Cave Bear in the much shorter interspace between the canine tooth and the third molar tooth counting from behind forwards; it differs likewise in having this interspace occupied by two small and simple-fanged premolars, completed in outline in *fig.* 23. The crown of the penultimate grinder is broader in proportion to its length or antero-posterior diameter. The difference in regard to the presence of the two first false molars must be allowed due weight, since the present Fen Bear has its grinders much worn, whilst the Cave Bear, with which it is compared, is a younger but full-grown specimen, with the tubercles of the grinding teeth entire, and the last molar tooth of the Fen Bear has a narrower posterior termination than in the Cave Bear. The Fen Bear differs also from the *Ursus priscus*, a smaller extinct species of Cave-haunting Bear, which retains the two first false molars, by their being in contact, which results from the narrower interspace between the canine and the third false molar, which interspace is relatively as wide in the *Ursus priscus* as in the *Ursus spelæus*, and a great proportion of this interspace divides the first from the second false molar in the *Ursus priscus*. This likewise cannot be a difference dependent on age or sex, for the jaw of the Fen Bear here described belonged to an individual absolutely larger than the *Ursus priscus*, with which it was compared; and, judging from the size of the canine teeth, the present specimen of the Fen Bear was probably an old male. The grinding surface of the molars prove it to have been also an older individual than the *Ursus priscus* with which it is compared, and to have attained that age when no difference could be expected to take place in the length of the interspaces of any of the teeth. In all the characters in which the upper jaw of the Fen Bear differs from that of the two species of Cave Bear with which it has been compared, it agrees with the *Ursus Arctos.*

In regard to the two varieties of existing European Bear, brown and black, held by some to be distinct species, the entire skull in the Woodwardian Museum shows that the most recent of the extinct British Bears, in its less convex forehead, and the greater length of the sagittal crest, resembled the Black Bear of Norway and Siberia, more than it did the Brown Bear of the Alps and Pyrenees.

As it may aid in the subsequent attempt to elucidate the true specific characters of the extinct Cave Bears (*Ursi spelæus* et *priscus*), as well as those of the existing *Ursus Arctos*, I shall add a few observations arising out of the comparison of the lower jaw of the Bear from the Manea Fen. The specimen, which is in the collection of Sir Philip Egerton, is the left ramus of the lower jaw.

It equals in length the largest specimen of the lower jaw of the *Ursus spelæus*, but differs from that species in the more simple form of the last premolar, or the fourth grinder, counting from behind forwards; for, whereas the Cave Bear has two distinct tubercles and a ridge developed from the base of the principal cone of that tooth, in the present species there is only the principal cone, as in the Black, Brown, and White Bears. The Bear of the Fen also differs from the *Ursus spelæus* in the shorter interspace between the last described molar and the canine, even when its lower jaw is compared with the lower jaw of a Cave Bear of less dimensions. The preceding interspace in the Fen Bear contains the sockets of two small spurious molars, each with a simple fang (given in outline in *fig.* 22), but there is no trace of these in the Cave Bear, save in very rare exceptions; and this difference cannot be the effect of age, because the lower jaw of the Fen Bear, which has the grinders moderately worn by mastication, is here compared with the jaw of a young and small *Ursus spelæus*, in which the tubercles of the grinding teeth are all entire.

The Fen Bear resembles the *Ursus priscus* in so far as the latter retains the first false molar, but differs in possessing the second, which is wanting in a younger specimen of the *Ursus priscus*; it differs also in the greater extent of the interspace between the canine and the third false molar; and, more importantly, in the form of that tooth, which in the *Ursus priscus* presents a second cusp on the inner side, and a little behind the first, which cusp is wholly wanting in the corresponding tooth of the Fen Bear. The ramus of the jaw is deeper, and the slope of the symphysis is more gradual.

These characters are illustrated in the comparative views of the dentition of the lower jaw of *Ursus Arctos*, *U. priscus*, and *U. spelæus*, *fig.* 35.

In all the particulars in which the Fen Bear differs from the two extinct species above cited, from the caverns, it agrees with the existing *Ursus Arctos*, and especially with the darker variety of Europe, from which it does not appear to differ in any well-marked specific character. The Grisly Bear of North America agrees with the Cave Bear (*Ursus spelæus*), and differs from the *Ursus Arctos* and the present British fossil representative of that species in the absence of the first two false molars and in the more complicated crown of the third false molar.

CARNIVORA. *URSIDÆ.*

Fig. 25.

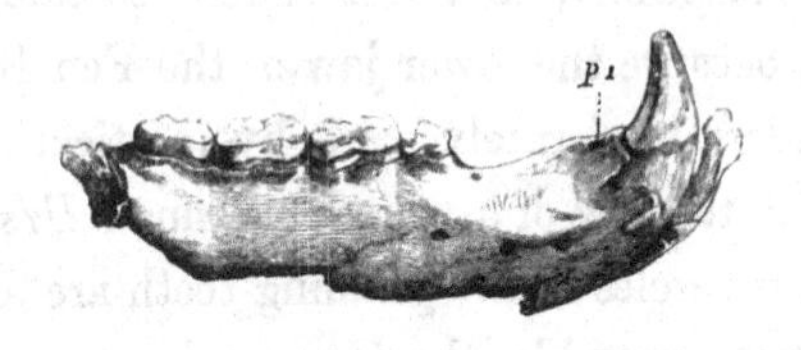

FOSSIL URSUS PRISCUS, ½ Nat. size. Kent's Hole.

Fig. 26.

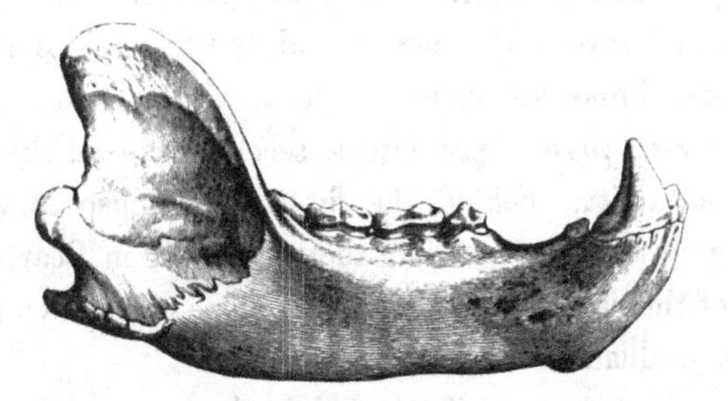

Recent *Brown Bear.—Alps.* ½ Nat. size.

URSUS PRISCUS.

Ursus priscus, GOLDFUSS, Nova Acta Academiæ Nat. Cur. t. x. pt. 2, p. 259.
Ours intermédiaire, CUVIER, Ossem. Fossiles, 4to., ed. 1823, tom. iv. p. 356.

THIS extinct species was established upon characters deduced from a nearly entire cranium, discovered in the celebrated bone-cave at Gailenreuth, and which is now in the British Museum. Though this specimen has evidently, by the abraded summits of the grinding teeth, belonged to an aged individual, this must have been less than the Bear from the Manea Fen or the great European Black Bear. The contour of the skull of the *Ursus priscus* is less elevated than in the Brown, or Alpine variety, and the flattened forehead passes into the nose with a less sensible concavity than in the skull of the Fen Bear (*fig.* 24). The coronoid process of the lower jaw is rather broader and higher, and the interval between the antepenultimate molar and the canine tooth is longer.

By the latter character, a very interesting fossil of a Bear, from the cavern called "Kent's Hole," near Torquay Devon, is referable to the *Ursus priscus*, heretofore only known from the German cave-depositaries of Ursine remains. The British fossil consists of a large proportion of a lower jaw, with the incisors, canines, and the entire series of molar teeth on both sides. The most perfect ramus is figured from the outside at Cut 25, and beneath it the entire right ramus of the lower jaw of the existing European species, for the illustration of the last cited character of the greater relative length of the interspace between the antepenultimate molar and the canine in the *U. priscus*. The persistent premolar in front of the antepenultimate molar is in place, and the socket of the first small single premolar is distinctly preserved in the fossil, thus manifesting a well marked character by which the *Ursus priscus* resembles the *Ursus Arctos*, and differs from the *Ursus spelæus;* in which, at least, that molar is most commonly wanting, and its socket obliterated. The trace of a socket of a second small single-fanged premolar is visible in the jaw from Kent's Hole near the large premolar, with which the series of grinding teeth commences, and, in the Gailenreuth specimen, the corresponding small premolar is retained in the upper jaw.

The absorbent condition of the fossil jaw from Kent's Hole hardly permits a doubt that it is of the same antiquity as the remains of the gigantic *Ursus spelæus*, found in the same cavern; Cuvier makes a like observation with respect to the *Ursus priscus* of the Gailenreuth cavern.

Joseph Whidbey, Esq., civil engineer, has recorded in the "Philosophical Transactions" his discovery in the limestone quarries at Oreston, near Plymouth, of a cavern containing fossil remains, described by Sir Everard Home as teeth and bones belonging to the Rhinoceros, Deer, and a species of Bear. This discovery is likewise noticed by Dr. Buckland in the "Reliquiæ Diluvianæ," p. 67. Mr. Whidbey says, "These bones were lately found in a cavern one foot high, eighteen feet wide, and twenty feet long, lying on a thin bed of dry clay at the bottom; the cavern was entirely surrounded by compact limestone rock, about eight feet above high-water mark, fifty-five feet below the surface of the rock, one hundred and seventy-four yards from the original face of the quarries, and about one hundred and twenty yards, in that direction, from the spot where the former bones" (those of a Rhinoceros) "were found in 1816." *

The remains of the Bear consist of

1. Left internal incisor.
2. Left upper canine, much worn by use.
3. Left lower canine.
4. Right lower canine.
5. Penultimate molar of the right side of the upper jaw.
6. Penultimate molar of the left side of the lower jaw.

The first three specimens correspond in size and form with the teeth of the great *Ursus spelæus*. The canine No. 4, though completely formed and showing marks of service, is smaller, and agrees in size with that of the *Ursus priscus*. This might possibly be a sexual difference. But

* Philos. Trans. 1821, p. 133.

the penultimate molars, which, by their much abraded tuberculate surface, indicate an aged individual, are not only smaller than the corresponding teeth in the *Ursus spelæus*, but have a shorter and broader crown and smaller fangs, agreeing in these characters with the *Ursus priscus*. Thus the caverns at Oreston, like those at Torquay and Gailenreuth, testify to the coexistence of two species of Bear, both apparently exterminated anterior to the historical period.

The next section will be devoted to the account of the largest of these species, and to the evidences of its former existence in England.

The geological relations of the freshwater deposit of eastern Norfolk, in which the jaw of the *Ursus spelæus* first to be noticed was found, is illustrated in the subjoined vignette, for which I am indebted to the kindness of Charles Lyell, Esq.

Fig. 27.

Runton Gap.

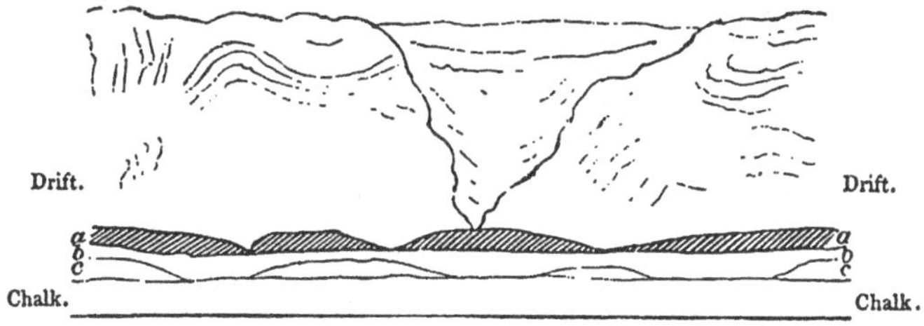

Drift. Drift.

a b c a b c

Chalk. Chalk.

a. Black earth with shells. } Freshwater.
b. Reddish sand. }
c. Norwich crag in patches.

CARNIVORA. *URSIDÆ.*

Fig. 28.

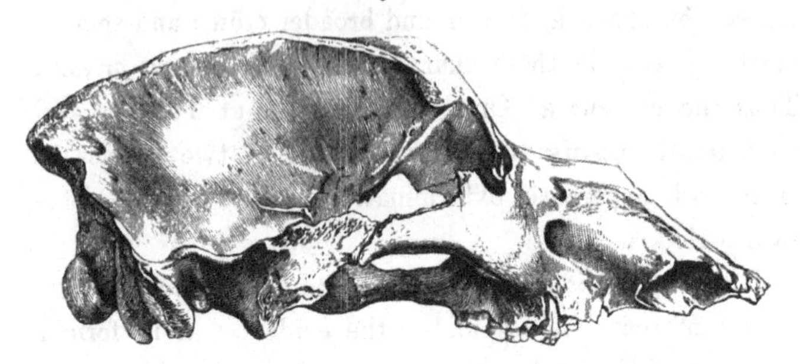

Fossil, $\frac{1}{8}$ Nat. size, Gailenreuth.

URSUS SPELÆUS. Great Cave Bear.

Ursus spelæus,	BLUMENBACH, CUVIER, Bulletin des Sciences, par la Soc. Philomath, No. 50, 1796. Annales du Muséum, tom. vii. Ossem. Fossiles, 4to. 1823, tom. iv. p. 345.
Ursus fornicatus magnus,	SCHMERLING, Recherches sur les Ossem. Fossiles découverts dans les Cavernes de la Province de Liège, 4to. 1833, p. 105.
Ursus arctoideus,	BLUMENBACH, CUVIER, loc. cit.
Ursus fornicatus minutus,	SCHMERLING, loc. cit.
Ursus planus,	OKEN.

Fossil Bear different from the White Bear, HUNTER, Phil. Trans. vol. lxxxiv. 1794.

JOHN HUNTER, who first instituted an anatomical comparison between the remains of extinct Bears and the bones of those of the present period, selected the White, or Polar Bear, for this purpose, as being the largest existing species with which he was acquainted, as well as that to which the fossils of gigantic Bears from the German caverns had been referred by Esper and other preceding writers. In regard to the cranium, Hunter* alludes, with philosophic caution, to the modifications of shape which are due to age in carnivorous animals, and he restricts himself to pointing

* Loc. cit. p. 419.

out the difference in the proportion of length to breadth in the skull of an old White Bear, and in that of the great Cave Bear; the individual skulls which he compared are still preserved in juxtaposition in the Museum of the College of Surgeons, as they were left by Hunter, when removed by death from the last and richest field of his extensive and various researches.

This difference in the proportions of the skull, though one of the most striking between the fossil and recent species of Bears, is not the only one. The last molar tooth of the upper jaw in the White Bear (*Ursus maritimus*) has a smaller antero-posterior diameter, and a narrower posterior termination. The interspace between the antepenultimate molar and the canine tooth presents the remains of two sockets, one near the molar, the other near the canine, which in young, but full-grown Polar Bears contain small and single-fanged premolars. The youngest specimens of Cave Bear which I have seen, exhibit no trace of either of these small premolars, or of their sockets; they doubtless existed in the fœtus, but normally were very soon lost; the exceptions are extremely few in which their traces are visible in the jaws of full-grown Cave Bears. The posterior palatal foramina are situated opposite the middle of the last molar tooth in the skull of the White Bear, but opposite the interspace between the penultimate and last molars in the skull of the Cave Bear. The zygomatic arches are wider and shorter, and the base of the zygomatic process behind the glenoid cavity is more horizontal in the White Bear than in the Cave Bear. The Grisly Bear (*Ursus ferox*), —a larger species than the White Bear, and unknown to Hunter,—agrees with the Cave Bear in the great proportional size of the last molar tooth, but the interspace between the antepenultimate grinder and the canine is relatively less than in the Cave Bear (*U. spelæus*), and it contains two small and simple premolars in specimens, which, from the worn state of the molar teeth, have belonged to older individuals than those Cave Bears whose skulls show no trace of premolars.

The superiority of size, and some other characters which distinguish the great *Ursus spelæus*, have been pointed out in the works of Rösenmuller, of Soemmerring, of Goldfuss, and of Cuvier: the most striking distinction is the convex elevation of the forehead, and the sudden sinking of the concave line, which leads forwards to the nasal bones. This character is well shown in the fine fossil cranium, from the cavern of Gailenreuth (*fig.* 28); which is introduced at the head of the present section in the absence of the opportunity of representing the same character in a British specimen of the skull of the *Ursus spelæus*.

The evidence of the former existence of this extinct species in England is derived from the lower jaw and other bones of the skeleton, especially the humerus and femur, and from teeth, either detached, or *in situ* in the lower jaw.

M. de Blainville, however, the latest author on the relations of recent and fossil Bears, concludes a detailed summary of the characters indicated by his predecessors in proof of the specific distinction of the *Ursus spelæus*, by the statement, that it differs in no respect in its osteology or dentition from the characters which he has found in the *Ursus Arctos*, and especially in the *Ursus ferox*.*

* "D'après ces différentes considérations, nous regardons comme presque hors de doute que les cranes de l'Ours fossile attribuês à *l'U. spelæus* proviennent d'individus adultes du sexe mâle les plus vigoureux, et ne constituent nullement une espèce. Et en effet, toutes les parties characteristiques que nous avons exposêes dans notre Ostéographie et dans notre Odontographie, ne présentent rien de dif-

If, however, the differences which have been pointed out in the upper jaw and teeth of the *Ursus spelæus*, as compared with the *Ursus maritimus* and *U. ferox*, should be deemed explicable on the influences of age, sex, and climate, no known extent of the operation of these causes can account for the differences which are observable in the dentition of the lower jaw, and in other characters derivable from the skeleton of the *U. spelæus*.

The lower jaw of the *Ursus spelæus* differs from that of the *Ursus maritimus* in the greater convexity of the inferior contour of the ramus of the jaw, in which latter circumstance it differs, though in a somewhat less degree, from the *Ursus ferox*, and from the Black Bear of Europe (*Ursus Arctos*).

The posterior molar tooth, in the lower jaw, is always broader in proportion to its antero-posterior diameter in the *Ursus spelæus* than in the *Ursus ferox*, and still more so than in the *Ursus Arctos*. The space between the canine and the series of the last four molar teeth is usually longer, and almost always edentulous and without any trace of the sockets of the small deciduous premolars: the first of the four persistent grinding teeth has a more complex crown than in the *Ursus priscus*, or in any existing species of Bear: besides the principal cusp there are two small tubercles on its inner side, and a ridge extending along the outer and back part of the base of the crown.

An entire right half or ramus of the lower jaw of a Bear, from the lacustrine formation near Bacton on the Norfolk coast,* presents all these distinctive characters of the *Ursus spelæus*; as, for example, the long and edentulous interval

férent de ce que nous trouvons dans l'*U. Arctos*, et surtout dans l'*U. Arctos ferox* de l'ouest de l'Amérique septentrionale."—*Ostéographie des Ours*, 4to. 1840, p. 57.

* This formation is shown at *a*, *fig.* 27.

between the canine tooth and the first of the series of four molars; the complicated crown of the first and smallest of these persistent teeth, and the superior breadth of the fourth molar as compared with that in the common and Grisly Bears.* The size of the Bacton fossil is not equal to that of the jaw of the largest specimens of Cave Bear, but it exceeds some of the jaws which have apparently belonged to young females of the *Ursus spelæus:* it measures ten inches three lines in length, and the length of the series of molars is three inches and a half. In the lower jaw of an *Ursus spelæus* from the Gailenreuth cavern, now in the British Museum, measuring eight inches nine lines in length, the series of four molars is three inches ten lines in length; in another jaw of the *Ursus spelæus* from the same locality measuring twelve inches in length, the series of molar teeth is also three inches ten lines in length; and these important and least varying instruments of digestion precisely correspond in number, size, and structure, with those in the shorter jaw.

In the *Ursus priscus*, and the largest specimens of Europæan, Polar, or Grisly Bears, the specific differences in the forms and proportions of the molar series of teeth are readily recognisable, although the total length of the jaw may exceed that of the jaws of the young, and probably female Spelæan Bears, which have acquired their adult dentition.

An idea of the formidable size which the old males of the *Ursus spelæus* attained in this country, may be estimated by the upper canine tooth, from the cave at Kirkdale, figured by Dr. Buckland,† and by the one here figured (*fig.* 29) from Kent's Hole, Torquay. It matches the canine teeth of the largest of the continental specimens of the *Ursus spelæus*, the size of which extinct Bear Cuvier says must have equalled that of a large Horse.

* *Fig.* 35, *b.* † *Reliquiæ Diluvianæ*, Pl. 6, fig. 1.

In the same bone-cave, near Torquay, has been found the anterior part of the lower jaw, with the canines of corresponding magnitude, of the *Ursus spelæus*, in which the small simple-fanged premolar close behind the canines has been retained on each side; and its crown has been flattened by attrition. A few exceptional instances of this retention of the teeth, which are commonly deciduous at an early period in the great Cave Bear, have been observed in lower jaws from the German and Belgian caverns.

Fig. 29.

Upper canine, fossil.
URSUS SPELÆUS.

The fossil *humerus*, or arm-bone (*fig.* 30), of a large bear from Kent's Hole, manifests all the characters of that bone in the *Ursus spelæus*, which appear to me to be as well marked as those distinguishing the humeri in any other two species of one genus.

Cuvier conceived that he had met with two very distinct forms of fossil humerus, belonging to equally gigantic extinct species of Cave Bears. He says,—

"We find two kinds of humeri, both belonging to Bears, and yet very different from each other; John Hunter has already represented them (Phil. Trans. 1794, pl. xx.); but since that time no author has insisted upon their difference.

Fig. 30.

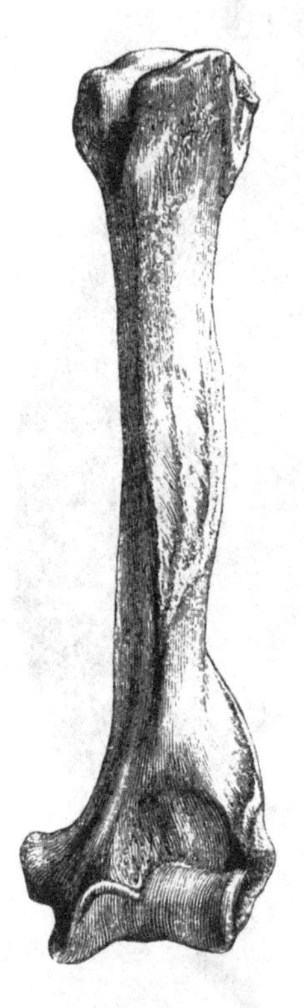

URSUS SPELÆUS, FOSSIL.

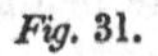

Fig. 31.

URSUS MARITIMUS, RECENT.

The second kind of humerus from these caves is known to me by a very perfect specimen in our Museum, by the figure in Hunter, and by a drawing of a part, including the lower three fourths of the bone, for which I am indebted to the late Adrien Camper. It differs remarkably from the preceding, by a hole pierced above the internal condyle for the passage of the ulnar artery."*

Whatever may be deemed the value of the character of the perforation of the inner condyle, I can affirm that it derives no accession from the other differences manifested by the figure in Hunter's memoir, which Cuvier supposed to be of a fossil Bear; that figure having been, in fact, taken from the imperforate humerus of an old Polar Bear, inserted in the plate (pl. xx. Phil. Trans. 1794), and placed above the figure of the true fossil humerus in order to illustrate the differences between the recent and fossil species. The bone of the Polar Bear was placed by Hunter in the same drawer with two humeri of the Cave Bear (*Ursus spelæus*), from Gailenreuth, which it exceeds in size, and which are the identical specimens alluded to in the following passage of Hunter's Memoir:—"These are two ossa humeri rather of less size than those of the recent White Bear." Hunter does not allude to any other differences, probably intending these to be illustrated by the figures. These figures, in fact, show that the humerus of the White Bear (*Ursus maritimus*, *fig.* 31) is broader at both extremities, and thicker in proportion to its length. The supinator ridge forms an angle instead of being continued downwards in a gentle convex curve; the internal condyle is much thicker and stronger, where it bounds the olecranal cavity, and it extends inwards

* "On trouve deux sortes d'humérus, tous deux appartenant à des Ours, et cependant fort différens l'un de l'autre, John Hunter les a déjà représentés (Phil. Trans. 1794, pl. xx.); mais depuis on n'a insisté dans aucun ouvrage sur leur différénce. La deuxième sorte d'humérus de ces cavernes, pl. xxv. fig. 4, 5, 6, et 7, m'est connue par un échantillon bien entier que notre Muséum possède, par la gravure de Hunter, et par le dessin que je dois à feu Adrien Camper d'une portion qui en comprenoit les trois quarts inférieurs. Elle diffère éminemment de la précédente par un trou percé au dessus du condyle interne pour le passage de l'artère cubitale. (*Voy. a*, fig. 4 et 5.)"—*Ossemens Fossiles*, 4to. 1823, tom. iv. p. 362.

to a greater distance from the articular surface; the del-toidal ridge reaches lower down in the White Bear; the antero-posterior diameter of the proximal third part of the bone of the White Bear exceeds in a marked degree that of the extinct species.

The decease of Hunter took place before the printing of his observations on the fossil cave-bones, and the individual to whom the task of superintending the printing was en-trusted, described both the figures of the humeri in the Plate, as belonging to the fossil species. Cuvier, who did not perceive the resemblance of the upper figure to the humerus of the White Bear, and who, therefore, did not recognise the mistake, avails himself of it to illustrate his opinions respecting the specific distinction of his *Ursus spelæus* and *U. arctoideus*.

Cuvier, in fact, possessed a fossil humerus of one of the great Cave Bears, the internal condyle of which was perforated as in the feline tribe, whilst other humeri were imperforate, and corresponded with the lower figure in Hunter's plate. But the perforated fossil humerus figured by Cuvier differs from that of the White Bear in the shorter deltoid ridge, the narrower proximal and distal extremities, the convex outline of the supinator ridge, and the inferior production of the inner condyle; in short, in all those characters by which the imperforate fossil humerus has been shown above to differ from that of the White Bear. Not any of the three fossil humeri in the Hunterian Collection have the perfora-tion of the internal condyle; and amongst the extremely numerous humeri of large Bears that have since been obtained from the bone-caves of Germany, not any have been found to present the perforation which Cuvier regards as the specific character of this bone in the *Ursus spelæus*;

it is most probably, therefore, as Dr. Schmerling* and Professor de Blainville† conjecture, an accidental anomaly. But the differential characters which both the imperforate and perforate humeri of the great Cave Bear present, when compared with those of any recent species, cannot be reconciled by the hypothesis, that these are merely degenerated descendants of the *Ursus spelæus.*

The nearly entire humerus of the bear from the Cave of Paviland‡ presents all the characters of that of the *Ursus spelæus* above described.

The ulna of the Cave Bear (*Ursus spelæus*), compared with one of the same length from the Polar Bear, is less straight, being more convex towards the radius; is thicker, particularly at the anterior part of the shaft; the ridge on the outside of the distal end of the bone is more produced; the styloid process is more pointed; and the concavity on the inner side of the proximal articular surface is deeper.

The ulna of the Bear from the freshwater deposit near Bacton (*fig.* 27, *a*), as well as a larger ulna from Kent's Hole, agree with that of the *Ursus spelæus* from the German caves.

The upper extremity of the radius of the Cave Bear, from a bone-cave in the Mendips, and the gnawed shaft and lower end of a radius from Kent's Hole, match the largest specimens from the German caverns in size, and equally demonstrate the oval form of the upper articular surface which rotates on the humerus and ulna, and the larger oblique oval surface at the distal end, which distinguish the radius of the great extinct Bear from the corresponding bone in the great feline animals.

The scapho-lunar bone, the os magnum with its charac-

* Loc. cit. p. 130. † Loc. cit. p. 71.

‡ Buckland, Reliquiæ Diluvianæ, p. 82.

teristic shallow surface* for the proximal tuberosity of the metacarpal bone of the index, and some of the metacarpal and phalangeal bones of the *Ursus spelæus* have been obtained from British bone-caves, as Kent's Hole, that at Paviland, and those in the Mendips.

Fig. 32. *Fig.* 33.

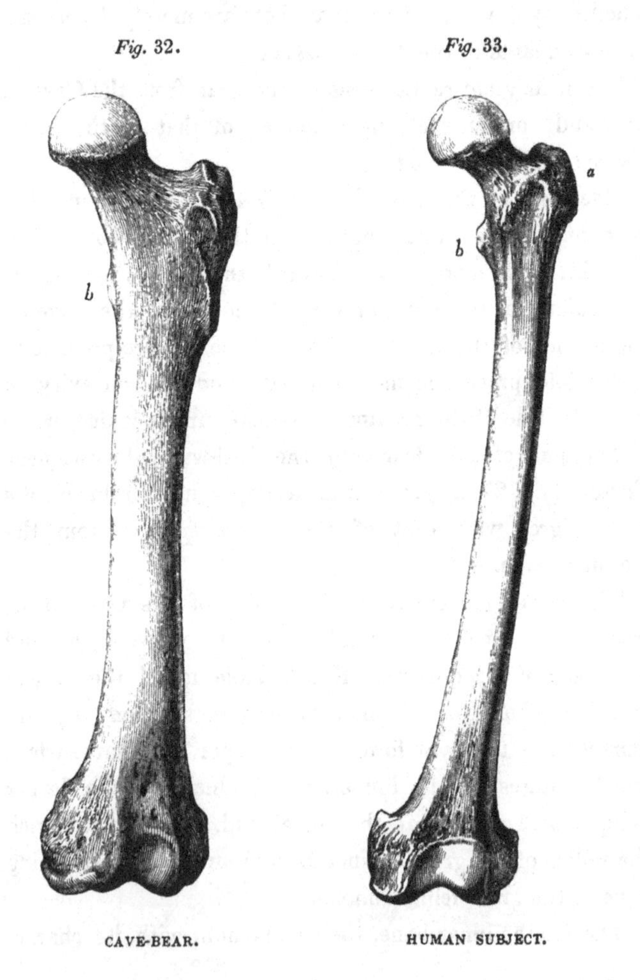

CAVE-BEAR. HUMAN SUBJECT.

* The corresponding surface is a deep depression in the os magnum of the *Felis spelæa*.

Of no other quadruped than the Bear is the femur more likely to be mistaken by the unpractised Anatomist for that of the human subject, especially the femur of the gigantic extinct species commonly found in caves: figures of the human thigh-bone (*fig.* 33) and of that of the *Ursus spelæus* (*fig.* 32), reduced to the same proportions, are, therefore, subjoined.

The bear's femur differs chiefly in its greater thickness compared with its length; in being straighter; in the much greater vertical extent of the large trochanter (*a*), and the less projection of the small trochanter (*b*), in the less oblique inflection of the neck of the bone, in the minor expansion of the distal condyles, and in the smaller size of the articular surface for the patella or knee-pan.

The difference between the femur of the *Ursus spelæus* and the femur of the *Ursus Arctos* and *Ursus ferox*, is analogous to that which has been pointed out in the humeri; the femur of the Grisly Bear being broader in proportion to its length, especially at its two extremities: it is owing to this breadth that the lesser trochanter is thrown wholly to the posterior surface of the bone, the inner margin being continued beyond it, whilst in the Cave Bear the lesser trochanter, though on the posterior surface of the bone, projects a little beyond the inner margin. At the distal end of the bone the tuberosity above the internal condyle, corresponding with that in the humerus, is larger and more prominent in the Grisly than in the Cave Bear: the same difference in the position of the lesser trochanter is presented by the White Bear as compared with the Cave Bear, and the extremities of the bone are relatively broader in the White Bear.

I have determined portions of the fossil leg-bones (tibiæ and fibulæ), entire ankle or tarsal bones, and bones of the hind-foot (metatarsals and phalanges), by their size and general ursine characters, to belong to the *Ursus spelæus*, from different Cave localities in England: but none of these bones have presented any well-marked modification of form by which they might be distinguished, in addition to their size, from the corresponding bones in the smaller extinct and existing species of Bear. But the coincidence of such appreciable modifications in the femur, ulna, and humerus, of the great Cave Bear, with those in the form and proportions of the head, and in the form and the relative size of certain teeth, offer as good grounds for the specific distinction of the *Ursus spelæus* as for that of the *Ursus maritimus*, or of any other existing species defined by Pallas and Cuvier, and admitted by the best modern zoologists.

The question which the Palæontologist ought to propose to himself in his first survey of the fossils of any particular district, is the value of the differential characters which such remains may present, as compared with those which distinguish the living species, according to the zoological systems and principles of the time being. It is true that the extent of the influence of external causes, operating through a vast series of ages, has not yet been determined; but this only renders it the more imperative to take cognizance of all modifications in fossils which, according to present knowledge, cannot be so explained.

To refuse to recognise such differences as have been pointed out in the skeleton of the great Cave Bear, because they may be accounted for by a hypothetical degeneration of the specific type, and thereupon to record the fossil species as the primæval state of the present *Ursus Arctos*, seems a voluntary abandonment of the most valuable instrument in ulterior endeavours to solve the higher and more general problems in zoology.

Observation has well determined the extent of modification which the skull of a carnivorous species may undergo according to age, to sex, to the free or the constrained exercise of its destructive weapons; and the relative size of the intermuscular crests, the relative strength of the zygomatic arches, and the proportions of the canines to the other teeth are well known to vary within certain limits.

But in the *Ursus spelæus* we have to account for the greater relative size and complexity of certain molar teeth; for the more extended diastemata, accompanying more lengthened jaws; for a premature loss of certain teeth and their sockets, without any predominating development of neighbouring canines to account for it; for narrower zygomata, with longer and higher parietal crests; for large frontal sinuses impressing a striking and readily recognisable feature upon the skull.

M. de Blainville has endeavoured to explain the last-cited modification, on the supposition that the primæval Bears had their frontal sinuses more developed in virtue of their respiring a fresher, drier, and more invigorating atmosphere than their less fortunate and degenerated descendants.* But we may question whether the flat-headed *Ursus ferox* has a less exposed locality or breathes a more humid and impure atmosphere on the rocky mountains in the far west of North America, than did the Cave Bears of the ancient German and British forests; and we may

* "L'intensité même de l'acte respiratoire dans les lieux plus découverts, où l'air est plus vif, plus sec, plus frais, développe tous les sinus que se trouvent sur le trajêt de l'air, et, des-lors, les frontaux sont dans ce cas aussi bien que tous ceux qui entourent les fosses nasales; des-lors aussi, par l'écartement des deux lames de l'os, le gonflement des fosses frontales, indépendantes et separées par un sillon."—De Blainville, *Ostéogr.*, *des Ours*, p. 36.

more than doubt that the cold and bracing sea-breezes inhaled by the still flatter-headed Polar Bear, should be less efficient in expanding the sinuses along the respiratory tract, than the musty air of the sepulchral retreats in which the Cave Bears slept of old.

Existing Bears, regarded as distinct species by modern zoologists, do in fact differ in the relative convexity of their forehead, and the flat-headed species, as the Polar and American Bears, are unquestionably not those which habitually respire the least pure and invigorating air. Instead, therefore, of speculating on the atmosphere as a physical cause of the inflation of the bony cells, it would be more profitable, if it were possible, to trace the relationship between the different degrees of development which the frontal sinuses may present in different species of Bears, and their peculiar habits and modes of life. We may thus, I think, see the reason why, in the piscivorous species of the Polar ice, the receptacles of air in the bones of the head are least developed, viz., to offer least resistance to its progress through the water when diving after its prey.

The opposite extreme in the condition of the frontal sinuses of the *Ursus spelæus*, may have had some corresponding relation to the habits of that gigantic extinct species.

From the great proportional size and more complicated tubercular surface of the posterior molar teeth, especially in the upper jaw, and from the greater complication on the crown of the smallest persistent molar in the lower jaw, one might be led to suppose that the *Ursus spelæus* fed more on vegetables than the Grisly Bear does. In which case it might be inferred from the slight traces of abrasion in the teeth of full-grown specimens, that the vegetable food, in whatever proportion it entered into their diet, was of a soft nature, as berries, or tender twigs or sprouts. The size and strength of the *Ursus spelæus*, and the huge canines with which its jaws were armed, would, however, enable it to cope with the large Ruminants and ordinary Pachyderms, its contemporaries in ancient Britain and on the Continent, and to successfully defend itself against the large Lion or Tiger, whose remains have been found in the same caverns.

In regard to such depositaries of fossil remains in this country it has been proved, chiefly by the researches of Dr. Buckland, that England differs very remarkably from the rest of Europe in the small number of its ancient bears, as compared with the hyænas; the proportionate numbers of *Ursus spelæus* and *Hyæna spelæa* being reversed in the island and on the continent. How far this difference depends on the accident of a discovery of retreats of the Hyæna in this country, which remain to be found on the Continent, or whether it is to be regarded as an indication of some geographical separation having existed at the remote period of these beasts of prey, analogous to that which now divides England from the Continent, may be determined by ulterior researches.

Having already discussed the question of the specific characters and relations of the extinct Bears of this country, I shall conclude by briefly indicating the chief localities in which their fossil remains have been discovered.

The tusk of a Bear, equalling in size that of the *Ursus spelæus*, discovered by Dr. Buckland in the celebrated hyæna-cave at Kirkdale in Yorkshire has been already cited. A few teeth of a feline animal, indicating a magnitude equal to the largest Bengal Tiger, were also found. The paucity of such remains is rendered more striking by the contrast, of the incalculable numbers of hyænas' teeth which the same cavern has furnished.

With respect to the larger Carnivora, Dr. Buckland has well observed that, "it is more probable that the Hyænas found their dead carcases, and dragged them to the den, than that they were ever joint tenants of the same cavern." It is, however, obvious, he adds, that they were all contemporaneous inhabitants of ancient Yorkshire.*

In the bone-cavern lately explored on Durdham Down, near Bristol, Mr. Stutchbury determined, amongst the remains of the Carnivorous animals, one Bear and eleven or twelve individual Hyænas.

In the cave at Paviland, in the lofty limestone cliff facing the sea on the coast of Glamorganshire, the following parts of a large species of Bear are enumerated by Dr. Buckland:—Many molar teeth; two canines; the symphysial end of two lower jaws, exhibiting the sockets of the incisor teeth and of the canines, the latter are more than three inches deep; a humerus nearly entire; many vertebræ; two ossa calcis; metacarpal and metatarsal bones.

At Oreston, on the coast of Devonshire, several caverns or cavernous fissures were discovered during the quarrying of the limestone rock for the construction of the breakwater at Plymouth. The first of these, described in the Philosophical Transactions for 1817, contained the bones of a species of *Rhinoceros;* in the second, a smaller cavern distant one hundred and twenty yards from the former, and described in the Philosophical Transactions for 1821, were found, associated with the tooth of a Rhinoceros and parts of a deer, some teeth and bones of a species of *Ursus.*

The fossils referable to the Bear here discovered, include a canine tooth, left side, lower jaw; a canine tooth, left side, upper jaw; the penultimate grinder, right side, upper

* Reliquiæ Diluvianæ, p. 35.

jaw; the penultimate grinder, left side, lower jaw; a portion of the sacrum; portions of two tibiæ; a portion of the ulna; a portion of the femur.

Those specimens, which from their smaller size and modifications of form, are referable to the *Ursus priscus*, have been already described; the remainder agree in size with the large *Ursus spelæus*, and I have been gratified in confirming, by a close examination of these specimens, the accuracy of the opinion which Cuvier, on analogical grounds, entertained of their nature.*

Perhaps the richest cave-depositary of the fossil bones of Bears hitherto found in England is that called Kent's Hole, near Torquay. The natural history, with a special account of the organic riches of this cave, will be given in the second volume of the "Reliquiæ Diluvianæ," which Dr. Buckland is now preparing for the press. It is to the assiduous researches of the late Rev. Mr. Mac Enery, that the discovery of the various and interesting fossils of this cave is principally due, and some of the rarest and most valuable of this gentleman's collection have been lately acquired by the British Museum. Among the Ursine fossils meriting especial notice, are portions of the skull and teeth of the *Ursus spelæus*, some of the latter equalling in size the largest specimens from the German caverns.

The anterior portion of a lower jaw, including the anchylosed symphysis, with two enormous canines, is likewise remarkable from the circumstance of its retaining a

* "Sir Everard Home assure qu'il y avoit des os d'ours dans cette caverne d'Oreston près Plymouth, d'où l'on en a tant retiré d'éléphans et de rhinocéros. Il y a trouvé une pénultième molaire supérieure, une inférieure qu'il déclare de l'ours brun ou noir, et plusieurs autres os qu'il croit en venir probablement aussi; expressions d'après lesquelles il semble qu'il ne les juge pas de nos espèces des cavernes. Ils me paroissent toutefois devoir venir de ces espèces-ci, d'autant que M. Buckland m'apprend y avoir découvert récemment des os d'hyènes et de loups."—*Ossem. Fossiles*, 4to., 1823, t. iv. p. 348.

small and simple-fanged premolar in the interspace, or diastema, between the canines and the double-fanged molars. Similar, but rare instances, from Continental caves, of this variety in the *Ursus spelæus*, have been noticed above.

Amongst the bones of the trunk and extremities of the *Ursus spelæus* from Kent's Hole, there occur remarkable examples of diseased action; a lumbar vertebra, for example, presents extensive exostosis from the under part and sides of the body; the distal extremity of a radius (*fig.* 34) exhibits an oblique fracture of that bone, in the attempt to heal which a new and irregular ossific mass has been deposited on the surface of the bone. Several bones and teeth of the Bear from Kent's Hole exhibit very decided marks of having been gnawed, most probably by a hyæna. One of the fragments of the lower jaw of a young Bear (*fig.* 36) shows the same interesting transitional state of dentition which has been discovered in fossils from the Continental Bear-caves. The point of the permanent canine (*l*) has just protruded from its socket, and the crown of the last molar (*m*) is hollow and without a fang.

Fig. 34.

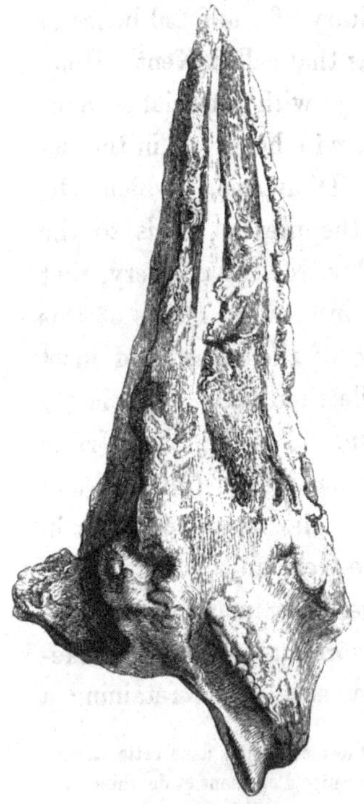

Fossil, ½ nat. size.

The unstratified drift and newest tertiary strata in several localities of England have yielded remains of large carnivorous quadrupeds, and among them those of the Bear.

In the valley of the Thames these deposits afford considerable quantities of brick-earth, and in working this material at Grays in Essex, and also at Whitstable, remains of a large species of *Ursus* have been discovered.

Mr. Brown of Stanway has obtained remains of a large species of Bear from the freshwater formations of Clacton, where they are associated with the Mammoth, Rhinoceros, and other large extinct quadrupeds. The lower jaw from the lacustrine beds near Bacton, in Norfolk, containing evidences of the Mammoth, Trogontherium, Palæospalax, and other extinct quadrupeds is referable, as has been already pointed out, to the *Ursus spelæus.*

In the newer pliocene fluviatile deposits traced by Mr. Strickland from Warwickshire into the valley of the Severn, near Tewkesbury, the remains of a Bear, which is regarded with great probability as one of the extinct species of *Ursus*, were discovered associated, as in the freshwater deposits in Essex, with remains of Hippopotamus, Rhinoceros, Mammoth, the great Aurochs, Wolf, and Deer.

The latest Ursine remains having any claim to be admitted into a record of British Fossils, are the entire skull and portions of the upper and lower jaws of the Bear from the Cambridgeshire Fen, and they belong to the existing European black variety of the *Ursus Arctos.*

The oldest fossil referable to the genus *Ursus* from British strata is the crown of a molar tooth, which was found associated with the teeth of a hog, and of a species of *Felis* as large as a Leopard, at Newbourn, near Woodbridge, Suffolk. Mr. Lyell, after examining the locality from which Mr. Colchester obtained these teeth, inclines to the belief that they came from the red crag. The Bear's tooth is the antepenultimate grinder of the right side, upper jaw; it is smaller than the corresponding tooth in the *Ursus spelæus*. Similarly coloured and triturated teeth of fishes have been procured in abundance from the same pit. According to this view, the fossil Bear in question belongs to the miocene strata.*

In conclusion, it may be stated on the evidence at present acquired, that the period of the existence of

Fig. 35.

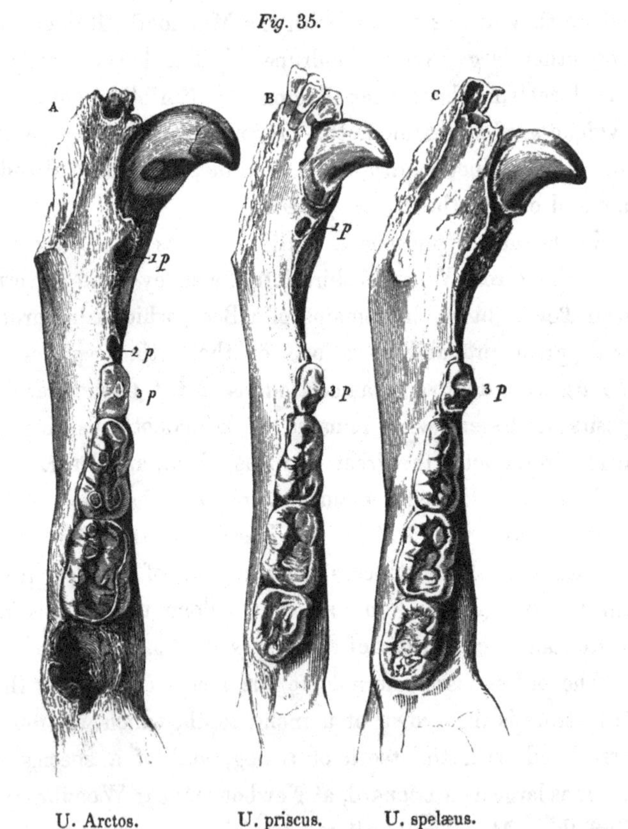

U. Arctos. U. priscus. U. spelæus.

the Ursine genus in this island has extended from the

* For the data respecting this view the reader is referred to Mr. Lyell's paper in the "Annals of Natural History." No. 23, pp. 187 and 188, 1839.

middle, or miocene tertiary formations, through the older and newer pliocene, and that the genus surviving, or under a new specific form reappearing after the epoch of the deposition and dispersion of those enormous, unstratified, superficial accumulations of marine and freshwater shingle and gravel, called drift and diluvium, has been continued during the formation of vast fens and turbaries upon the present surface of the island, and until the multiplication and advancement of the human race introduced a new cause of extermination, under the powerful influence of which the Bear was finally swept away from the indigenous Fauna of Great Britain.

The adjoining figures illustrate the characters derivable from the lower jaw and its dentition, of three of the species by which the genus has been represented in England, during the different periods above cited. C, *fig.* 35, is the jaw of the extinct *Ursus spelæus*, from the Norfolk pliocene; it shows the complex premolar (3 *p*) and the long toothless interval between it and the canine: B is the jaw of the *Ursus priscus* of the post-pliocene epoch, in which the interval is shorter and retains the first small premolar (1 *p*): A is the jaw of the *Ursus Arctos* from the Cambridge fen, in which the shorter interval retains two small premolars, and the third (3 *p*) has a more simple crown.

With the present experience of physiologists as to the range of variety of which a specific form is susceptible, through the long continued operation of external influences, we cannot attribute the anatomical differences which have been pointed out in the fossil teeth and bones of Bears derived from the above-cited series of formations, to varieties of one species produced by such accidental causes. On the contrary, those Bears which existed anterior to the present condition of the surface of the British Islands must be referred to two species distinct from any now known, and which have disappeared altogether from the face of the earth. Moreover, the two extinct species alluded to, called *Ursus spelæus* and *Ursus priscus*, have not come after each other, as they themselves have been succeeded by the *Ursus Arctos* in later times, but their fossil remains are found associated together in the caves of Britain, as in those of the Continent. This is a circumstance which of itself weighs against the hypothesis, that the present European Bears are the degenerate descendants of the huge Spelæan species.

The *Ursus priscus* scarcely differs less than the *Ursus Arctos* from the *Ursus spelæus*, yet it is as ancient a species as the more formidable one, and has equally suffered from causes of extinction which we are at present unable fully to understand.

On the other hand, we may, by the study of British fossils alone, avoid the error of the opposite extreme of multiplying nominal species, if, guided by the known laws that regulate the range of deviation from a true specific type, we make due allowance for diversities of age and sex in a carnivorous and combative quadruped like the Bear; and we thus distinguish from the *Ursus priscus*, or the *Ursus Arctos*, the fossil remains of young, though adult, individuals, and those of the females of the great *Ursus spelæus*, which have given rise to the nominal species, *Ursus arctoideus* and *Ursus planus*.

Fig. 36.

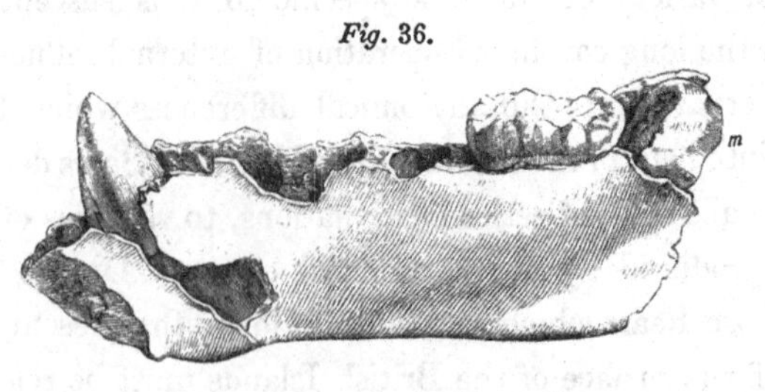

Young Ursus spelæus. Kent's Hole.

Fig. 37.

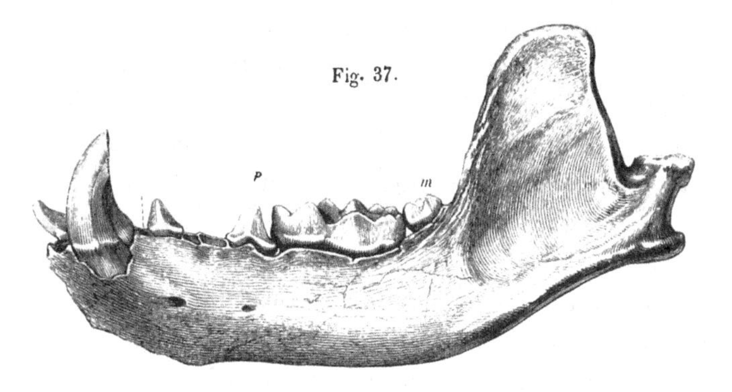

Nat. size, fossil, Kent's Hole.

MELES TAXUS. Badger.

Meles vulgaris fossilis, H. VON MEYER, Palæologica, 1832, p. 47.
Blaireau fossile, SCHMERLING, Ossem. Foss. de Liège, tom. i. p. 158.

WHILST some of the larger species of Bear have yielded to the influence of the last general physical changes which the surface of the earth has undergone, and the entire genus has been blotted out of the indigenous Fauna of Great Britain by the hostility of man, a comparatively weak and diminutive species of the Ursine family has survived both causes of extirpation. The remains of a Badger, not distinguishable from the existing British species, have been discovered in the caves at Torquay and Berry Head, Devonshire, in juxtaposition with the bones of the extinct Mammalia, and manifesting precisely the same mineral condition, so that no reasonable doubt can be entertained of their equal antiquity with the Spelæan Bear, Hyæna, and Tiger. Bones of the Badger, as might be expected from its habits of burrowing and concealment, have likewise been found in the dark recesses of caves, which were evidently, by their chemical condition, the remains of animals recently introduced into such localities, but these are readily distinguishable from the true absorbent fossils.

The most perfect fossil specimen from British localities is alluded to by M. de Blainville,* on the authority of Mr. Mac Enery, as having been found in Kent's Hole. It is now preserved in the British Museum, and, with the obliging permission of Mr. König, has been figured for the illustration of the present section (*fig.* 37). It is an entire ramus of the lower jaw, with all the teeth *in situ* except two of the incisors and the second premolar. It corresponds precisely in size and shape, and in the forms and proportions of the several kinds of teeth, with the existing male Badger. The last premolar (*p*) answering to the carnassial or sectorial tooth in the typical *Carnivora*, has the same large size and complicated crown, and the first true molar (*m*) which terminates the series, has the same diminutive size as in the common Badger.

We may conclude, therefore, that the food, like the dentition, of the diminutive plantigrade associate of the gigantic Cave Bear and Hyæna, must have been the same as that of its existing descendant; and that it must have owed its safety from the formidable contemporary beasts of prey, to the same cautious concealment and nocturnal habits which still continue to preserve the harmless species, amidst the more numerous and dangerous class of enemies which has arisen from the increasing population of a civilized country.

Fossil remains of the Badger have been discovered in the cave at Berry Head, Devon. They have been obtained on the Continent, hitherto, exclusively from cave localities.

* Ostéographie de Sub-ursus, p. 47.

M. de Blainville* cites amongst the fossil bones of the *Meles taxus*, a portion of a lower jaw from the grotto of Avison, in the department of the Gironde. But the most perfect collection of the remains of the Badger is that described by Dr. Schmerling,† in his account of the fossils of the caverns near Liege; it is there expressly asserted that the bones of the *Meles* were found under the same circumstances and in the same fossilized state as those of the *Ursus spelæus*, and other extinct quadrupeds from the same caverns; and, after a detailed comparison of the fossils with the bones of the recent Badger, the historian of the Belgian bone-caves affirms their specific identity.

A fossil skull of a Badger in the Museum of the Philosophical Institution at York, would seem to carry the antiquity of the *Meles taxus* to a higher point than the Cave epoch, and as far back as any species of the Ursine genus has been traced. The specimen is stated to have been obtained from the red crag at Newbourn, in Suffolk, and Professor Phillips has assured me, that it has the same mineralized condition and general appearance which characterise the ordinary and recognised fossils of that miocene formation. Should this specimen prove authentic, the *Meles taxus* is the oldest known species of Mammal now living on the face of the earth.

My friend Mr. Bell‡ has pleaded the cause of the poor persecuted Badger, on the ground of its harmless nature and innocuous habits; the genuine sportsman will, doubtless, receive favourably the additional claim to his forbearance and protection which the Badger derives from its ancient descent.

* Loc. cit. p. 47. † Loc. cit. p. 158. ‡ British Quadrupeds, p. 123.

CARNIVORA. *MUSTELIDÆ,*

Fig. 38.

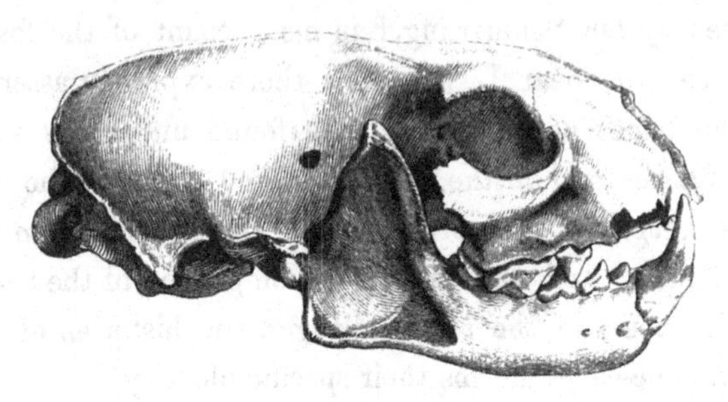

Fossil, nat. size. Cave.

PUTORIUS VULGARIS. Pole-cat.

Espèce de la grandeur du Putois,	CUVIER, Oss. Foss. iv. p. 467.
Mustela antiqua,	H. VON MEYER, Palæologica, 1832, p. 54.
Mustela Putorius Fossilis,	SCHMERLING, Ossem. Foss. de Liège 1833, tom. ii. pl. i. figs. 1 and 2.

FOSSIL remains of Mammalia much smaller than the Badger have been discovered in bone-caves and raised beaches in England, and give equal proof of their contemporaneity with the extinct species of the Cave epoch: demonstrating that the Pole-cat and Stoat, for example, have survived, as species, those changes, or catastrophes, during which their gigantic contemporaries perished; and that they have either been continued down to the present time by uninterrupted descent from generation to generation, without any appreciable alteration of their specific character; or, that, having perished with the *Ursus spelæus*, they have again been introduced at a subsequent epoch under the same specific form.

The fossil remains of a small Carnivore of the Weasel-family (*Mustelidæ*), of the same size as the common Pole-cat, were first noticed by Cuvier, in his Memoir "On the Bones of Carnivora, associated with those of Bears in Hungary and Germany," published in the "Annales du Muséum," for 1807,* and subsequently reproduced in the successive editions of his great work on extinct animals.

The remains in question were a few bones of the trunk and extremities: one of the vertebræ, the antepenultimate dorsal, differed from that in the common Pole-cat, and resembled that in the Cape species, called Zorille, in its greater breadth compared with its length; an approximation which Cuvier recognised with much interest, seeing that the bones of the Cave Hyæna resembled most closely those of the existing spotted Hyæna of the Cape.† The other remains, however, of the little fossil Carnivore bore a closer resemblance to the bones of the common Pole-cat, which induced Cuvier to leave its affinities doubtful, and to forbear adducing the Cave Polecat in support of a once favourite idea, that it was in Southern Africa that we should look for the existing quadrupeds most nearly allied to those extinct species recognised by bones found in the Caves of Europe.

The entire skull (*fig.* 38), discovered in association with larger extinct quadrupeds in the bone-cave recently explored at Berry Head, Devon, by the Rev. Mr. Lyte, affords decisive

* Tom. ix. p. 437.

† "La vertèbre dorsale est moins longue et plus grosse que dans le *putois:* elle ressemble à celle du *zorille*, et ce rapprochement me frappa d'abord singulièrement, vu que les os d'hyène de ces cavernes ressemblent aussi beaucoup à ceux d' hyène tachetée, qui vient du Cap comme le *zorille*."—*Ossem. Foss.* ed. cit. p. 467.

M. de Blainville affirms, upon a re-examination of the fossils described by Cuvier, that the dorsal vertebra is the last of that series in the Martin-cat (*Fouine*) and belongs to a distinct animal from the pelvic bone and caudal vertebræ, both which he refers, with Cuvier, to the Pole-cat.—*Ostéographie de Mustela*, p. 57.

evidence against its identity with the Zorille (*Zorilla capensis*), and in favour of its specific relations to the common Pole-cat (*Putorius vulgaris*). The fossil slightly exceeds in size the recent skulls of this species, and the canine teeth of the fossil are relatively larger, but the correspondence in every proportion and in the relative position of each process, foramen, and suture, is so close, that the above specified differences must be referred to the characteristics of a large and vigorous male animal. The last tooth—the tubercular or first true molar, *m*, *fig.* 39—of both upper and lower jaws is, indeed, rather smaller in the fossil than in the recent skulls: but I find in these that it varies in size more than the other molar teeth do. The specimen figured in M. de Blainville's "Ostéographie, Mustela," Pl. xiii., exhibits the variety in which the tubercular grinders are large.

The differences observable in the dentition of the fossil *Putorius* of the Caves, and in that of the Cape Zorille, are much more decisive. The canines are considerably smaller in the Zorille: the sectorial, or penultimate teeth—the last of the premolars—are smaller, and that of the lower jaw has a broader crown in the Zorille: the first two small premolars of the upper jaw are further apart, and the corresponding teeth of the lower jaw have the hinder margin of the compressed crown notched, forming two hinder tubercles instead of one as in the *Putorius vulgaris* and in the fossil under consideration.

An almost entire skull of a Pole-cat, in the usual condition of fossil remains of extinct quadrupeds, has been found in one of the raised beaches near Plymouth.

The fossil remains of *Putorius* alluded to by M. de Blainville, as cited by M. Keferstein from the "Reliquiæ Diluvianæ," belong exclusively to the smaller species, the subject of the next section.

Dr. Schmerling has figured the entire skull of a fossil Pole-cat from one of the caves in Belgium, and, like that from Berry Head, it is identical with the existing *Putorius vulgaris*.

In the subjoined cuts of the teeth of the last-cited fossil, *i* indicates the alveoli of the three incisors of the upper jaw; *l* the socket of the great laniary or canine tooth; *p m* the teeth included between the dotted lines, and called premolars, the largest being the 'carnassière' of the French Anatomists, or the 'sectorial tooth'; *m* is the true molar, or tubercular tooth, the only one of that kind developed in the present family of Carnivora.

Fig. 39.

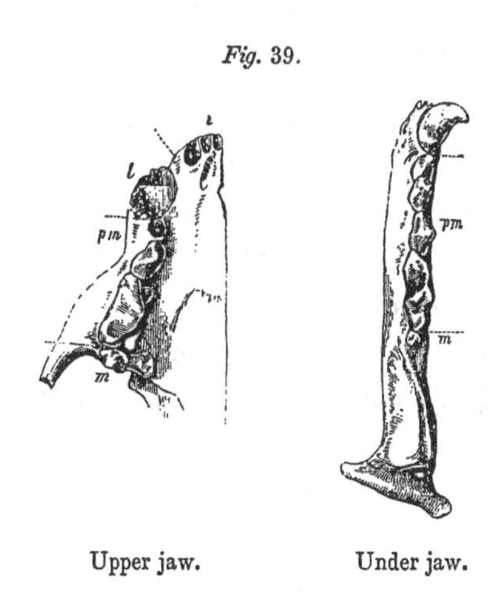

Upper jaw. Under jaw.

Fossil Pole-cat, nat. size.

CARNIVORA. *MUSTELIDÆ*

Fig. 40.

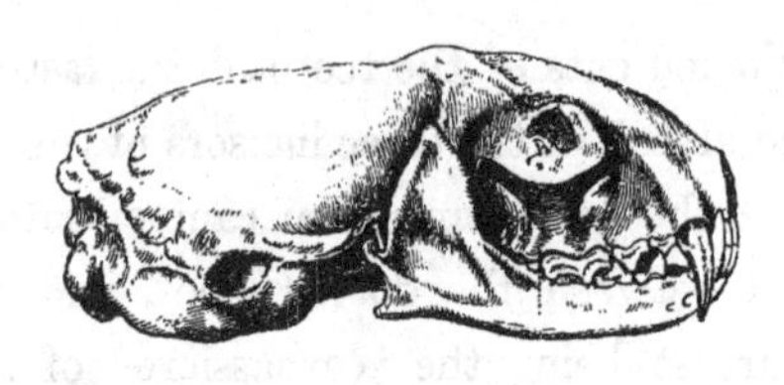

Fossil, nat. size. Cave.

PUTORIUS ERMINEUS. Stoat.

Fossil Weasel,	BUCKLAND, Reliquiæ Diluvianæ, pp. 18, 73. Pl. vi. figs. 28, 29. Pl. xxiii. figs. 11, 12, 13.
Belette commune,	CUVIER, Oss. Foss. iv. p. 475.
Putorius vulgaris,	OWEN. Report of Brit. Association, 1842.

THE most instructive fossil of the ancient British Ermine was discovered by Mr. Bartlett of Plymouth in the Bone-cave at Berry Head, and is now in the British Museum. It is a remarkably perfect skull, with the lower jaw cemented by stalactitic matter in its natural position (*fig.* 40); the specimen is absorbent from the loss of animal matter, and slightly stained red by the ferruginous deposit of the waters which percolated the limestone fissures. The zygomatic arches are broken, but the teeth are unusually complete, the incisors of the upper jaw, and the long, slender, and sharp canines in both jaws being entire.

The size of this skull, and a slight superiority of breadth in proportion to its length, indicate it to have belonged to the larger species of Weasel, called the Stoat or Ermine (*Putorius ermineus*). Fig. 41, shows a small exostosis or bony tumour on the right os frontis of the specimen.

Should the entire skeleton or the whole series of caudal vertebræ of the same individual ever be found in a fossil state, they would yield more decisive evidence in respect of the two existing British species, since the Stoat has seventeen vertebræ in the tail, and the common Weasel but fifteen.

A less entire skull (*fig.* 42), which, by its size, must also be referred to the larger Weasel, (*Putorius ermineus,*) discovered by Mr. Mac Enery in Kent's Hole, and having all the fossilized characters of the extinct mammals of that rich natural mausoleum, is now also in the British Museum. In this skull the thin cranial bones are broken away: the lower jaw is lost, but the upper molar teeth are preserved *in situ*.

The specimen is cited by M. de Blainville, from a figure of it communicated to him by Mr. Mac Enery, as appertaining without any doubt to the common Weasel* (*Belette*). As there is no appreciable difference in the dentition of the Ermine and common Weasel, the question cannot be satisfactorily determined; but, if the present specimen belong to the *Putorius vulgaris*, it indicates an individual of unusually large size.

Dr. Buckland first made known the fact that the Weasel had been associated with the extinct Hyæna, a few jaws and teeth of this small vermineous carnivore having been found fossil in the celebrated cave at Kirkdale. Two of these teeth, the sectorial premolar and the tuberculate true molar of the upper jaw, are figured by the author of the "Reliquiæ Diluvianæ," and they are pronounced by Cuvier to be exactly like those teeth in the common existing species: they, however, equally resemble those of the Ermine. The lower jaw from the Kirkdale cavern, figured

* Loc. cit. p. 59.

in Pl. xxiii. fig. 13, of the "Reliquiæ Diluvianæ," fully equals in size that of the largest *Putorius ermineus*, and exceeds the fossil jaw figured by Dr. Schmerling, in his work on the Fossils of the Belgian Caverns.

Further evidence of the antiquity of the Weasel is adduced by Dr. Buckland, on the authority of Mr. Clift, from marks of nibbling by the incisor and canine teeth of a small quadruped, of the size of a Weasel, on the ulna of a Wolf and the tibia of a Horse, found fossil in one of the caves at Oreston: and the author of the "Reliquiæ Diluvianæ" observes, with his usual acumen, that, "the weasel's teeth must have made their impressions on the bones of the wolf and horse before they were buried in diluvial mud." The account which Mr. Bell has given, in his History of the existing Quadrupeds of Britain, of the food and habits of the Weasel, is, however, scarcely reconcileable with the idea of its applying its slender acuminate teeth to the act of gnawing bones, and we shall be justified, therefore, in requiring further evidence before admitting the *Putorius vulgaris* into the catalogue of British Fossils, as the associate of the extinct Mammalia of the Oreston caves.

Fig. 41. *Fig.* 42.

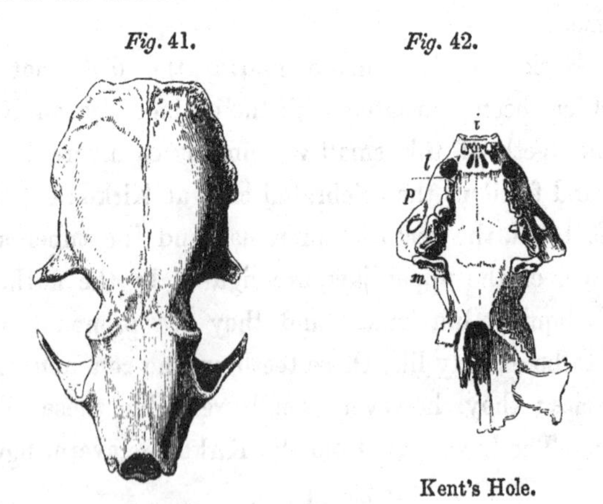

Kent's Hole.

Fossil Stoats, nat. size.

CARNIVORA. *MUSTELIDÆ.*

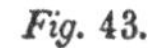

Fig. 43.

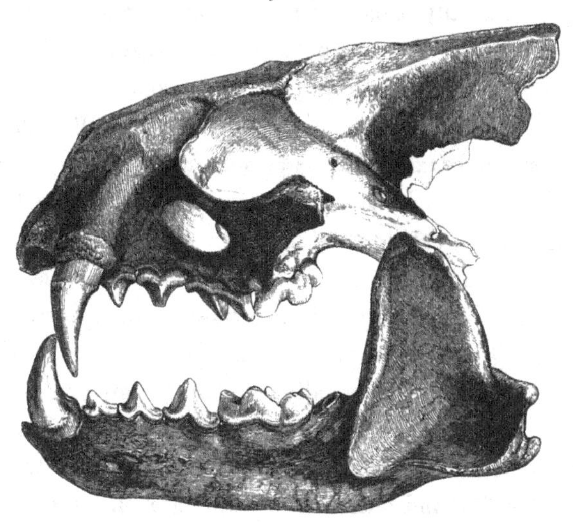

Nat. size. Fens, Cambridgeshire.

LUTRA VULGARIS. Common Otter.

Mustela Lutra,	M. DE SERRES and DUBRUEIL, Mem. du Muséum, tom. xviii. p. 334. Pl. 17, fig. 14 and 15.
Lutra antiqua,	H. V. MEYER, Palæologica, p. 55.
Loutre,	CROIZET and JOBERT, Ossem. Foss. du Puy-de-Dome, p. 89.
Potamotherium Valletonii,	Geoffroy St. Hilaire, cited in De Blainville's Ostéographie.
Lutra vulgaris,	OWEN, Report of Brit. Association, 1842.

THE fen-lands of Cambridgeshire, as I am informed by my friend Professor Sedgwick, to whom I am also indebted for the opportunity of describing and figuring the subject of the above engraving (*fig.* 43), are chiefly composed of turf-bog, occasionally alternating with marl and clay containing fresh-water shells of living species ; they commence a little below Cambridge and are irregularly expanded on both sides of the river, as far as the sea-coast. In Littleport-fen, below Ely, those marsh lands are of very wide extent, and are gradually blended with the great marshes of the Bedford level. The turf-bogs are of irregular thickness, varying from two or three feet, up to fourteen or fifteen feet, and rest either immediately upon the gault, Kimmeridge clay, and Oxford clay, or more rarely upon the thin beds of gravel which have been partially drifted over these great horizontal argillaceous deposits. In all the fens under cultivation the turf-bog is cut through in various places to get at the subjacent clay, which is now commonly used as a top dressing for the corn-land: in digging for this clay blackened bones are occasionally found immediately under the bog, and, therefore, either resting on the marly surface of the Kimmeridge and Oxford clays or on the surface of the thin layers of drifted and finely comminuted gravel, composed of flints from the chalk escarpment, and of pebbles from the green sands and oolites. On such a bed, beneath about ten feet of peat-bog, the fractured skull and lower jaw, with a few other bones, of the Otter, were found associated with the antlers of a Roe-buck.

They presented the same blackened colour and increased specific gravity that characterise the bones of the Bear, Wolf, Wild Boar, and Beaver, which have been found under similar circumstances, and, like these animals, which now no longer exist in England, the Otter in question must have lived before the fen-lands began to accumulate.

The jaws which preserve their series of teeth nearly complete, exhibit the characteristic dentition of the Otter; the incisors (*fig.* 44, *i*) are wanting: the canines (*l*) are shorter than those of the Fox, narrower than those of the Badger, larger and relatively thicker than those of the Martin-cat, and might, therefore, be recognised if found detached; *p*, indicates the premolar teeth, of which the first in the upper jaw, which is absent in the Pole-cat and Weasel, retains its characteristic place on the inner side of the canine: the sectorial premolar *s*, has its inner lobe much more developed than in Putorius, and the tubercular molar is relatively larger. Similar modifications of these teeth distinguish the dentition of the lower jaw of the Otter, which agrees in the number and kind of teeth with that of the Pole-cat. The increased grinding surface relates to the inferior and coarser nature of the animal diet of the Otter, the back teeth being thus adapted for crushing the bones of the fishes before they are swallowed. The fossil Otter of the Cambridge-fens, if the specimens may be termed fossil, does not, like the Otter of the caves at Lunel-vieil, surpass the existing individuals in strength or size: the cranium was, in fact, somewhat less than that of the old male Otter with which I compared it.*

A portion of the lower jaw of an Otter, from the Norwich crag at Southwold, and the characteristically bent humerus from the same formation near Aldborough, which Mr. Lyell has proved to be partly of fluviatile origin, carry the date of the *Lutra vulgaris* in England, as far back as the older pliocene period.

I have hitherto met with no fossil remains of the Otter in the newer pliocene fresh-water deposits of England, and the amphibious habits and cautious concealment of the Otter prevent any surprise at the absence of its remains in those ossiferous caves which have served as retreats to the larger extinct Carnivora, and which have yielded so many valuable evidences of the antediluvian inhabitants of Great Britain.

* M. de Blainville cites this spelæan Otter as the *Lutra antiqua*; but M. Marcel de Serres expressly states, in respect to the most characteristic fossil bone,— "Notre maxillaire se distingue donc uniquement par sa force et ses proportions de celui de la Loutre commune." Mem. du Muséum, tom. xviii. p. 337.

Whatever revolutions the surface of this island has subsequently undergone, the Otter, as a species, still survives. The Beaver was its associate in the ancient rivers of England, from the pliocene era down to the historical period, when this large herbivorous rodent finally disappeared, according to some Naturalists, in the year 1188. The intimate dependence of the Beaver upon the bark of trees for food, and upon the wood for the fabrication of its dams and dwelling-place, makes its extirpation a very obvious consequence of the clearance of the forests of England.

The Otter, independent of the vegetation which, of yore, overshadowed the rivers which it haunted, burrowing in their banks for shelter and concealment and preying exclusively on their scaly inhabitants, has been comparatively little affected by the changes which cultivation has produced in the lands through which those streams now flow.

Fig. 44.

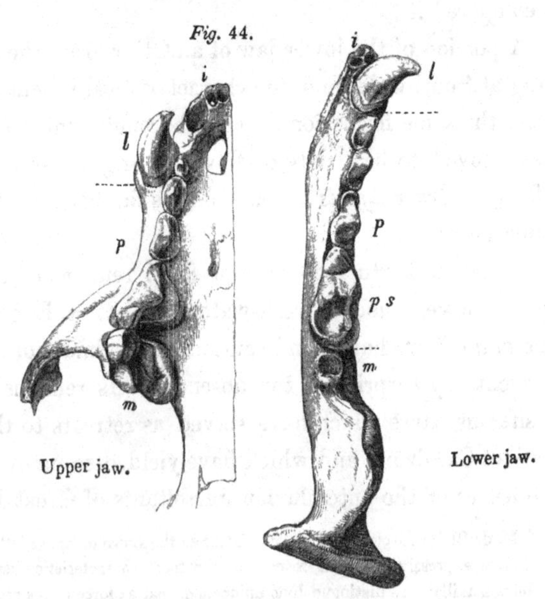

Otter, nat. size, Fens, Cambridgeshire.

Fig. 45.

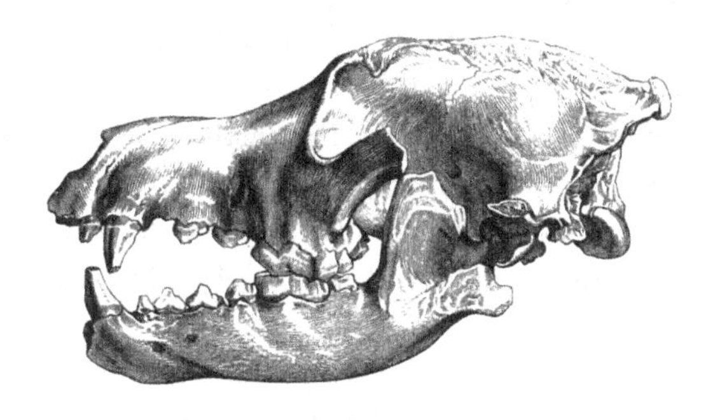

Fossil Wolf, ¼ nat. size, Kent's Hole.

CANIS LUPUS. Wolf.

Loup fossile,	Cuvier, Annales des Muséum, tom. x. p. 432; Ossem. Foss. tom. iv. p. 458.
Wolf,	Clift, Philos. Trans. 1823, p. 90. Pl. viii. and xii.
"	Buckland, Reliq. Diluv. pp. 18, 75, 89.
Canis spelæus,	Goldfuss, Nova Acta, Nat. Curios, t. xi. pt. ii. p. 451.
Loup fossile,	Schmerling, Ossem. Foss. de Liège, t. ii. Pl. ii. iii. and iv.
Canis Lupus,	Owen, Report of Brit. Association, 1842.

The fossilized state of bones and teeth of the Wolf discovered in caves, and their association with remains of extinct species of Mammalia found in the same state and position, carry back the date of the existence of this Carnivore in great Britain to the period anterior to the deposition and dispersion of the superficial drift. At a subsequent period, when evidence of the state of the British Fauna can be derived from historical records, we find the Wolf amongst the earliest animals which are thus noticed. In Ireland, a species continued to exist until the year

1710; in Scotland, to the year 1680; in England, it was extirpated at a much earlier period.

The first mention of the enduring remains of a large species of *Canis*, indicating the antiquity of this genus in England, is made in the "Account of the Assemblage of fossil Teeth and Bones in the Cave at Kirkdale in Yorkshire," by Dr. Buckland, published in the Transactions of the Royal Society for 1822: this was followed in the succeeding year by a paper, containing a description of more numerous and perfect fossil remains of a Wolf, by Mr. Clift, in the same Transactions.

The remains of the large species of *Canis* discovered in the Kirkdale Cavern were singularly scanty as contrasted with the prodigious number of fossil teeth and bones of the genus *Hyæna*, much fewer, indeed, than was originally supposed, Cuvier having pointed out that some of the teeth at first referred to the Wolf, were the deciduous teeth of young Hyænas. In the "Reliquiæ Diluvianæ," Dr. Buckland says, "Of the Wolf, I do not recollect that I have seen more than one large molar tooth." This is figured in Plate XIII. (*fig.* 5 and 6); it is the carnassial, or sectorial tooth of the right side of the lower jaw, and offers no character by which the Wolf can be distinguished from the larger varieties of the Dog.

At Paviland, on the coast of Glamorganshire, in one of the caves called Goat's Hole, facing the sea, in the front of a lofty cliff of limestone, which rises more than one hundred feet perpendicularly above the mouth of the caves, and below them slopes, at an angle of about 40°, to the water's edge, presenting a bluff and rugged shore to the waves, there were found, associated with remains of the extinct Mammoth, Rhinoceros, and Hyæna, the following fossils of a species of *Canis*, the size of a Wolf;—one lower jaw, one heel-bone, (calcaneum,) and several bones of the foot (metatarsals). These parts of the skeleton in the Wolf, are not distinguishable from those in the larger varieties of the Dog; and, since in the same cave the left side of a human skeleton was found, under a cover of six inches of earth, whilst a modern breccia has been formed, consisting of earth cemented by stalagmite, and containing shells of edible mollusks and birds' bones of existing species, the analogical probability that the canine remains were those of a Wolf is not so great as in the case of the fossils from Kirkdale.

In the enormous quarry at Oreston, near Plymouth, produced by the removal of an entire hill of limestone for the construction of the breakwater, there is an artificial cliff, ninety-three feet above high-water mark, the face of which is perforated and intersected by large irregular cracks and cavities, which are more or less filled up with loam, sand, or stalactite. These apertures are sections of fissures and caverns that have been laid open in working away the body of the rock, and are disposed in it after the manner of chimney flues in a wall.* The most remarkable of these cavernous fissures have been successively described by Mr. Whidbey, the engineer of the breakwater, in the Philosophical Transactions for 1817, 1821, and 1823. The vignette (*fig.* 50) is copied from one of the illustrations of the latest of those memoirs. In the gallery, or cavern, marked E, were found several bones and teeth of a species of *Canis*, identical in size and other characters with those from the caves of Kirkdale and Paviland, and not distinguishable from those of the recent Wolf. The chief of these remains, with the associated fossils, and those from neigh-

* Buckland, "Reliquiæ Diluvianæ," p. 68.

bouring cavernous fissures, are described and figured by Mr. Clift.*

Of the fossils from the Oreston Caverns, which I have personally examined, the following are referable to a Wolf or large species of *Canis*—

The left side of the lower jaw, with the entire series of teeth.

Four less entire rami of the lower jaw, with various proportions of the dental series: one of these is from a young, but nearly full-grown animal, and is remarkable for the evidence of disease, probably the consequence of injury inflicted by the bite of a stronger animal: the jaw is enlarged by exostosis and ulcerated near the angle, which is perforated, at *a*, by the ulceration consequent upon an abscess, or sinus, which has eaten through the bone. It has been figured by Mr. Clift, and the subjoined cut gives a reduced view of this singular example of antediluvian disease.

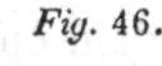

Fig. 46.

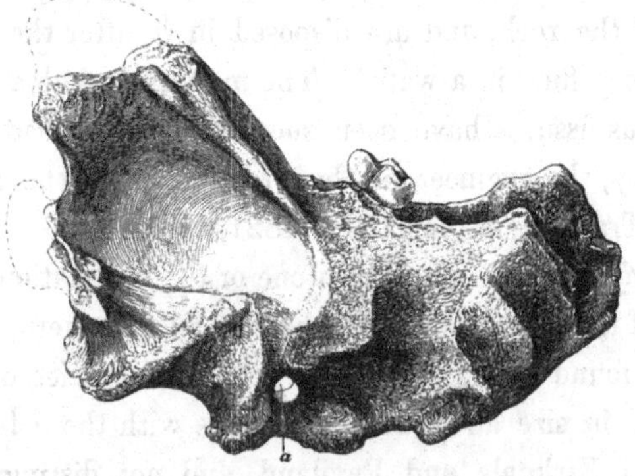

Diseased jaw, fossil Wolf, Oreston.

Besides the jaws there were collected detached specimens of nearly all the teeth of both upper and lower jaws;

* Philos. Trans., 1823, p. 78.

three fractured cervical vertebræ; one fractured dorsal vertebra; one fractured lumbar vertebra; two shafts of right humeri; a left humerus, wanting the head, or upper end; portions of three ulnæ, one of which exhibits the marks of having been gnawed by a small quadruped, and is alluded to at p. 118: a portion of the right radius; two metacarpal bones; a phalanx of the fourth toe of the right fore-foot; the left femur; the lower end of the left tibia; three metatarsal bones; the proximal phalanx of the second toe of the left hind-foot.

All the specimens are absorbent and stick to the tongue, from the loss of their original animal matter. They were found firmly imbedded in stiff clay: some of the bones which were on or near the surface of the clay, were coated by a thin crust of stalagmite; and they adhered so firmly to the clay, that many were broken by the workmen in separating them from it.

The above bones constitute but a small proportion of the fossil remains that were obtained from the Oreston caverns. In the oblique fissure, A and B, (*fig.* 50,) about forty feet above the bottom of the quarry, Mr. Whidbey had collected fifteen large maund-baskets full of bones, skulls, horns, and teeth, before the arrival of Dr. Buckland, who says, "In the upper parts of the cavity from which they were taken, we saw appearances of as many more, still undisturbed, and forming a mass which entirely blocked it up, to an extent which we could not then ascertain,"* In a collection subsequently made by Joseph Cottle, Esq., of Bristol, five jaws of the Wolf or large Dog, and five detached teeth of the same species were included.

Dr. Buckland, who examined the cavernous fissures at

* "Reliquiæ Diluvianæ," p. 71.

Oreston in company with Mr. Warburton, M.P., the present President of the Geological Society, states—

"The bones appeared to us to have been washed down from above, at the same time with the mud and fragments of limestone through which they are dispersed, and to have been lodged wherever there was a ledge or cavity sufficiently capacious to receive them; they were entirely without order, and not in entire skeletons; occasionally fractured, but not rolled; apparently drifted but to a short distance from the spot in which the animals died; they seem to agree in all their circumstances with the osseous breccia of Gibraltar, excepting the accident of their being less firmly cemented by stalagmitic infiltrations through their earthy matrix, and, consequently, being more decayed; they do not appear, like those at Kirkdale, to bear marks of having been gnawed or fractured by the teeth of hyænas, nor is there any reason to believe them to have been introduced by the agency of these animals."*

In respect to all the fossils referable to the genus *Canis*, which were submitted to Mr. Clift's inspection, the closest and most careful comparisons demonstrated a perfect agreement of the jaw-bones, in size, in form, and in the arrangement of the teeth, with those of a full-grown recent Wolf. "The os humeri," Mr. Clift says, "is perfectly similar, and has the rounded aperture through its lower extremity to receive the curved process of the olecranon."† This character is shown at *a*, in the fossil figured in cut 47. Nevertheless, the experience of comparative anatomists teaches that the teeth and bones of the existing Wolf, referred to in the foregoing comparisons, are not distinguishable from those of the larger varieties of the Dog, and

* "Reliquiæ Diluvianæ," p. 73.

† Philos. Trans. 1823, p. 97.

my own observations have uniformly led me to the same conclusions.

Cuvier,* premising that the accurate Daubenton, who seems first to have instituted the comparison, had expressed how difficult it was to distinguish the skeleton of a Wolf from that of a "Matin," (Wolf-dog or Irish Greyhound,) or a shepherd's dog of the same size,† proceeds to say that, more interested than Daubenton in discovering such distinguishing characters, he had long laboured for that especial object, comparing carefully the skulls of many individuals of those races of Dogs with the skulls of Wolves. He limits his observations, however, to the points of difference which had attracted Daubenton's notice, observing that the Wolf has the triangular

Fig. 47.

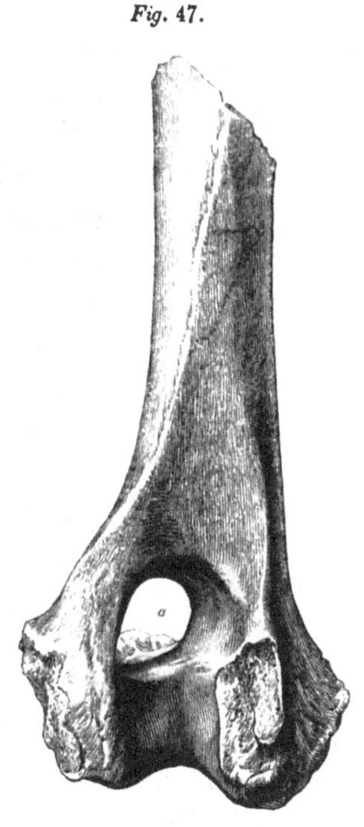

Fossil humerus of Wolf.

* Ossem. Fossiles, tom. iv. p. 458.

† I do not find in the excellent description of the Wolf in Buffon's "Histoire Naturelle," (4to. 1758, tom. vii. p. 53) the expression which Cuvier cites. Daubenton says, that the skeleton of the Wolf perfectly resembles that of the Dog in the number and position of the bones and teeth : the only appreciable difference being in the figure of certain bones, and in the size of the teeth and claws. The bony crests prolonged from the back part of the skull are longer in the Wolf than in the Matin. The teeth, especially the canines, are larger, and all the bones are rather stronger (un peu plus gros). The anterior part of the sternum is less curved upward than in the dog. Daubenton also alludes to an accidental anchylosis of the last lumbar vertebra to the right iliac bone in one skeleton of a Wolf examined by him ; p. 64. I cannot, however, appreciate any difference in the curvature of the sternum in the skeleton of an Arctic Wolf, as compared with a Newfoundland Dog.

part of the forehead behind the orbits a little narrower and flatter, the occipito-sagittal crest longer and loftier, and the teeth, especially the canines, proportionally larger.

Fig. 48.

Canine, nat. size.

Figure 48 shows one of the fossil canines from Oreston, of the natural size. But, adds Cuvier, these shades of difference are so slight, that they are frequently more marked between two individual dogs or between two wolves; and that he can hardly avoid concluding as Daubenton had done, that the Dog and the Wolf are of the same species.*

M. de Blainville, who gives the result of a very elaborate and detailed comparison of the teeth and bones of the Wolf and Dog in his "Ostéographie," concludes by invalidating even the slight shades of distinction admitted by Daubenton and Cuvier in the configuration of the cranium, and cites the wild races of the Dog, and especially the Dingo of New Holland, as indistinguishable from the Wolf by the cranial characters which his predecessors had pointed out.

* Cuvier, loc. cit. p. 458. He does not cite the work containing this alleged opinion of Daubenton. In the great "Histoire Naturelle," the words of the painstaking coadjutor of the eloquent Buffon, are "Plus j'ai observé les chiens et les loups, soit à l'extérieur, soit à l'intérieure, plus je les ai comparés les uns aux autres, tant les mâles que les femelles, plus j'aurois été porté à conclure de la ressemblance qui est dans leur conformation, qu'ils sont de la même espèce, si M. Buffon n'avoit tenté inutilement de faire accoupler le chien avec la louve." Tom. vii. p. 54. The success of the experiment which Daubenton seems to have thought essential to establish the conclusion of the specific identity of the Wolf and Dog was subsequently obtained by John Hunter, who carried the experiment a step further in regard to the hybrid offspring. See his "Observations tending to show that the Wolf, Jackall, and Dog, are all of the same species." Hunter's Works, Palmer's Edition, vol. iv. and my note at p. 324.

In regard to the Dingo, M. de Blainville's observation is accurate in respect of the configuration of the skull and the relative capacity of the cerebral cavity: the skull of this wild species of *Canis* is, however, always smaller than that of the Wolf, in so far as the entire animal is less. And it might be contended that the Dingo was a variety of the Wolf rather than of the Dog.

However this may be, the cranial characters of the Wolf pointed out by Cuvier are good and available in its determination when compared with those of a Dog of equal size, and a cranium, therefore, was the most desirable fossil for the resolution of the question of the nature of the ancient species of *Canis*, associated in Great Britain with spelæan Bears and Hyænas.

The rich cavernous depositary of the Mammalian remains of that epoch, called Kent's Hole, has afforded, thanks to the persevering explorations of Mr. Mac Enery, the desired evidence, viz., an almost entire skull with the teeth (*fig.* 45).

This specimen exactly equals in size the skull of a fine male Arctic Wolf, has the same flat and narrow triangular frontal space, an equally developed occipito-sagittal crest, and as large canines. The only differences worth mentioning, which a close comparison has yielded, are, that the antepenultimate or sectorial molar is a little larger in the fossil, and the lower border of the jaw rather more convex.

Fig. 49.

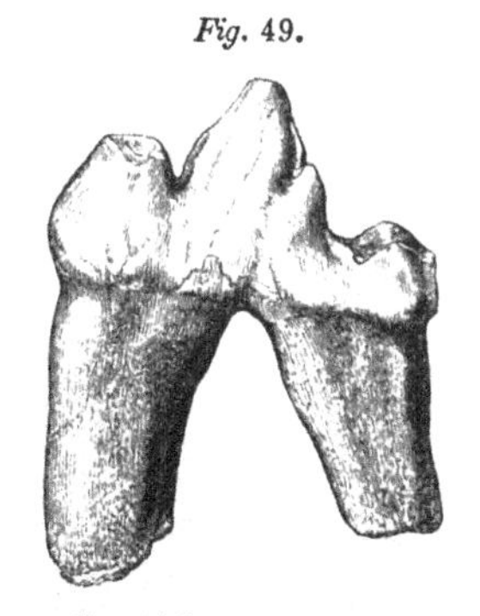

Sectorial molar, nat. size.

The latter character is not, however, appreciable in the Oreston fossils, and the sectorial molar varies as much in size in different individuals of the Dog. Fig. 49 gives the natural size of this tooth in the fossil Wolf of the Oreston cavern. Other more important points of concordance between the skull from Kent's Hole, and those of the existing Wolf leave no reasonable ground for doubting their specific identity; and the Naturalist who does not admit that the Dog and the Wolf are of the same species, and who might be disposed to question the reference of the British Fossils described in the present section to the Wolf, must in that case resort to the hypothesis, that there formerly existed in England a wild variety of Dog having the low and contracted forehead of the Wolf, and which had become extinct before the records of the human race.

The conclusion, however, to which my comparison of the fossil and recent bones of the large *Canidæ* have led me is, that the Wolves which our ancestors extirpated, were of the same species as those which, at a much more remote period, left their bones in the limestone caverns by the side of the extinct Bears and Hyænas.

Fig. 50.

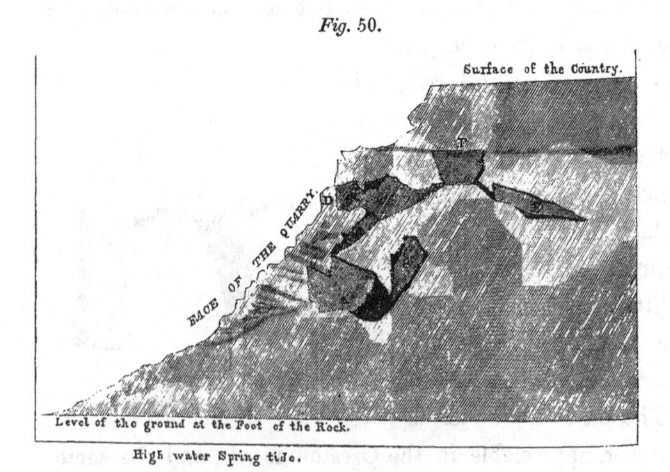

Section of the Caves at Oreston.

CANIS FAMILIARIS. Dog.

Recognizable remains of the Dog have been, in fact, obtained from Bone-caves. Dr. Schmerling* has described and figured an almost entire skull, two right rami of lower jaws, a humerus, ulna, radius, and some smaller bones indicating two varieties of the domestic dog, notably differing in size from each other, as well as from the Wolf and Fox, whose bones, with those of the Bear and Hyæna, occur in the same cavern.

The canine remains in question, are too small for the Wolf and too large for the Fox, and the conclusion that they belonged to the Dog, is admitted by M. de Blainville to be proved by the frontal elevation of the skull, which exceeds that in the Wolf.

The skull of a small variety of Dog with the latter characteristic well developed, was submitted to Mr. Clift by a person who had obtained it from an English Bone-cave: it had belonged, in Mr. Clift's opinion, to a small bull-dog or large pug: and it was not in the same absorbent state, as the true cave fossils.

Possibly the bones of the Dog described by Dr. Schmerling, may have been in the same comparatively recent state in which the Human remains of the Belgian caverns, attributed, together with those of the Dog, to the epoch of the extinct species, were proved to be by Dr. Buckland.

* Recherches sur les Ossemens Fossiles découverts dans les Cavernes de Liège, tom. ii. pl. ii.

CARNIVORA. *CANIDÆ.*

Fig. 51.

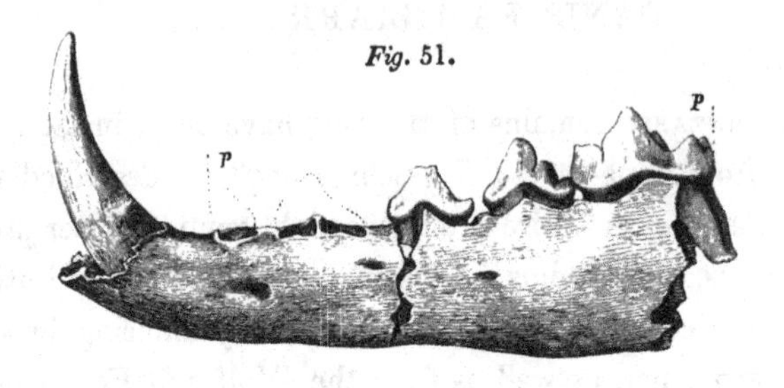

Fossil, nat. size, Kent's Hole.

VULPES VULGARIS. Common Fox.

Renard fossile,	CUVIER, Ann. du Mus. ix. p. 435; Ossem. Foss. iv. p. 461.
Canis spelæus minor,	WAGNER, Kast. Archiv. fur Natur, xv. p. 17.
Fox,	BUCKLAND, Reliquiæ Diluvianæ, p. 15.
Canis vulpes,	OWEN, Report of British Association, 1842.

IN entering upon the consideration of the fossil remains of a species of *Canidæ* of the size of the Fox, we cease to encounter those difficulties which beset the investigation of the fossils of the larger species, discussed in the preceding sections.

No Naturalist or Comparative Anatomist has ever had recourse to the Fox for the primary source of any of the domestic races of Dogs, and their specific identity has never been maintained. The varieties of Dog which have degenerated to the size of the Fox usually exhibit, in an exaggerated degree, those characters which distinguish the skeleton of the Dog from that of the Fox. The known wild varieties of Dog differ, besides, in their superior size; and the resemblance of the Fox to the Wolf and the Jackall, in the opinion of Mr. Bell,* is scarcely sufficient even to constitute it a species of the same generic group.

The skull of the Fox is narrower, and contracts more rapidly anterior to the orbits; the forehead is more contracted and flatter than in the Wolf or Jackall. The exoccipital mastoid process is longer, the orbito-frontal process is shorter, the upper margin of the squamo-temporal bone is straighter and the zygomatic arch is broader and more open than in the Wolf or Dog.

The scapula indicates an approach to the Feline tribe, in its longer coronoid process and its bifid acromion. The clavicle is more developed. The bones of the extremities, especially of the feet, are more slender than in the Jackall, and still more so than in the Dog or Wolf. In regard to the teeth, the canines (*fig.* 52) are relatively more slender and more curved than in the Wolf, Dog, or Jackall, and the upper true or tubercular molars, like those in the Jackall, are relatively to the carnassial tooth, larger than in the Wolf and Dog.

Fig. 52.

Canine, Fox. Oreston.

With these grounds for determining the small fossils of the genus *Canis*, one may unhesitatingly concur with Mr. Mac Enery, in referring to the Fox the right ramus of the lower jaw discovered by him in Kent's Hole, so superficially situated, indeed, as might justify the suspicion of its recent introduction.

The remains of the Fox from the same cavern, now in the British Museum, present, however, precisely the same fossilized state as the bones of the Spelæan Bear and

* "British Quadrupeds," p. 255.

Hyæna. One of these fossils, the anterior half of the left ramus of the lower jaw, is figured at the commencement of the present section; it retains the canine and the last three of the series of five premolars. A second fossil, (*fig.* 53,) consisting of the hinder half of the same ramus of the lower jaw of another individual, retains the last premolar or sectorial tooth, *p*, and the first tubercular molar, *m*.

Mr. Whidbey obtained from the gallery E, of the Oreston cavern, (*fig.* 50,) which yielded the bones of the Wolf, several fossil remains of the Fox, of which I have identified the following:—

Two canine teeth of the lower jaw.

A cervical vertebra.

A dorsal vertebra.

The shaft of a humerus.

A portion of the shaft of a femur.

The two latter fossils are relatively more slender than in the Jackall. Some of the above remains are noticed by Mr. Clift, in his Paper in the Philosophical Transactions, before quoted, and all are, as he describes, "equally fragile and absorbent with those of the other animals."*

Although, from the habits of concealment of the Fox, its bones might be expected to be found in caves and cavernous fissures more commonly than those of the Dog or Wolf, yet the testimony of Mr. Whidbey is adverse to the hypothesis of the recent introduction of the above-mentioned fossils into the Oreston caverns. With respect to them, he writes, "These, I think, will be the last bones I shall send you from these caves, as they are now nearly worked out. The cave B," (*fig.* 50,) "terminated near where it was first seen; the head of it was closed over with a body of limestone. The joints of the

* Loc. cit. p. 96.

rocks were not so close but that water might drop down into the cave, and, about these points, some stalactites were found in small pieces. I have not seen anything to encourage the idea that the cavern had a communication with the surface since the Flood; the present state of the quarries shows nothing like it."*

* Philos. Trans. 1823, p. 96.

Fig. 53.

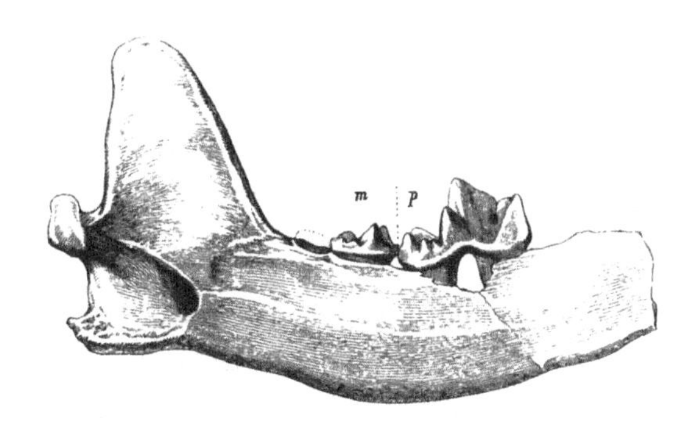

Fossil Fox, nat. size, Kent's Hole.

CARNIVORA. *VIVERRIDÆ.*

Fig. 54.

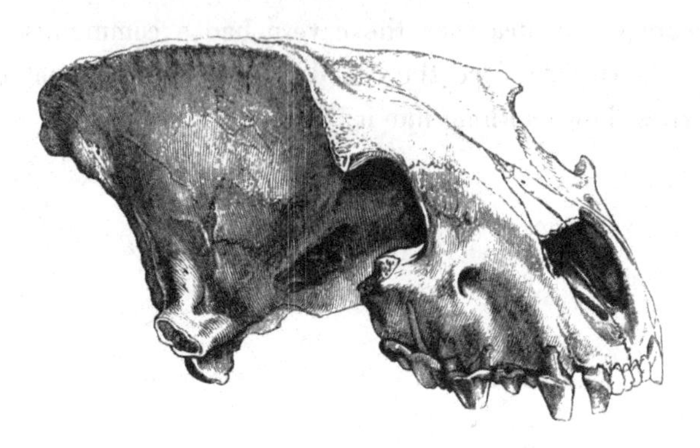

Fossil, ¼ nat. size, Kent's Hole.

HYÆNA SPELÆA. Cave Hyæna.

Hyène fossile,	CUVIER, Ann. du Muséum, tom. vi. p. 127.
Hyæna spelæa,	GOLDFUSS, Die Umgebungen von Muggendorf, 1810, 12mo. p. 280.
Fossil Hyæna,	BUCKLAND, Reliquiæ Diluvianæ, passim.
Hyæna spelæa,	OWEN, Report of British Association, 1842.

THE Hyæna is the largest and most aberrant of that tribe of Carnivorous quadrupeds of which the Genet and Civet-cats may be regarded as the type, and it makes the nearest approach to the Feline genus in its dentition. But its habits are less destructive; it seeks the dead carcase rather than a living prey, and does not disdain carrion; in this respect, bearing the same analogy to the Lion and Leopard, that the Vulture does to the Eagle and Falcon. With the number of incisors $\frac{6}{6}$, and canines $\frac{1-1}{1-1}$, common to the Carnivora, the Hyæna has four molars in each ramus of the lower jaw; animals of the Cat kind having three molars, and those of the Dog kind seven, in the same bone.

The four lower molars all belong to the spurious series, the last being the sectorial tooth. The upper jaw of the Hyæna has five molars on each side, a small tubercular true molar terminating the series, which includes four, instead of three premolars as in the genus *Felis*.

The most characteristic modification in the dentition of the Hyæna is the strong conical shape of the second and third premolars, in both upper and lower jaws, the base of the cone being belted by a strong ridge, which defends the subjacent gum.* This form of tooth is especially adapted for gnawing and breaking bones, and the whole cranium is modified by the enormous developement of the muscles which work the jaws and teeth in this operation. The tooth of the Hyæna most liable to be mistaken for one of a large Feline animal, is the sectorial or last molar

Fig. 55.

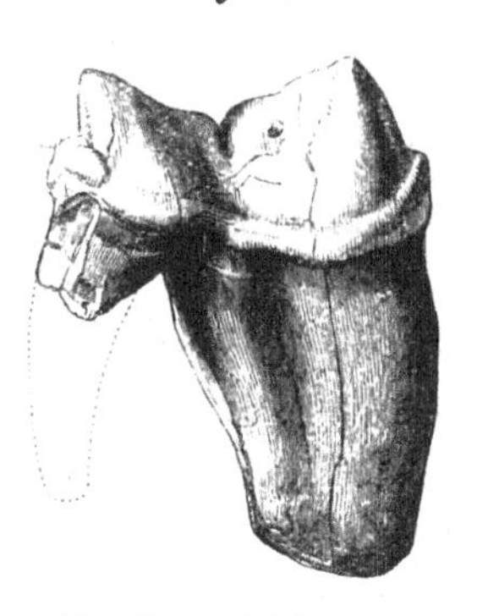

Nat. size, Kirkdale Cavern.

of the lower jaw (*fig.* 55); it is distinguished by the presence of two small tubercles, one at each end of the base of the crown.

* An eminent civil engineer, to whom I once showed the jaw of a Hyæna, said, that, if he wanted a model for the form of a hammer best adapted for breaking stones for roads, he should take the strong, conical, ridged tooth of that animal.

The existing species of Hyæna are confined to the warmer climates. The striped Hyæna (*Hyæna vulgaris*) abounds in Abyssinia and Nubia, and extends through the adjacent parts of Africa and Asia. The spotted Hyæna, (*Hyæna crocuta*,) and a rarer species, the *Hyæna villosa* of Smith, inhabit the Cape of Good Hope. The extinct species, to which the present section refers, resembled more the spotted than the striped Hyæna, but was a much larger and more formidable animal than either. This lost species was first determined by Cuvier, by the comparison of fossil remains from Continental localities, which proved it to have abounded in that ancient world of which his immortal works have stamped him as peculiarly the naturalist. We find the Hyæna, says Cuvier, not only in the same caverns which contain so many fossil bones of Bears, but also in the unstratified drift, (*terrains d'alluvion*,) where the remains of the Elephants are interred.

The discovery of the *Hyæna spelæa*, as a British fossil, is due to Dr. Buckland, in whose graphic and philosophical language the circumstances of the discovery, and the deductions of the habits of the living animals, will be here principally narrated.

In the summer of 1821, the workmen quarrying the slope of a limestone rock at Kirkdale, in the vale of Pickering, intersected the mouth of a long hole, or cavern, closed externally with rubbish and overgrown with grass and bushes. Nearly thirty feet of the outer extremity of the cave was removed before it was visited by Dr. Buckland, who found its entrance a hole in the perpendicular face of the quarry, about three feet high and five broad, as represented in the vignette (*fig.* 60). The cave is about twenty feet below the incumbent field, and extends about two hundred and fifty feet into the interior of the hill, expanding and contracting itself irregularly from two to seven feet in breadth, and two to fourteen feet in height. "Both the roof and floor, for many yards from the entrance, are composed of regular horizontal strata of limestone, uninterrupted by the slightest appearance of fissure, fracture, or stony rubbish of any kind; but, farther in, the roof and sides become irregularly arched, presenting a very rugged and grotesque appearance, and being studded with pendent and roundish masses of chert and stalactite; the bottom of the cavern is visible only near the entrance, and its irregularities, though apparently not great, have been filled up throughout, to a nearly level surface, by the introduction of a bed of mud or loamy sediment.

"There is no alternation of mud with any repeated beds of stalactite, but simply a partial deposit of the latter on the floor beneath it; and it was chiefly in the lower part of the earthy sediment, and in the stalagmitic matter beneath it, that the animal remains were found; there was nowhere any black earth or admixture of animal matter, except an infinity of extremely minute particles of undecomposed bone. In the whole extent of the cave, only a very few large bones have been discovered that are tolerably perfect; most of them are broken into small angular fragments and chips, the greater part of which lay separately in the mud, whilst others were wholly or partially invested with stalagmite; and others, again, mixed with masses of still smaller fragments, and cemented by stalagmite, so as to form an osseous breccia. In some few places, where the mud was shallow and the heaps of teeth and bones considerable, parts of the latter were elevated some inches above the surface of the mud and its stalagmitic crust, and the upper ends of the bones thus projecting, like the legs of pigeons through a pie crust, into the void space above, have become thinly covered with stalagmitic drippings, whilst their lower extremities have no such incrustation, and have simply the mud adhering to them in which they have been imbedded; an horizontal crust of stalagmite, about an inch thick, crosses the middle of these bones, and retains them firmly in the position they occupied at the bottom of the cave. A large flat plate of stalagmite, corresponding, in all respects, with the above description, and containing three long bones, fixed so as to form almost a right angle with the plane of the stalagmite, is in the collection of the Rev. Mr. Smith, of Kirby Moorside. The same gentleman has also, among many other valuable specimens, a fragment of the thigh-bone of an Elephant, which is the largest I have seen from this cave.

"The effect of the loam and stalagmite in preserving the bones from decomposition, by protecting them from all access of atmospheric air, has been very remarkable; some that had lain uncovered in the cave for a long time before the introduction of the loam, were in various stages of decomposition, but, even in these, the further progress of decay appears to have been arrested as soon as they became covered with it, and, in the greater number, little or no destruction of their form, and scarcely any of their substance, has taken place. I have found, on immersing fragments of these bones in an acid, till the phosphate and carbonate of lime were removed, that nearly the whole of their original gelatine has been preserved. Analogous cases of animal remains preserved from decay by the protection of similar diluvial mud, occur on the coast of Essex, near Walton, and at Lawford, near Rugby, in Warwickshire; here the bones of the same species of Elephant,

Rhinoceros, and other diluvial animals, occur in a state of freshness and perfection, even exceeding that of those in the cave at Kirkdale, and from a similar cause, viz., their having been guarded from the access of atmospheric air, or the percolation of water, by the argillaceous matrix in which they have been imbedded, whilst other bones, that have lain the same length of time in diluvial sand or gravel, and been subject to the constant percolation of water, have lost their compactness and strength, and great part of their gelatine, and are often ready to fall to pieces on the slightest touch, and this where the beds of clay and gravel in question alternate in the same quarry, as at Lawford.

"The bottom of the cave, on first removing the mud, was found to be strewed all over, like a dog-kennel, from one end to the other, with hundreds of teeth and bones, or, rather, broken and splintered fragments of bones, of all the animals above enumerated; they were found in greatest quantity near its mouth, simply because its area in this part was most capacious; those of the larger animals, Elephant, Rhinoceros, &c., were found co-extensively with all the rest, even in the inmost and smallest recesses. Scarcely a single bone has escaped fracture, with the exception of the astragalus, and other hard and solid bones of the tarsus and carpus joints, and those of the feet. On some of the bones, marks may be traced which, on applying one to the other, appear exactly to fit the form of the canine teeth of the Hyæna that occur in the cave. The Hyæna's bones have been broken, and apparently gnawed equally with those of the other animals. Heaps of small splinters, and highly comminuted, yet angular fragments of bone, mixed with teeth of all the varieties of animals above enumerated, lay in the bottom of the den, occasionally adhering together by stalagmite, and forming, as has been before mentioned, an osseous breccia. Many insulated fragments, also, are wholly or partially enveloped with stalagmite, both externally and internally. Not one skull is to be found entire; and it is so rare to find a large bone of any kind that has not been more or less broken, that there is no hope of obtaining materials for the construction of a single limb, and still less of an entire skeleton. The jawbones also, even of the Hyænas, are broken to pieces like the rest, and, in the case of all the animals, the number of teeth and of solid bones of the tarsus and carpus, is more than twenty times as great as could have been supplied by the individuals whose other bones we find mixed with them."*

Fragments of jaws were by no means common, but Dr. Buckland observed about forty which belonged to the *Hyæna spelæa*. The greatest number of the teeth are those of the Hyænas and the Ruminant animals.

Dr. Buckland says, "Mr. Gibson alone collected more than three hundred canine teeth of the Hyæna, which, at least, must have belonged to seventy-five individuals, and, adding to these the canine teeth I have seen in other collections, I cannot calculate the total number of Hyænas, of which there is evidence, at less than two hundred or three hundred.

"The only remains that have been found of the Tiger species are two large canine teeth and a few molar teeth, exceeding in size those of the largest Lion or Bengal Tiger. There is one tusk only of a Bear, which exactly resembles those of the extinct *Ursus spelæus* of the caves of Germany.

"In many of the most highly preserved specimens of teeth and bones, there is a curious circumstance, which,

* "Reliquiæ Diluvianæ," p. 15.

before I visited Kirkdale, had convinced me of the existence of the den, viz., a partial polish, and wearing away to a considerable depth of one side only; many straight fragments of the larger bones have one entire side, or the fractured edges of one side, rubbed down and worn completely smooth, whilst the opposite side and ends of the same bones are sharp and untouched, in the same manner as the upper portions of pitching stones in the street become rounded and polished, whilst their lower parts retain the exact form and angles which they possessed when first laid down. This can only be explained by referring the partial destruction of the solid bone to friction from the continual treading of the Hyænas, and rubbing of their skin on the side that lay uppermost in the bottom of the den."

In the adjoining cut, (fig. 56,) of the section of the Kirkdale cave, before the mud had been disturbed, A is a stratum of mud, covering the floor of the cave to the depth of one foot, and concealing the bones; B, stalagmite, incrusting some of the bones, and formed before the mud was introduced; C C, stalagmite formed since the introduction of the mud, aud spreading horizontally over its surface; D, insulated stalagmite on the surface of the mud; E E, stalactites hanging from the roof above the stalagmites.

Fig. 56.

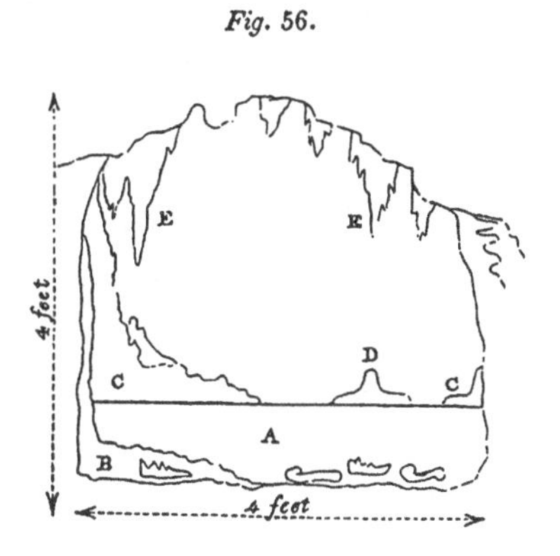

Section of Kirkdale cave, from the "Reliquiæ Diluvianæ."

Dr. Buckland justly inferred, from the facts which his persevering researches elicited, and particularly from the comminuted state and apparently gnawed condition of the bones, that the cave at Kirkdale had been, during a long succession of years, inhabited as a den by Hyænas, and that they dragged into its recesses the other animal bodies whose remains are found mixed indiscriminately with their own. "This conjecture," he states, "is rendered almost certain by the discovery I made, of many small balls of the solid calcareous excrement of an animal that had fed on bones, resembling the substance known in the old *Materia Medica* by the name of 'album græcum;' its external form is that of a sphere irregularly compressed, as in the fæces of sheep, and varying from half an inch to an inch and half in diameter; its colour is yellowish white; its fracture is usually earthy and compact, resembling steatite, and sometimes granular; when compact, it is interspersed with small cellular cavities, and, in some of the balls, there are undigested minute fragments of the enamel of teeth.

"It was, at first sight, recognized by the keeper of the menagerie at Exeter Change, as resembling, both in form and appearance, the fæces of the Spotted, or Cape Hyæna, which he stated to be greedy of bones beyond all other beasts under his care.

"This information I owe to Dr. Wollaston, who has also made an analysis of the substance under discussion, and finds it to be composed of the ingredients that might be expected in fæcal matter derived from bones, viz., phosphate of lime, carbonate of lime, and a very small proportion of the triple phosphate of ammonia and magnesia; it retains no animal matter, and its originally earthy nature and affinity to bone will account for its perfect state of preservation."

The force of this evidence, the most conclusive that could be added to the previously ascertained facts, has been attempted to be invalidated by subsequent statements,* founded, however, on imperfect observation of the habits of living Hyænas, which statements later and better testimony has disproved.† The best informed naturalists fully concur in the truth of the picture which Dr. Buckland has given of the habits of the recent species.

"The strength of the Hyæna's jaw is such that, in attacking a dog, he begins by biting off his leg at a single snap. The capacity of his teeth for such an operation is sufficiently obvious from simple inspection; and consistent with this strength of teeth and jaw is the state of the muscles of his neck, being so full and strong that, in early times, this animal was fabled to have but one cervical vertebra. They live by day in dens, and seek their prey by night, having large prominent eyes, adapted, like those of the rat and mouse, for seeing in the dark. To animals of such a class, our cave at Kirkdale would afford a most convenient habitation, and the circumstances we find developed in it are entirely consistent with the habits above enumerated."

Cuvier emphatically sanctions this happy application of the natural history of the Hyæna to elucidate the phenomena of the Kirkdale cavern, which, he says, might seem

* Wernerian Transactions, vol. i. p. 385.

† That of Colonel Sykes, quoted in the Edinburgh Philosophical Journal, vol. xii. p. 315. M. de Blainville, in the course of his argument against the conclusions of Dr. Buckland, adopted by Cuvier, says, in reference to the inference deduced by Dr. Buckland from the minute fragments of the enamel of teeth, which he detected unaltered in the Coprolites of the Kirkdale cavern: "But I have yet to learn that any animal feeds upon teeth, and can even digest them; so that this peculiarity might afford an additional objection against the opinion of M. Buckland, that the bones of Mammifers, found in great quantity in the cave of Kirkdale; with those of Hyænas, had been brought there by them, and not at all by inundations." — "Or, je ne connois encore aucune animal qui se nourrisse de dents et puisse même les digérer; en sorte que cette particularité pourrait être une objection de plus à exposer contre l'opinion de M. Buckland, que les os de Mammifères trouvés en grande quantité dans la caverne de Kirkdale, avec ceux

to have been described in the striking quotation which Dr. Buckland has given from Busbec:—

"Sepulchra suffodit, extrahitque cadavera portatque ad suam speluncam, juxta quam videre est ingentem cumulum ossium humanorum veterinariorum et reliquorum omne genus animalium."

No heaps of bones, however, were found on the outside of the Kirkdale cave, as described by Busbec on the outside of the Hyænas' dens in Anatolia; such evidences of the old inhabitants must have been dispersed by the geological force which has left more conspicuous and lasting traces of its operation in the Vale of Pickering.

It is a most interesting and remarkable fact, and one of which Dr. Buckland has ably availed himself in support of his explanation of the causes of the accumulation of the fossil bones in the cave at Kirkdale, that the remains of the Hyænas which occur in the unstratified drift or diluvial gravel show no marks of gnawing or fracture, like those on the cave bones. An entire under jaw, a radius, and ulna, of a very old and large Hyæna, which were associated with the remains of the Mammoth and extinct two-horned Rhinoceros, at Lawford, near Rugby, were in the highest possible state of preservation, and, "supplied, says Dr. Buckland, the only link that was deficient to complete the evidence I wanted to establish the Hyæna's

d'Hyènes, y ont eté apportés par elles, et nullement par inondations."—Ostéographie des Hyènes, p. 76. Since, however, it is incontestable that Hyænas devour and digest the bones of other Mammals, we must suppose them capable of digesting the dentine and cement of teeth, which substances form so large a proportion of those organs, and are so closely similar to bone in physical and chemical properties. With regard to the enamel in the Coprolites, Dr. Buckland expressly states that it was undigested. No doubt, Hyænas do not feed upon teeth; but to render such an objection valid against Dr. Buckland's interpretation of the coprolitic fossils of Kirkdale, it ought to be shewn that the modern Hyænas, before they proceed to crunch the head of a deer or sheep, are careful to extract all the teeth.

den at Kirkdale." . . . . . "The Hyæna at Lawford appears, from its position in the diluvial clay, to have been one that perished by the inundation that extirpated the race, as well as the Elephant, Rhinoceros, and other tribes that lie buried with it; and, consequently, as it could have had no survivors to devour its bones, we should on this hypothesis expect to find them entire, as they are actually found in the specimens before us."*

With them were found some small bones of the foot, apparently of the same individual Hyæna; and subsequently an almost entire cranium was found in the same superficial deposit, but at some distance from the lower jaw, which, however, fitted so well the glenoid articular cavities in the cranium, as to make it highly probable that it belonged to the same individual. The teeth in the upper maxillary bones of that skull shewed, by the extent to which they had been worn down, the same advanced age as those in the lower jaw. The socket of the small tubercular, or fifth molar tooth, is preserved on each side of this rare and beautiful cranium, illustrating the character first observed by M. de Blainville,† in a fragment of the upper jaw of a *Hyæna spelæa* from a Continental locality, now in the Parisian Museum; viz. the small size and rounded form of the fifth or tubercular molar, the socket of which is shown at *m*, in fig. 57.

Fig. 57.

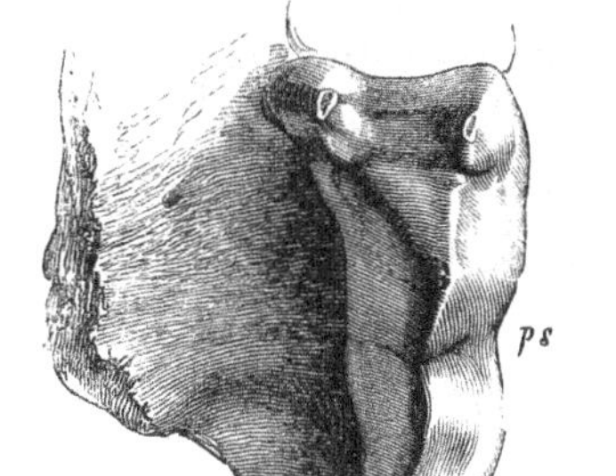

Upper sectorial molar, *p s*, and socket of tubercular molar, nat. size, *Hyæna spelæa*.

* Loc. cit. p. 27.

† Ostéographie des Hyènes, 4to., p. 62.

By its small size, this tooth confirms the deductions from other anatomical characters of the closer affinity of the extinct Hyæna to the spotted than to the striped species of the present day; and, in its rounded form, M. de Blainville sees a confirmation of the specific distinctness of the *Hyæna spelæa* from the *Hyæna crocuta*, in which the small tubercular molar has a subtriquetral crown. The skull of the *Hyæna crocuta* now before me manifests another distinction in the double fang by which the small tubercular molar is implanted in the jaw, whilst that of the *Hyæna spelæa* was inserted, as M. de Blainville remarks, by a single fang.

Baron Cuvier has particularly cited the discovery of the Hyæna's remains in the diluvium at Lawford, near Rugby, as a proof of that Carnivore having been associated in England, as on the Continent, with the Rhinoceros, Mammoth, and other great extinct Pachyderms of the unstratified drift formations.

Several instances of the same nature have been subsequently brought to light. Mr. Murchison, in his great work, "The Silurian System," notices the association of the *Hyæna* with the Rhinoceros in a fissure of the Aymestry limestone, constituting one of the vertical joints of the rock, which had been irregularly opened out by ancient disturbance of the beds, and subsequently filled by the drift and detritus of the superficies. "These jointed rocks form the eastern side of a deep comb, the higher parts of which are occupied by the upper Ludlow rock; the lower by the Aymestry limestone, which, where it contains the bones, is about forty feet above the little brook that waters the valley. In extracting the limestone for use, these fissures were perceived to be filled with calcareo-argillaceous cement of a whitish colour, like hardened mortar, in which remains of animals have been from time to time detected, including stags' horns, and bones of great size. In clearing away the limestone, a large part of the principal fissure has been obliterated, and most of the bones first discovered have been lost. Through the zeal, however, of Dr. Lloyd and Mr. Duppa Lloyd, other remains have recently been collected, which sufficiently prove the character of the accumulation; for not only have bones of deer and ox been found, but also a perfect tooth of a *Hyæna* and the femur of a *Rhinoceros*, together with several small bones which have not been determined."*

The specimen of the *Hyæna spelæa* obtained by Mr. Brown of Stanway, from the till which forms part of the beach at Walton Naze on the Essex coast, is not only more characteristic of the extinct species, consisting as it does of nearly the whole of the left ramus of the lower jaw, but the satisfaction of a personal examination of it has been afforded me, with the permission of taking the subjoined (fig. 58). The four molars remain in the jaw, which also includes the symphysis and the socket of the canine (*l*), showing that this tooth had equalled that of the largest fossil Hyænas from the cave depositaries. The crowns of the posterior molars are much abraded, especially the last, *p s*; and the exposed parts of the fangs shew them to

Fig. 58.

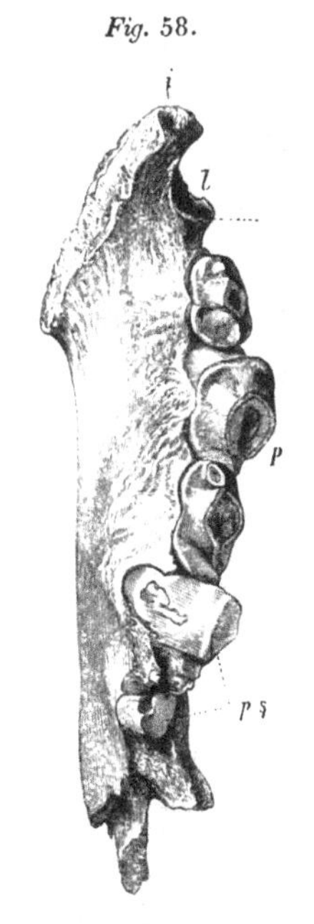

Lower jaw, *Hyæna spelæa*, Walton, ½ nat. size.

* Silurian System, p. 553.

be encrusted with a thick coat of cæmentum, as happens in aged quadrupeds.

The Hyæna is associated in the till at Walton with remains of the spelæan Bear and Tiger, the Mammoth, Rhinoceros, Hippopotamus, and other Mammalia of the extinct Fauna of the newest tertiary and drift periods.

Remains of the *Hyæna spelæa* occur, similarly associated with extinct Pachyderms, in the brick-earth at Erith, which contains extinct and recent fresh-water shells; and in a corresponding formation, constituting a superficial deposit, and filling rents that traverse the limestone called Kentish Rag, near Maidstone.

But the most perfect and abundant fossils of the extinct Hyæna have been discovered under circumstances similar to those in which the species was first determined to have belonged to the extinct Fauna of this island, viz. in limestone caves and fissures.

The cavernous fissures of the limestone quarries at Oreston yielded several specimens of the *Hyæna spelæa*, among which Mr. Clift distinguished at least five or six individuals of various ages; some of them equalling the largest of those found at Kirkdale in 1820. The posterior part of a skull appeared to Mr. Clift of uncommon magnitude, measuring twice as much from every determinate point to another, as a recent full-grown Hyæna's skull.*

This specimen has accordingly been referred by some Palæontologists to the *Hyæna spelæa major* of Goldfuss, which M. de Blainville regards with much reason as a variety of the common extinct spelæan species; merely adding, in reference to the Oreston specimen, a remark which calls for more precise dimensions of the specimens compared. The only recent skull with which

* Philosophical Transactions, 1823, p. 87, pl. xi.

Mr. Clift could compare the Oreston fossil in 1822, belonged to a small individual of the striped species (*Hyæna vulgaris*), there being no cranium of the *Hyæna crocuta* in the Hunterian Museum at that period. The following are the dimensions of the Oreston fossil, compared with the skull of the Spotted Hyæna.

	Hyæna spelæa.		*Hyæna crocuta.*	
	In.	Lines.	In.	Lines.
From the summit of the occipital crest to the posterior border of the glenoid cavity	6	0	4	10
To the upper border of the foramen magnum	3	4	2	10
Greatest breadth of occiput	5	0	3	10

From these dimensions it will be seen, that the largest of the Hyænas from Oreston did not surpass in size the existing Spotted Hyæna of the Cape, more than did the individuals of the extinct species that have been discovered at Kirkdale and Lawford.

In the portion of the cranium from Oreston, the convolutions of the brain have left deep impressions upon the inner surface, and the bony tentorium which divided the cerebrum from the cerebellum is well shewn; the air-sinuses are seen to have extended from the frontal to the occipital region beneath the sagittal crest; and to their intervention between the outer and vitreous tables of the skull is due the survival, by one of the old cave Hyænas of Muggendorf, of an extensive fracture (*a*, *a*,) which well illustrates the healing processes in bone.

I have subjoined an original figure of this unique specimen, now in the British Museum, and which has been described by Sœmmering* and Cuvier.† This example

* Nova Acta Acad. Nat. Cur. xiv. p. 1. tab. 1 et 2. *H. fossilis ex antro Muggendorfiano, cujus crista terribili morsu læsa et sanata.*

† "It is," says Cuvier, "that of an old Hyæna, which had probably received a violent bite across the occipital crest, either from one of its own species, or from one of the great lions or tigers that coexisted in the same localities, and whose

of ancient natural surgery is well paralleled by the healed fractures in the skull of the *Mylodon robustus*, described

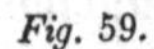

Fig. 59.

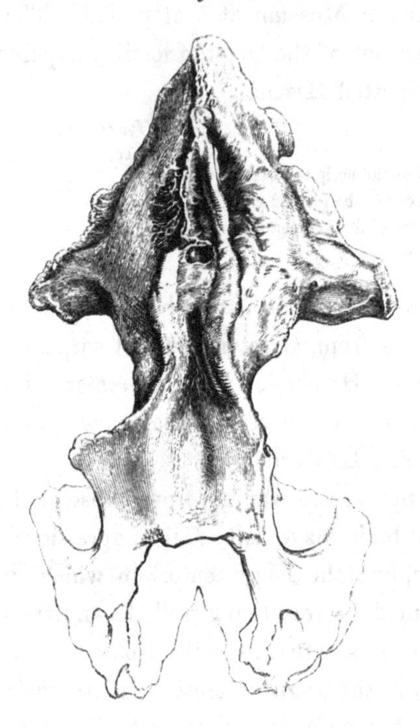

Fractured and healed skull of an extinct Hyæna, ⅓ nat. size.

and figured in my memoir on that gigantic leaf-eating quadruped of South America.* The Oreston cranium differs from that of the *Hyæna crocuta*, not only in its superior size, but in the absolutely smaller interspace between the occipital condyle and the occipito-mastoid process, and in the relatively greater extent of the posterior plate of the glenoid cavity.

bones are found in the same caverns."—*Ossem. Fossiles*, vol. iv. p. 399, pl. xxx. fig. 6 and 7.

* Description of the skeleton of an extinct gigantic Sloth, &c., 4to. Van Voorst, 1842.

Another fragment of a skull from the same locality indicates a younger *Hyæna spelæa*, by the smaller size of the sagittal and occipital crests, and the limited extent of the frontal sinuses, which are not continued backwards beyond the frontal bones.

A left ramus of the lower jaw of the *Hyæna spelæa* from Oreston, corresponding in size with the larger fragment of the skull, differs from the *Hyæna crocuta* in the greater relative breadth of the posterior ridge of the second premolar tooth. Mr. Clift* has figured a portion of the lower jaw of a young Hyæna, "in which remain one of the shedding teeth, and two permanent ones, which had not sufficiently advanced in their growth to have protruded through the gum, but are still enclosed within their alveolar cavities."

The Oreston specimens of the Cave Hyæna were found in the fissure marked B, fig. 50, p. 132.

Two canine teeth, much worn, of the *Hyæna spelæa* are recorded by Dr. Buckland to have been discovered, associated with the remains of the *Mammoth* and *Rhinoceros*, in the cave of Crawley rocks, near Swansea; these fossils are preserved in the collection of Miss Talbot at Penrice Castle.

In the same collection is preserved the lower extremity of the left humerus of the *Hyæna spelæa* from the cave at Paviland, on the coast of Glamorganshire, noticed above at p. 124.

Numerous and highly characteristic specimens of the *Hyæna spelæa* have been obtained from the caves and fissures of the mountain limestone at Bleadon, and near Hutton and Banwell in the Mendip Hills.

Remains of eleven or twelve Hyænas were discovered by

* Loc. cit. pl. x. fig. 7.

Mr. Stutchbury in the cavernous fissure lately opened on Durdham Down near Bristol. These remains were associated, as usual, with those of the Bear and Wolf, of a large Bovine animal, of the Rhinoceros, Hippopotamus, and young Mammoths. The bones were all detached and broken into small bits; and the proportion of teeth and horns to the other parts of the body greatly preponderated. In reference to the possible modes of the accumulation of these remains, Mr. Stutchbury argues that "the first method by floods is excluded, because, as in all diluvial accumulations, there would have been a mixture of rolled stones of various kinds. If the animals had fallen into the fissure, whole skeletons, or at least all the bones of a single individual, would have been entombed. But, so far from this being the case, the receptacle would not contain a number approaching to that of the animals whose remains are here found. On the other hand, the theory that the cave was the den of Hyænas, is consistent with all the observed facts. The habits of these animals to tear up putrid carcases, to carry off portions to their dens in rocks, to crush with violent force the bones of their prey, the gnawed and splintered condition of the bones, are circumstances which render the last-adduced theory highly probable, and worthy to be assumed as the true one." . . . . . "By comparison of the teeth of the fossil Hyæna with those of recent animals, their enormous size was strikingly shewn; those of the Hyæna proved it to have been larger than the largest known species of tiger." Mr. Stutchbury does not give the admeasurements.

The skull from the bone-cave called Kent's Hole near Torquay, figured at the head of the present section, (cut 54,) measures fourteen inches in total length, and exhibits the dental characters, and the strong intermuscular ridges of the formidable spelæan Hyæna in great perfection.

Subjoined (fig. 60) is a great proportion of the lower jaw, with the dentition complete, excepting the incisors: it corresponds with the abovecited cranium in size.

Fig. 60.

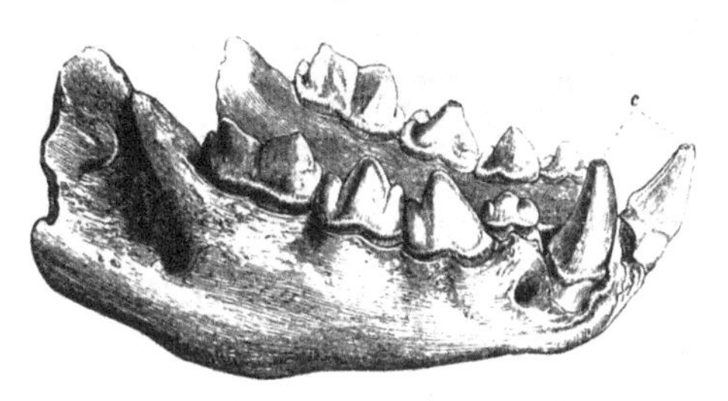

Lower jaw and teeth, *Hyæna spelæa*, ⅓ nat. size. Kent's Hole.

Several characteristic specimens of the *Hyæna spelæa* from this cavern are preserved in the collection of Dr. Buckland; and some very interesting ones were obtained for the British Museum, at the sale of the collection of the late Mr. Mac Enery. Among these is the anterior part of the lower jaw, shewing a malposition of the second permanent premolar on the left side; the corresponding deciduous tooth is retained, worn down to the stumps, and its successor projects, external to it, from the outer side of the jaw. Here, as in the Kirkdale and Oreston caves, the jaw of a young Hyæna was found, which shews the deciduous and permanent teeth (fig. 61). The point of the permanent canine has just begun to protrude from the socket; the three deciduous molars are retained, the last having the form of the sectorial tooth: these are succeeded and displaced by the first three molars of the adult, which have the conical form: the permanent sectorial tooth, *s*, is developed behind these, and rises behind the deciduous sectorial, which it does not displace; it is developed earlier than the anterior permanent molares.

Fig. 61.

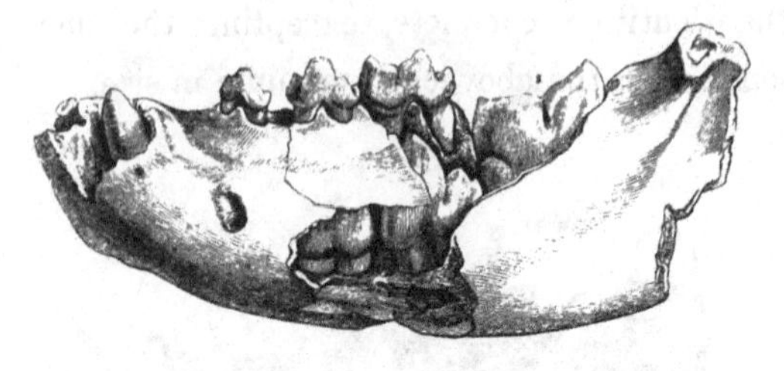

Lower jaw of young *Hyæna spelæa*, ½ nat. size. Kent's Hole.

A great proportion of the skeleton of the *Hyæna spelæa* has now been recovered from the different localities of that extinct species in England. The larger bones of the extremities found in Kent's Hole are fractured, as in the Kirkdale cave; but the smaller bones, as the astragalus, calcaneum, metacarpals, and metatarsals, are, for the most part, remarkably perfect. They differ from their analogues in the skeleton of the *Hyæna crocuta* chiefly in their larger and more robust proportions: the scapula appears to be rather narrower in proportion to its articular extremity; the deltoid crest of the humerus is longer and stronger.

In the numerous specimens of the fossil Hyæna from British localities, which I have examined and compared in public and private collections, I have not hitherto detected any characters indicative of a species distinct from the *Hyæna spelæa;* the differences observed have been those only of size and dental development, depending on diversity of sex and age. Of that fossil species which is more nearly allied to the Striped Hyæna (*Hyæna Monspessulana*, Christol), no trace has presented itself to my notice. It appears to have been confined to the middle of France, Languedoc, and Italy. Fossil remains of the Hyæna have been discovered by MM. Baker and Durand* in the tertiary strata of the Sewalik Hills; and, what is

* Journal of the Bengal Asiatic Society, vol. iv. 1835, p. 569, pl. 46.

more remarkable, the Hyæna was represented in the ancient Fauna of South America by a species which its discoverer, Dr. Lund, has termed *Hyæna neogæa.**

The following are some judicious remarks, by Sir Henry de la Beche, on the mode of observation to be pursued in the exploration of caverns in search of fossil remains. "An observer, after entering a cavern, may again return from it without the slightest suspicion that it is ossiferous, and yet the cave contain the abundant remains of animals. Many in our own country, which have furnished hundreds of bones and teeth of various mammiferous creatures to those who properly searched for them, have been visited from time immemorial by numbers who never observed a trace of such exuviæ. Caverns are far more abundant in limestone rocks than in others; and hence the frequent occurrence of stalactitical and stalagmitical matter in ossiferous caves, which often masks the organic riches beneath it." .... "When an observer discovers bones in a cavern, he should pay particular attention to their mode of occurrence. Let him make a complete section of the stalagmite, mud, silt, sands, or gravel, as the case may be, noting the depth of each different bed, and carefully abstract specimens from each before fragments of it become mingled with the others. He must be careful to mark whether different kinds of bones or teeth occur in particular beds, or are all mingled together. He should also make different sections of the cave at various points, particularly noting where or in what directions it may communicate with the surface, for caverns frequently lead to the surface in other places than their entrances, such places being filled with fallen rubbish. An observer should be particularly careful in ascertaining the

* Blik paa Brasiliens Dyreverden, &c., in the Transactions of the Royal Academy of Copenhagen, vol. vii. 1841, pp 93, 94.

general external conditions of the situation where the cavern occurs; noting whether it was ever probable that it was concealed by gravel or angular fragments of rock, which, having been subsequently removed by natural or artificial causes, a free entrance into the cavern was obtained."

"If a cavern has remained open to the surface during long periods up to the present time, it may have been tenanted first by creatures now extinct, and subsequently by those now existing; and hence their various remains may be detected in it, sometimes mixed, at others in beds above each other. Consequently, the remains of Man and his works may be discovered in such places, as has been the case, particularly in the South of France."*

The subjoined vignette gives a view of the mouth of the cave at Kirkdale, in the face of the quarry near the brow of a low hill, from the engraving given by Dr. Buckland in the "Reliquiæ Diluvianæ," pl. 2. fig. 1.

* How to observe, p. 182.

Fig. 62.

Mouth of the cave at Kirkdale.

CARNIVORA. *FELIDÆ.*

Fig. 63.

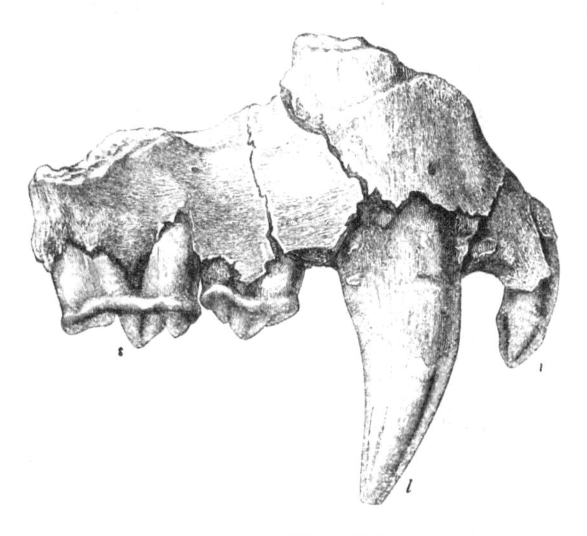

½ nat. size. Kent's Hole.

FELIS SPELÆA. Great Cave Tiger.

Lion,	ESPER and ROSENMULLER, Beschreibung der Zoolithen in den Gailenreuter Höhlen.
Animal du genre du Tigre ou du Lion,	CUVIER, Annales du Muséum, ix. 1806, p. 429.
Fossil animal of the Tiger-kind,	BUCKLAND, Reliquiæ Diluvianæ, pp. 17, 72, 261.
Felis spelæa,	GOLDFUSS, Die Umgebungen von Muggendorf, 1818.
	CUVIER, Ossem. Foss. Ed. 1823. iv. p. 449.
Felis spelæa,	OWEN, Report of British Association, 1842.

IT is too commonly supposed that the Lion, the Tiger, and the Jaguar are animals peculiarly adapted to a tropical climate. The genus *Felis* is, however, represented by species in high northern latitudes,* and in all the intermediate countries to the equator; and there is no genus of Mammalia in which the unity of organization is more closely maintained, and in which, therefore, we find so little ground in the structure of a species, though it may most abound at the present day in the tropics, for inferring its special adaptation to a warm climate. A more influential, and, indeed, the chief cause or condition of the prevalence of the larger feline animals in any given locality, is the abundance of the vegetable-feeding animals in a state of nature, with the accompanying thickets or deserts unfrequented by man. The Indian Tiger follows the herds of Antelope and Deer in the lofty Himalayan chain, to the verge of perpetual snow. The same species also passes that great mountain barrier, and extends its ravages, with the Leopard, the Panther, and the Cheetah, into Bocharia, to the Altaic chain, and into Siberia as far as the fiftieth degree of latitude; preying principally, according to Pallas, on the wild Horses and Asses.†

It need not, therefore, excite surprise that indications should have been discovered, in the fossil relics of the ancient Mammalian population of Europe, of a large feline animal, the contemporary of the Mammoth, of the tichorrhine Rhinoceros, and of the gigantic Cave Bear and Hyæna, and the slayer of the Oxen, Deer, and equine quadrupeds that so abounded during the same epoch.

These indications were first discovered in the bone caves of Germany; and Cuvier, in his usual masterly review of the materials which were accessible up to the period of his Memoir on the Cave *Carnivora* in the Annales du Muséum for 1806, concludes that the most charac-

* "Lynx boreale frigus non timet," Pallas, Zoographia Rosso-Asiatica, i. p. 13.

† *Ib.* pp. 7—19.

teristic of the fossils of the great feline animal could be referred neither to the existing Lion or Lioness, nor to the Tiger, still less to the Leopard or Panther; but that it more resembled, in the curvature of the lower border of the under-jaw, the Jaguar.

M. Goldfuss, having subsequently obtained an almost entire fossil cranium of the large extinct feline animal, described it under the name of *Felis spelæa*;* which name Cuvier adopted in the later edition of his great work,† adding to the distinctions which Goldfuss had pointed out between the fossil and the skulls of the existing Felines, including the Jaguar, that the suborbital foramen appeared to be smaller, and placed further from the margin of the orbit than in the existing Lion or Tiger. Although in the uniform and gentle curve of the upper contour of the fossil skull, it resembles more that of the Leopard than any of the larger Felines, Cuvier subsequently speaks of the extinct species as "a Lion or Tiger."

There is a constant and well-marked character, of which Cuvier appears not to have been aware, by which the skulls of the existing Lion and Tiger may be distinguished from one another; it consists in the prolongation backwards, in the Lion, of the nasal processes of the maxillary bones to the same transverse line which is attained by the upper ends of the nasal bones; whilst, in the Tiger, the nasal processes of the maxillary bones never extend nearer to the transverse line attained by the upper ends of the nasal bones than one-third of an inch, and sometimes fall short of it by two-thirds of an inch, where they terminate by an obtuse or truncated extremity, whilst in the Lion they are pointed.‡ It is very desirable that this character should be deter-

* Nova Acta Acad. Nat. Cur. tom. x. pt. ii. p. 489, tab. 45.

† Ossemens Fossiles, vol. iv. 1823, p. 449.

‡ See Proceedings of the Zoological Society, January, 1834.

mined, if possible, in the Continental specimens of the skulls of the *Felis spelæa*. If the nasal processes of the superior maxillary bones do not extend as far backwards as the nasal bones, it may be concluded that the species was not a Lion; but, as the shorter processes of the superior maxillary bones are present in the skull of the Jaguar and Leopard, as well as the Tiger, the approximation of the fossil to the striped or the spotted species of the genus *Felis* will depend upon other characters.

The most characteristic British fossil of the great spelæan Tiger, as it will, for convenience' sake, be here termed, is a considerable proportion of the right upper jaw, with the external incisor, the canine, and the second and third premolars *in situ;* the first and the tubercular molar being lost; fig. 63. The length of the fragment is six inches; the length of the canine tooth is five inches; the circumference of the base of its crown three inches and a half. These dimensions equal those of the same parts in the largest African Lion or Bengal Tiger; but it would seem, both from Continental fossils, and some that have been found in British fresh-water strata, that the spelæan Tiger had more powerful limbs and larger paws, as will be perceived from the following table of admeasurements.

	Felis spelæa.		*Felis tigris.*	
	In.	Lines.	In.	Lines.
Length of the first left metacarpal bone	4	6	3	9
first left metatarsal bone	5	3	4	3
second left metatarsal bone	5	9	4	6
third left metatarsal bone	6	0	4	7
fourth left metatarsal bone	5	6	4	1
proximal phalanx of the second toe left hind-foot	2	5	1	10

The teeth of the *Felis spelæa*, which Dr. Buckland determines among the fossils of the Kirkdale cave, exceeded in size those of the largest Lion or Bengal Tiger, but not in the same degree as the bones of the feet; nor have any fossil crania or teeth of the same surpassing magnitude as is displayed by those bones been yet discovered in the Continental caverns.

Fig. 64.

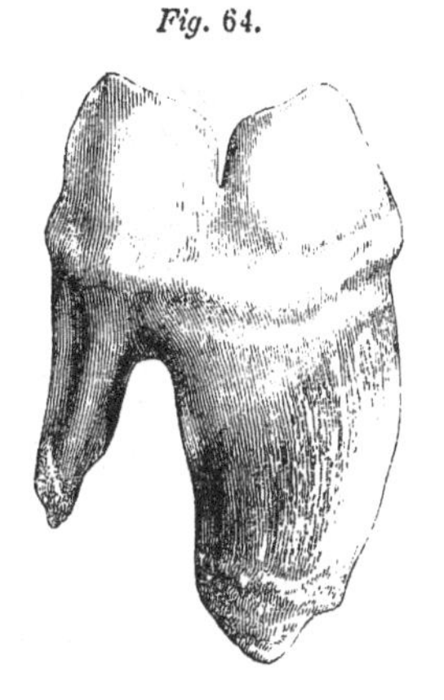

Sectorial molar. Nat. size. Kirkdale.

Four canine teeth and four sectorial molars of the lower jaw (fig. 64) are enumerated by Dr. Buckland, to whom we owe the first announcement of the *Felis spelæa* as a British fossil, amongst the specimens from the cave at Kirkdale.* With respect to the great Tigers indicated by these fossils, which were extremely rare as compared with the Hyænas, Dr. Buckland thinks it "more probable that the Hyænas found their dead carcasses, and dragged them to the den, than that they were ever joint tenants of the same cavern."†

A metatarsal bone, the third of the right hind-foot, from the Kirkdale cavern, surpasses a little in thickness, but not in length, the corresponding bone in a large Bengal Tiger; it may have belonged to a young, or a female, of the *Felis spelæa*.

Two canine teeth of the *Felis spelæa* were obtained from the cavernous fissures at Oreston: one of these belonging to the upper jaw measures three inches and three quarters in length, and both are inferior in size to the canine from Kent's Hole (fig. 65); but, like it, they present the two characteristic longitudinal indentations upon the crown *a c*: they may have belonged to a small female of the spelæan Tiger.

From the paucity of the remains of the *Felis spelæa* in

* Reliquiæ Diluvianæ, pp. 17 and 261.

† *Ib.* p. 35.

the cave of Kent's Hole, and the occurrence there of gnawed bones of Rhinoceros, Mammoth, and Horse, it is not improbable that they may have belonged to individuals whose carcasses were introduced, as Dr. Buckland conjectures those of Kirkdale to have been, by the agency of the *Hyæna spelæa*. The canine tooth (fig. 65) is rather smaller than the one in the portion of the upper jaw; but, from the thickly coated and solidified fang, *a b*, this tooth must have belonged to an old Tiger. M. de Blainville has figured a second and third molar tooth of the *Felis spelæa* from Kent's Hole, on the authority of Mr. Mac Enery.*

Fig. 65.

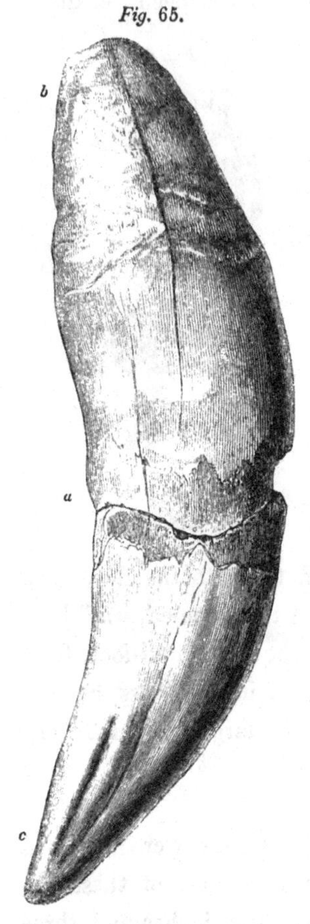

Canine of Felis spelæa.
Kent's Hole.

Fossil remains of the *Felis spelæa* have been obtained from the caves at Sandford Hill, Hutton, Banwell, and Bleadon: the most characteristic of these is in the possession of the Rev. D. Williams of Bleadon, Somerset.

* M. de Blainville frequently cites a "Description of the Cavern of Kent's Hole, Devonshire," which he supposes to have been published by Mr. Mac Enery, but which he regrets that he has not been able to procure. I have been assured by Dr. Buckland that Mr. Mac Enery never published such a work; and it is most probable that the drawings, or lithographic impressions, shewn by Mr. Mac Enery to Professor De Blainville, were those designed to illustrate the forthcoming second volume of the *Reliquiæ Diluvianæ*.

Like the *Hyæna spelæa*, the remains of the great extinct Tiger are not confined to ossiferous caverns, but occur in the superficial unstratified deposits. Portions of both upper and lower jaws, with parts of the rest of the skeleton, were discovered in 1829, together with remains of the Mammoth, Rhinoceros, Ox, Stag, and Horse, in a marl-pit near North Cliff, Yorkshire.* The pit is situated on the eastern boundary of the red marl, where that stratum approaches the low lias hills which skirt the south-western side of the Wolds. The section of the pit yielded the following strata :—

	Ft.	In.
Black sand	0	9
Yellow sand	1	6
White gravel, consisting of small pebbles of chalk, and angular fragments of flint, with a few pieces of *Gryphæa incurva*, and fewer pebbles of sandstone	2	6
Blue marl, irregularly penetrated by the gravel . .	5	0
Commencement of a blacker marl.		

This had been dug to the depth of ten feet, and here the greater part of the fossil bones were found. The horns of the Ox and the jaws of the spelæan Tiger lay near the bottom of the excavation: the antler of the Stag, the thigh-bone of the Mammoth, and one of the leg-bones of the Rhinoceros, lay low in the upper marl. The bones occupied a space of about twenty yards in length, and eight in width.

The following are the parts of the *Felis spelæa*, from the above deposit, now preserved in the Museum of the Yorkshire Philosophical Society: A fragment of the upper jaw, containing the second and the great sectorial premolar teeth: a lower jaw, with the part of the ascending rami and articular processes broken away; it measures from

* The circumstances attending the discovery of these bones are narrated by the Rev. W. Vernon, F.R.S., in the Philosophical Magazine for 1829, vol. vi. p. 225.

the fore-part of the canine to the end of the molar series five inches and a half; the depth of the jaw below the last molar tooth is two inches; the canine tooth is four inches and a half in length: the proximal end of a radius, the articular head of which exceeds by one-fourth that of the largest Lion or Tiger: the head of the femur; one metacarpal bone measuring four inches, nine lines in length; and two metatarsals, one belonging to the second toe, measuring five inches, five lines,—the other belonging to the middle toe, and measuring six inches in length: these bones indicate paws as large as those of the great Gailenreuth Tiger, whose admeasurements have been previously given.

The black marl contained abundance of fresh-water shells, amongst which Professor Phillips found *Limnæas* and *Planorbis* to predominate; and he determined that all the species discovered in this marl, twelve in number, agreed in every respect, even in their accidental variations, with the same species now existing in Yorkshire.

Professor Sedgwick,* who expresses well-deserved thanks to Mr. Vernon, for the zeal with which he investigated, and the fidelity with which he described, the excavation of the several regular deposits at North Cliff, containing the Mammalian and Testaceous fossils, observes that:— "Phenomena like these have a tenfold interest, binding the present order of things to that of older periods, in which the existing forms of animated nature seem one after another to disappear."

* Anniversary Address to the Zoological Society, 1830.

CARNIVORA. *FELIDÆ.*

Fig. 66.

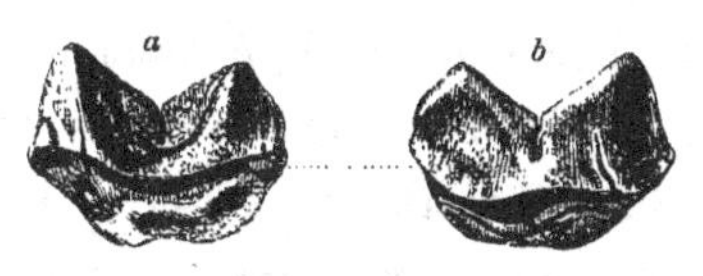

Left lower sectorial tooth of *Felis pardoides: a*, view from the inside ; *b*, view from the outside. Red Crag.

FELIS PARDOIDES. Owen.

The discovery of the fossil indicative of the present species is due to Mr. Wm. Colchester of Ipswich, who pointed out to Mr. Lyell, in June 1839, in his collection of fossil teeth from the Red Crag of Newbourn near Woodbridge, one tooth which differed greatly from the rest, and which they both suspected to belong to a carnivorous Mammal. The tooth being submitted to my inspection, I found, on comparison, that it agreed in size and shape with the posterior or sectorial molar of the left side of the lower jaw of the Leopard (*Felis leopardus*, Linn.); affording, if not proof of specific identity,—which, from the close correspondence in every character, save size, that pervades the dental formula of the different species of *Felis*, cannot be affirmed on the evidence of a single tooth,—at least sufficient indication that a feline animal as large as a Leopard existed at the geological epoch indicated by the formation in which it was found. The tooth in question (*fig.* 66), is imperfect: the enamelled crown is preserved, but the base has lost the fangs, and has the appearance of having been worn and polished after the fangs had been broken short off. The two compressed pointed lobes of the crown are more nearly equal in size than in the *Felis pardus.*

Mr. Lyell rightly states, that "this fossil resembles in colour that of many of the accompanying teeth of fishes, most of which belong to different species of the Shark family, with which the palatal bones of the *Myliobates*, a kind of Skate, are intermixed. It is deserving of remark, that in a great portion of the Shark's teeth, the softer or bony portion at the base has been worn away, more or less entirely, as if by attrition; while the upper part, or that covered by enamel, has suffered but slightly. In a word, they seem to have been subjected to the same mechanical action as the tooth of the Leopard."

"Newbourn is a village on the west side of the estuary of the Deben, and about six miles S.W. from Woodbridge. In the large pit of red crag at the northern extremity of the village (Mr. Wolton's pit), the crag presents its ordinary character of a purely marine deposit, containing the usual shells in great part comminuted. But the horizontal strata are traversed to the depth of about thirty feet by numerous fissures, which are from a few inches to a foot or more in width, and are filled principally with the detritus of red crag, in which numerous fragments of shells are still preserved. Some of these rents terminate downwards, coming to a point, with no signs of fracture below. As at present our information simply extends to the fact that the Leopard's tooth was picked up together with those of fishes in this pit, it might be suggested that the Mammalian relic was possibly derived from the contents of one of the fissures, the filling of which was an event certainly posterior, and perhaps long subsequent, to the era of the deposition of the crag."*

In the collection of the Rev. Edward Moore, of Bealings, near Woodbridge, the tooth of the Bear, noticed at p. 105,

* Annals of Natural History, vol. iv. 1840, p. 186.

and teeth of a species of Hog and of Deer, are preserved, from the same locality as the tooth of the pard-like Feline; and Mr. Lyell, judging from their appearance, inclines to the opinion that they are all of the age of the red crag. "They seem," he says, "to have undergone precisely the same process of trituration, and to have been impregnated with the same colouring matter, as some of the associated bones and teeth of fishes which we know to have been derived from the regular strata of the red crag."* The probability of the *Felis pardoides* being a veritable fossil of the red crag brings to mind the examples of the same genus in strata of equal antiquity, in the great *Felis aphanista*, Kaup, and the *Felis antediluviana*, Kaup, which is a species of the size of the Newbourn fossil: both *Felis aphanista* and *F. antediluviana* were discovered by Dr. Kaup † associated with Dinotheriums and Mastodons in the miocene sand at Epplesheim.

MM. Croizet and Jobert ‡ have also discovered in the tertiary strata of Auvergne, in the neighbourhood of Pardines, a fossil Cat, *Felis pardinensis*, about the size of the Leopard.

* Annals of Natural History, vol. iv. 1840, p. 188.
† Ossem. Foss. du Muséum de Darmstadt, pt. ii.
‡ Ossem. Foss. du Puy-de-Dôme, p. 201.

CARNIVORA. *FELIDÆ.*

Fig. 67.

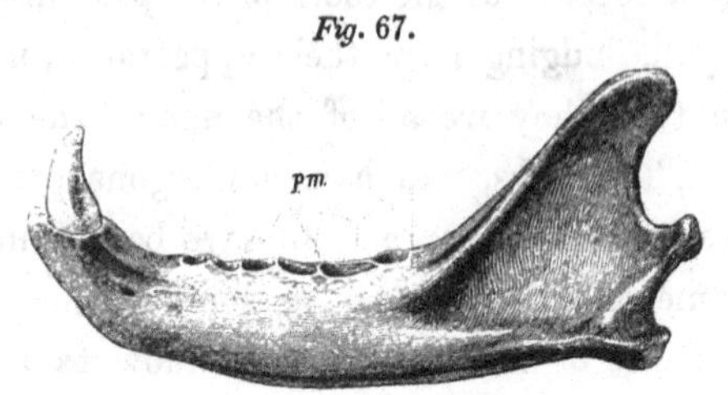

Nat. size. Grays, Essex.

FELIS CATUS. Wild Cat.

Felis ferus, SERRES, Recherches sur les Ossemens des cavernes de Lunel-Veil, 4to, 1839, p. 119.

FOSSIL remains of a feline animal about the size of the Wild Cat were first noticed by Dr. Schmerling in his description of the Caverns in the Province of Liege, where they were found in tolerable abundance. He assigns the right ramus of a lower jaw, which exceeds by a few lines the specimen figured above, to a species or variety which he calls *Felis Catus magna ;* and the greater proportion of the fossils, which include some entire skulls, to the *Felis Catus minuta.* These, however, do not vary from the standard of the existing Wild Cat more than the varieties due to age or sex are now observed to do.

MM. Marcel de Serres, Dubreuil, and Jean-Jean, have enumerated a considerable collection of bones of the Wild Cat discovered in the caverns of Lunel-Veil.*

The most authentic specimens of the *Felis Catus*, in relation to their antiquity, which appear yet to have been obtained from British localities, are the right ramus of the lower jaw, retaining the canine tooth, discovered in the brick-earth at Grays, Essex, and a corresponding part of

* Loc. cit.

the lower jaw, almost identical in size and shape, but retaining the three molar teeth, from the cave of Kent's Hole, Torquay.

The Essex jaw of the Wild Cat, which was found in the same deposit that has yielded so many remains of the Mammoth, was in the usual condition of the bones of that period. And the specimen from Kent's Hole, now in the British Museum, precisely accords, in colour and chemical composition, with the fossils of the extinct quadrupeds from the same cave. The outlines of the premolar teeth preserved in this jaw are added above the corresponding empty sockets of the jaw figured, with which they quite agree in size; and both are undistinguishable from the analogous parts of the still existing species of Wild Cat. We seem, therefore, here to have another instance of the survival, by a smaller and weaker species, of those geological changes which have been accompanied by the extirpation of the larger and more formidable animals of the same genus.

Our household Cat is probably a domesticated variety of the same species which was contemporary with the spelæan Bear, Hyæna, and Tiger. It appears, at least from an observation recorded by M. de Blainville, that grimalkin cannot be the descendant of the Egyptian Cat, as M. Temminck supposed. The first deciduous inferior molar tooth of the *Felis maniculata* has a relatively thicker crown, and is supported by three roots; whilst the corresponding tooth in both the domestic and wild Cats of Europe has a thinner crown and two roots. The tail of the domestic Cat is more tapering, and a little longer than in the wild Cat, but the extent to which this part is shewn, by a curious propagated variety of tail-less Cat, to be susceptible of modification, ought to warn us against inferring specific distinction from slight differences in the proportions of the tail.

CARNIVORA. *FELIDÆ.*

Fig. 68.

Machairodus megantereon, ½ nat. size. Auvergne.

Genus. MACHAIRODUS.

Ursus cultridens,	CUVIER, Ossemens Fossiles, 4to. 1824, vol. v. pt. ii. p. 517.
Ursus trepanodon,	NESTI, Lettera terza dei alcune ossa fossili non peranco descritte, al S. Prof. Paolo Savi, 8vo. Pisa, 1826.
Ursus cultridens Issiodorensis,	CROIZET ET JOBERT, Ossemens fossiles du Puy-de-Dôme, p. 200, 1828.
Felis cultridens,	BRAVARD, Monographie de deux Felis d'Auvergne, p. 143, tab. iii. fig. 10, 13, 1828.
Machairodus,	KAUP, Description d'Ossemens Fossiles du Muséum de Darmstadt, 2de Cahier, 1833.
Steneodon,	CROIZET, cited by Geoffroy in 'Revue Encyclopédique,' tom. lix. 1833.
Ursus cultridens,	OWEN, Report of British Association, 1842.

THE most remarkable of all the fossil teeth of large carnivorous Mammalia that have been hitherto discovered are the long, curved, compressed canines, the crowns of which, with finely serrated margins, more nearly resemble those of the teeth of the extinct Argenton Crocodile, or of the more ancient reptiles called *Megalosaurus* and *Cladyodon*, than the canine teeth of any known existing carnivorous Mammal.

These fossil falciform teeth have been found in the newer tertiary deposits in Italy, in Germany, in France, and in this country, for the most part singly and detached, and always very rare. They were first noticed in 1824 * by Cuvier, to whom the specimens discovered in the Val D'Arno were exhibited by Professor Nesti; and, from evidence relative to their association with the remains of a species of *Ursus*, Cuvier was induced to refer them to that genus, under the specific name of *Ursus cultridens*.

The first description of these large falciform canines is due to Professor Nesti, according to M. de Blainville, who cites his "Lettera terza dei alcune ossa fossili non peranco descritte, al Sign. Prof. Paolo Savi, Pisa, 1826." Cuvier makes mention of one of these teeth in the Cabinet of Fossils at Darmstadt, which, from a drawing transmitted to him by M. Schleyermacher, seemed to resemble in every respect the falciform teeth found in Tuscany.

Amongst the rich collection of fossils discovered, principally by the Rev. Mr. Mac Enery, in the bone-cave of Kent's Hole near Torquay, Devon, two canines were recognized by Dr. Buckland as very similar to those of Italy and Germany, on which Cuvier's species "*Ursus cultridens*" had been founded.

M. Bravard, however, having observed in parts of a fossil cranium of a large species of *Felis*, indications of an unusually long and compressed canine tooth, in the form of the socket of the upper canine, and the deep depression for the reception of its crown on the outside of the lower jaw,

* Supplement to the "Ossemens Fossiles," 4to. 1824, vol. v. pt. ii. p. 517.

when the mouth was closed, conjecturally restored the lost canine by one having the peculiar proportions of those previously referred to the *Ursus cultridens*.

Dr. Kaup, on the other hand, in his excellent illustrations of the fossils from Epplesheim in the Darmstadt Collection, lays stress on the obvious differences which the falciform canines present, as compared with the known Bears and feline animals; pointing out, in his comparison of them with the latter, that the compressed canines had neither the grooves nor the two ridges which characterize the canines in the genus *Felis*, and that no carnivorous quadruped had the enamelled crown of the canine so long, or its concave edge so serrated. The Darmstadt Professor dwells on the resemblance in these respects between the falciform canines in question, and the teeth of the *Megalosaurus*; and concludes by proposing to form a distinct genus, *Machairodus*, for the extinct species to which these singular teeth belonged.*

The author of the article *Machairodus* in the Penny Cyclopædia has cited my reasons for rejecting the idea of the Saurian nature of that genus; the proof of its belonging to the Mammalian class being afforded by the specimen figured at *b*, p. 244, vol. xiv. of that valuable work, "which shews that the tooth was originally lodged in a socket, and not anchylosed to the substance of the jaw, and that the fang was contracted and solidified by the progressive diminution of a temporary formative pulp, and did not terminate in an open conical cavity, like the teeth of all known Saurians, which are lodged in sockets." The article concludes by the remark, that "we are not without existing Ruminants with very long canine teeth in the upper jaw, with serrations on their edges, though not so

* Description d'Ossem. Foss. de Darmstadt, 4to. 2de cahier, p. 28.

broad in proportion as those of *Machairodus*." No bones, however, of any large Ruminant had ever been detected so associated with the teeth of the *Machairodus* as to countenance the supposition that they had formed the defensive weapons of a large hornless extinct species allied to the Musk-deer; whilst, on the other hand, the discovery in Kent's Hole of the external upper incisor (fig. 70), having its sharp edges as strongly serrated as in the great falciform canines, left little doubt that they appertained to the same species, and afforded corresponding proof of its carnivorous character.

The real affinities of the problematical *Machairodus* have at length been decided by M. Bravard's discovery of the skull of his *Felis megantereon*, retaining the falciform canine *in situ*, "armée encore de sa dent falciforme."* A reduced outline copy of the drawing of this interesting fossil, transmitted by M. Bravard to M. de Blainville, is placed at the head of the present section (fig. 68): the original had not been seen by the Parisian Professor; but, from a rapid inspection of a plaster model, the cranium seemed to him to bear a great resemblance to that of the Panther.

The Comparative Anatomist is prepared the more readily to accept this announcement, from the fact that the modification of the lower jaw, upon which M. Bravard had previously been led to found his new species of *Felis* (*F. megantereon*), is precisely such as would best accord with an unusually elongated form of canine: the modification in question consists of a sudden and considerable increase in the vertical diameter or depth of the symphysial part of the lower jaw; whilst a depression on the outer side, between the canine and the first molar, indicates the

* See De Blainville's Ostéographie, *Felis*, p. 140.

part which received the long descending crown of the upper canine when the mouth was closed.

The *Felis megantereon* of M. Bravard is much too small, of course, for the great falciform canines of the *Machairodus cultridens*, some of which measure nine inches in length, following the outer curve. I have had the satisfaction of obtaining the same kind of evidence of the feline affinities of the *Machairodus* from an inspection of the fossil remains discovered by Messrs. Falconer and Cautley in the tertiary deposits of the Sewalik mountain range, and transmitted by Captain Cautley to the British Museum. A portion of the left side of the upper jaw of a young *Machairodus*, with apparently the first or deciduous dentition, exhibits the characteristic elongated, compressed, and finely serrated canine *in situ:* the extremity of the crown is broken off, but the tooth evidently bore the same proportion to the molar series as does the canine of the *Felis megantereon* of Bravard.

The molar series in the Sewalik *Machairodus* included, in an extent of one inch and a half, three teeth: the first, which is simple, single-fanged, and very small, is indicated by the socket: the second, measuring eight lines in the antero-posterior diameter, is the carnassial or sectorial tooth; its crown is more compressed, its trenchant margins sharper, and the inner tubercle less developed than in the normal species of *Felis:* the socket of the third or tubercular molar is behind, or in a line with the sectorial tooth, as in the milk-teeth of the Lion. What remains of the crown of the canine indicates its great length: the breadth of its base is five lines; it is much compressed; the inner surface is flat, and both edges are finely but distinctly serrated. Like the larger canines of *Machairodus*, the outer convex side of the tooth is devoid of the two linear impressions which characterize the canine in the typical Felines; but, in the somewhat aberrant *Felis jubata*, these impressions are also obsolete.

A portion of the lower jaw of a larger *Machairodus*, from the Sewalik range, shews the beginning of the characteristic downward extension of the symphysis, and the depression on the outside of the ramus for the lodgment of the long upper canine. The molar series, which consists, as in the typical Felines, of three premolars, the last being the sectorial tooth, has a longitudinal extent of two inches; the second molar slightly overlaps the third, which has an antero-posterior extent of eleven lines. This portion of jaw indicates a species of *Machairodus* as large as the Jaguar: it most probably belongs to an adult of the same species as the one indicated by the instructive portion of the upper jaw.

MACHAIRODUS LATIDENS.

In this island, anterior to the deposition of the drift, there was associated with the great extinct Tiger, Bear, and Hyæna of the caves, in the destructive task of controlling the numbers of the richly developed order of the herbivorous Mammalia, a feline animal as large as the Tiger, and, to judge by its instruments of destruction, of greater ferocity.

In this extinct animal, as in the *Machairodus cultridens* of the Val d'Arno, and the *Mach. megantereon* of Auvergne, the canines curved backwards, in form like a pruning-knife, having the greater part of the compressed crown provided with a double-cutting edge of serrated enamel; that on the concave margin being continued to the base; the convex margin becoming thicker there, like the back of a knife, to give strength; and the power of the tooth being farther increased by the expansion of its sides. Thus, as in the *Megalosaurus*, each movement of the jaw with

Fig. 69.

Canine of Machairodus latidens, nat. size. Kent's Hole.

a tooth thus formed combined the power of the knife and saw; whilst the apex, in making the first incision, acted like the two-edged point of a sabre. The backward curvature of the full-grown teeth enabled them to retain, like barbs, the prey whose quivering flesh they penetrated. Three of these canine teeth, of one of which a side-view, and a view looking upon the concave edge, (fig. 69,) are subjoined, were discovered by the Rev. Mr. Mac Enery in Kent's Hole, Torquay, and were recognized by Dr. Buckland as bearing a close resemblance to the canines of the *Ursus cultridens* of the Val d'Arno. Professor Nesti, to whom Dr. Buckland transmitted casts of these teeth, recognized the same resemblance, but noticed their proportionally greáter breadth. The cast of one of the largest of the canines of the *Machairodus cultridens* from the Val d'Arno, presented to me by Mr. Pentland, measures eight inches and a half in length along the anterior curve, and one inch and a half in breadth at the base of the crown. The largest of the canines of the *Machairodus* from Kent's Hole measures six inches along the anterior curve, and one inch two lines across the base of the crown: the English specimens are also thinner or more compressed in proportion to their breadth, especially at the anterior part of the crown *a c*, which is sharper than in the *Mach. cultridens*.

These differences are so constant and well-marked as to establish the specific distinctness of the large British sabre-toothed Feline animal; for which, therefore, I propose the name of *Machairodus latidens*. the more important and prominent characters of the canine teeth, which this species has in common with the *Mach. cultridens* and *Mach. megantereon*, as well as with that from the Sewalik tertiary sand, fully justifying their separation from the typical *Felidæ*, in which family they form a well-marked and most interesting subgenus; and, to this, Dr. Kaup's name of *Machairodus*, although proposed under another view of the affinities of the fossils in question, may be most conveniently applied.

Fig. 70.

Incisor of Machairodus, nat. size. Kent's Hole.

The right external incisive tooth (fig. 70) strongly indicates, by the serration of the anterior and posterior margins of the crown, that it belonged to the same species as the falciform canines, and closely conforms in other respects with the external incisors of the existing Feline animals. Assuming it to belong to the *Machairodus latidens*, it proves this species to have relatively larger external incisors than any of the existing Felines, or than the *Mach. megantereon.* The obtuse consolidated fang, thickly coated by cement, which this incisor, like the canine, possesses, proves both kinds of teeth to have belonged to an aged animal.

Hitherto, no parts of the skeleton have been found in England so associated with the characteristic teeth of the *Machairodus* as to throw any additional light on the organization of this once formidable beast of prey. A comparison of fig. 69 with fig. 67 will shew that the *Machairodus latidens* must have equalled, or nearly equalled, in bulk the spelæan Tiger; and we can scarcely doubt, from its remains being found with those of the previously described large extinct Carnivora in the same recent tertiary deposits in India, Italy, Germany, and France, as well as in the caves of England, that it was their contemporary.

When we are informed that, in some districts of India, entire villages have been depopulated by the destructive incursions of a single species of large Feline animal, the Tiger, it is hardly conceivable that Man, in an early and rude condition of society, could have resisted the attacks of the more formidable Tiger, Bear, and Machairodus of the cave epoch. And this consideration may lead us the more readily to receive the negative evidence of the absence of well-authenticated human fossil remains, and to conclude that Man did not exist in the land which was ravaged simultaneously by three such formidable Carnivora, aided in their work of destruction by troops of savage Hyænas.

RODENTIA. *CASTORIDÆ.*

Fig. 71.

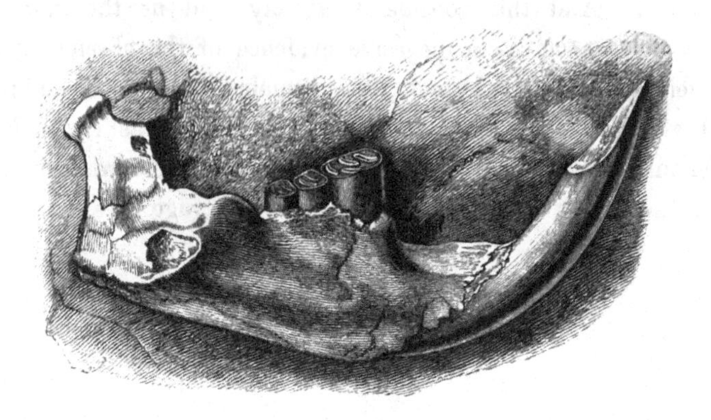

Lower jaw, ½ nat. size. Bacton.

TROGONTHERIUM CUVIERI. Cuvier's Gigantic Beaver.

Trogontherium Cuvieri, FISCHER, Mémoires de la Société des Naturalistes de Moscou, tom. ii. p. 250.
Castor trogontherium, CUVIER, Ossemens Fossiles, tom. v. pt. 1. p. 59.

THE discovery of the remains of a Rodent animal indicating an extinct sub-generic type in that order, and a species nearly allied to, but much exceeding in size the Beaver, which is now the largest of the indigenous Rodents of Europe, is due to M. Gothelf de Fischer, who has thus interpreted the characters of a fossil cranium from the sandy borders of the Sea of Azof.

Cuvier, to whom M. Fischer had transmitted a figure of this fossil, perceived its close resemblance to that of the Beaver, but observed, that it was about one-fifth larger than the European species, which itself surpasses in size the Beaver of North America. The length of the Siberian skull, from the occipital ridge to the most convex part of the incisors, was seven inches three lines. The chief difference which Cuvier recognized in the drawing was in the proportion of the last molar tooth of the upper jaw, which was longer, instead of being, as in the Beaver, shorter than the rest.

The first indication which presented itself to me of the *Trogontherium* as a British fossil, was from a fine specimen of the incisor of the lower jaw in John Hunter's Collection of Organic remains in whose manuscript catalogue it is described as "a long cutter of the Scalpris-dentata, or Glires genus, from Walker's Cliff, Norfolk." This tooth measures five inches and a half in length, and must have exceeded six inches when perfect, but it has suffered mutilation at both ends.

The chisel-crowned incisor in the lower jaw of the *Trogontherium* (fig. 71) measures seven inches, following the outer curve from the root to the abraded summit. This magnificent relic of the gigantic Beaver, which is now in the British Museum, was discovered by the Rev. Mr. Green, of Bacton, in that interesting lacustrine formation, with the submerged forest, which is noticed at p. 25: it was taken out of the bed of reddish sand which, at Ostend, has been spread immediately over the chalk. The incisive tooth is longer and stronger in proportion than in the existing Beavers, and doubtless operated with proportional effect upon the members of that ancient forest when they were green and flourishing. The projection of the crown, or exposed part of the incisor, is such, that the distance between its summit and the anterior border of the first molar is as great as from this part to the articular condyle; whilst, in the existing European Beaver, the diastema between the summit of the incisor and the first molar is little more than one-third of the extent from the incisor to the articular condyle. The lower incisor of the *Trogontherium* not only differs in both absolute and relative size, but also in shape. The anterior, or outer enamelled part of the tooth, is more convex, and, in a transverse section of the tooth (fig. 73, *a*), describes a semicircle. The inner, or mesial surface of the incisor, which in the existing and fossil Beavers is flat, is concave in the Trogontherium, as is also the outer surface of the tooth.

A well-marked sub-generic distinction, viz. the disproportionate size of the anterior molar, is well shewn in the figure above-cited, but I shall more particularly notice it in connection with the instructive portion of the lower jaw of the Trogontherium in the collection of Charles Lyell, Esq.,

Fig. 72.

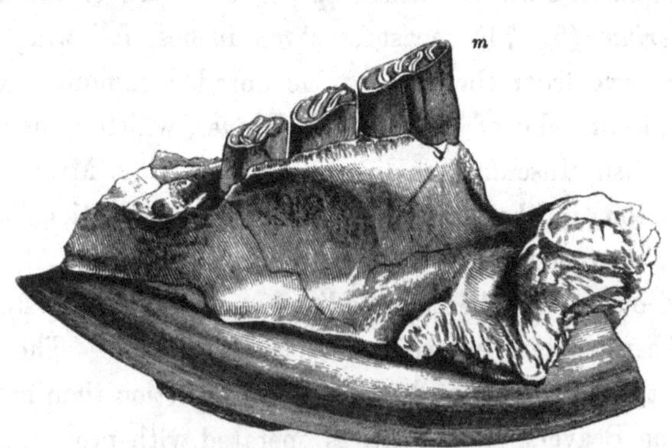

Trogontherium Cuvieri, nat. size. Fresh-water beds, Cromer.

by whose kind permission the specimen is here figured and described (figs. 72 and 73). It was discovered by Mr. Lyell in the blue clay which rests upon the Norwich crag at

Cromer. The stratum of clay is eight feet thick, and contains pyrites: its upper part is at about high water-mark, and it forms the beach. Here, or *in situ* in the blue clay, were discovered bones of the Mammoth, Rhinoceros, Ox, Horse, and Deer.

The Trogontherian relic is a portion of the right ramus of the lower jaw, containing half the root of the great incisor, and the three anterior molar teeth *in situ*. The proportions of these teeth differ conspicuously from those in the Beavers, both European and American. The antero-posterior and transverse diameters of the first grinder exceed by one-third those of the second grinder: both the second and third molars are smaller in proportion to the incisor than in the Beaver; and the socket of the fourth tooth shews this to have had a longer antero-posterior diameter than the third grinder, which is the reverse of the proportions of these teeth in the genus *Castor*. The grinding surface of the first molar, *m*, which alone bears the same proportion to the incisor as in the Beaver, has the same number and general direction of enamel-folds, viz. four,—three continued from the inner side of the tooth, and one from the outer side,—and this extends further into the substance of the tooth than in the Beaver, at an age when the molars are as much worn down as in the present specimen. The two succeeding grinders of the *Trogontherium* differ in a more marked degree from the corresponding teeth in the genus *Castor*, having but two inflected folds of enamel, one from the outer, the other from the inner side of the tooth; the latter also being relatively longer than the single inner fold of enamel in the Beaver's grinders, all of which retain the three outer folds of enamel.

The opportunity of instituting these comparisons is the more valuable, since M. Fischer has not entered into the details of the structure of the teeth in the upper jaw, which Cuvier has figured from the drawing transmitted to him,* and since neither of these anatomists appear to have had the opportunity of observing the dental characters of the lower jaw, in which they are probably best marked.

Cuvier even affirms that "the teeth, and all the forms of the head, bear the characters of a Beaver;" and proceeds, in the first edition of the "Ossemens Fossiles," to say that "it could not be distinguished from the head of the adult Beaver of Canada, if the fossil were not one-fourth larger. However, as it is not certain that we possess the skulls of those existing Beavers that attain the largest size; and since the Beaver formerly inhabited, and still, perhaps, inhabits the shores of the Euxine; since, also, nearly all the borders of the Sea of Azof are but vast alluvial formations,—I think one ought to know precisely the matrix of the skull in question before deciding whether it belonged to an extinct animal."† In the second edition of his great work, Cuvier modified his expressions, observing that in the drawing of the Siberian fossil the post-orbital process of the frontal bone has a somewhat different position from that of the Beaver, and that the temporal fossa seems scarcely to have exceeded the orbit in length; but he concludes, as in the first edition, by affirming that there can be no doubt respecting the *genus* of the animal, and that, until more certainty was acquired of its specific distinction, it might be provisionally named *Castor Trogontherium*.‡

The well-marked differences which the English fossils have demonstrated, not only in the proportions, but in the form

* Ossemens Fossiles, 1823, tom. v. pt. 1. pl. iii. fig. 11 and 12.

† Ossem. Fossiles, Ed. 1812, vol. iv. Rongeurs Fossiles, p. 4.

‡ Ed. 1823, vol. v. pt. 1. p. 59.

and structure of the teeth of the *Trogontherium*, will, I trust, be allowed to yield the same grounds for its sub-generic distinction, as have been proposed or accepted by the best modern Zoologists for the subdivisions of the same value in the rest of the Rodent Order.

Fig. 73.

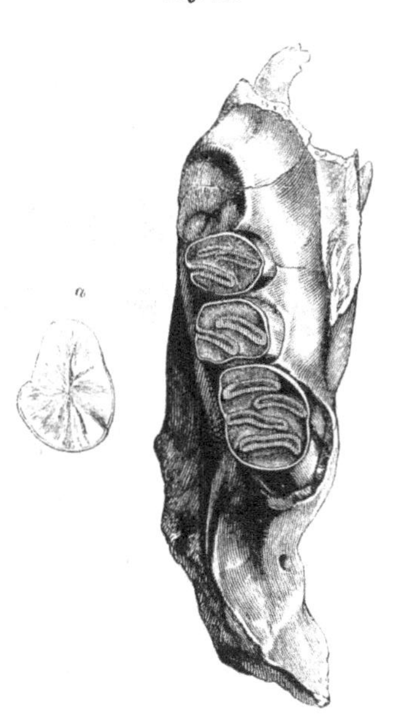

Portion of the lower jaw of the Trogontherium, shewing the grinding surface of the teeth. Nat. size. Cromer.

RODENTIA. *CASTORIDÆ.*

Fig. 74.

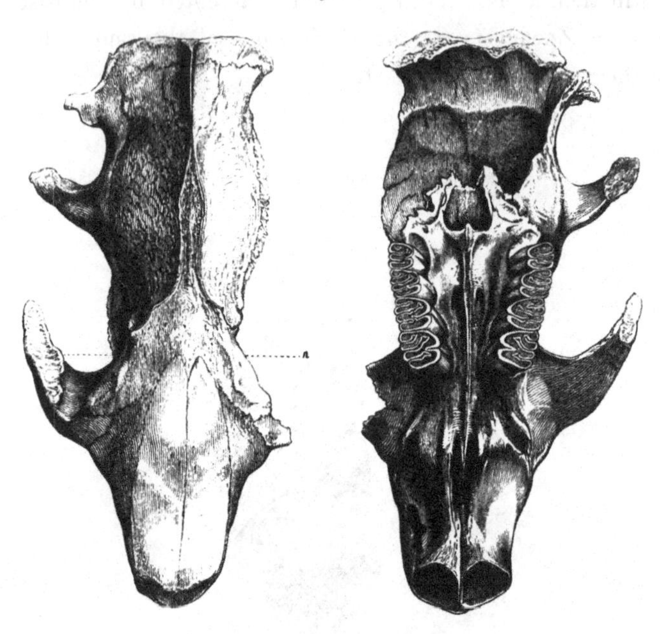

$\frac{2}{3}$ nat. size. Fens, Cambridgeshire.

CASTOR EUROPÆUS. European Beaver.

Castor des tourbières, CUVIER, Ossem. Fossiles, v. pt. 1. p. 55.
Castor fossilis, GOLDFUSS, Nova Acta Acad. Nat. Cur. tom. xi. pt. 2. p. 488.
Trogontherium Werneri, FISCHER, Mémoires de la Société des Naturalistes de Moscou, tom. ii. p. 250.

THAT the present European Beaver is not the degenerate descendant of the great Trogontherium, is proved, not only by the differences in the dental structure pointed out in the preceding section, but likewise by the fact that Beavers, in no respect differing in size or anatomical characters from the *Castor Europæus* of the present day, co-existed with the *Trogontherium*. Remains of the Beaver have been thus discovered by Mr. Green in the same fossilized condition, and under circumstances indicative of equal antiquity with the extinct Mammoth, in the lacustrine formations at Bacton. And M. Fischer, on his part, having received remains of a Beaver from near the Lake of Rostoff, in the department of Jarosslow, designated the species from the resemblance of the skull to that of the larger, and previously discovered Rodent, *Trogontherium Werneri*. Cuvier, however, to whom a drawing of the cranium had been transmitted, pronounced it to belong incontestably to the Beaver; it had the same dimensions, the same crests, and the same depressions as the skull of the *Castor Europæus*, with which it accorded in the smallest particulars. The contemporaneity of the beds in which this Beaver's skull and that of the Trogontherium were found in Russia, is not, however, so well ascertained as in the case of the Norfolk fossils referable to the *Castoridæ*. Remains of the Beaver (*Castor Europæus*), from the beach at Mundesley on the Norwich coast, are preserved in the collection of Miss H. Gurney, of North Repps Cottage, near Cromer; they are most probably from the fresh-water formation, and the Beaver may have been the inhabitant of that small river, which, Mr. Lyell imagines, "may have entered here, bringing down drift-wood, fresh-water shells, mud, and sand."* Mr. Woodward notices the occurrence of fossil remains of the Beaver in the cliffs at Mundesley, and in the oyster-bed at Happisburg, Norfolk, in his Geology of that county. Mr. Lyell submitted to my inspection some years ago a portion of the characteristic femur of the Beaver, from the fluvio-marine crag at Thorpe, in Suffolk.

* Philosophical Magazine, May, 1840, p. 253.

A portion of an incisor of the under-jaw of a Beaver, now in the Museum of the Geological Society of London, was found by the President, H. Warburton, Esq., M.P., in the fluvio-marine crag at Sizewell Gap, near Southwold, Norfolk. This formation has yielded remains, not only of the Rhinoceros and Mammoth, but also of the Mastodon, which carries the antiquity of the *Castor Europæus* far back into the tertiary period. Remains of the Beaver have been found associated with those of the Mammoth, Hippopotamus, Rhinoceros, Hyæna, and other extinct Mammalia, in the pleistocene fresh-water or drift formations of the Val d'Arno; and remains of both *Trogontherium* * and *Castor* † were found fossil by Dr. Schmerling in the ossiferous caverns in the neighbourhood of Liege. I have not yet obtained knowledge of any fossils of the Beaver family having been discovered in the bone caves of this country.

But the most common situation in which the remains of the Beaver are found in this island, as on the Continent, is the turbary, peat-bog, or moss-pit.

The earliest notice of such a discovery in this country is contained in a letter, dated February 24, 1757, from Dr. John Collet to the Bishop of Ossory, F.R.S., which is printed in the Philosophical Transactions for the year 1757, p. 109. It contains an account of the peat-pit near Newbury in Berkshire, and includes in the list of organic remains, "A great many horns, heads, and bones of several kinds of Deer, the horns of the Antelope, the heads and tusks of Boars, the heads of Beavers, &c.;" the author concludes by stating, "I have been told that some human

* Schmerling, Ossem. Foss. des Cavernes de Liège, tom. ii. pl. xxi. fig. 23, 24, 25.

† Ib. pl. xxi. fig. 40, 41.

bones have been found, but I never saw any of these myself, though I have of all the others. But I am assured that all these things are generally found at the bottom of the peat, or very near it. And, indeed, it is always very proper to be well and faithfully informed of the exact depth and place where anything of these kinds is found; whether it is in the earth above the peat, or in the clob, or in the true peat, or at the bottom of it, which will greatly assist us in forming a just judgment of the real antiquity of the things that are found, or at least of the time they have lain there."

This desirable kind of information I have been enabled to obtain, through Mr. Purdoe of Islington, a zealous collector of fossil remains, in relation to remains of jaws and teeth of the *Castor Europæus*, which were found twenty feet below the present surface in the Newbury peat valley. The section of the valley at this part disclosed, first, two feet of alluvium, then eight feet of a shell-marl, next ten feet of peat, then a second deposit of shell-marl, containing fresh-water shells of existing species; and in this stratum the Beaver's bones were found, associated with remains of the Wild Boar, Roebuck, Goat, Deer, and Wolf. The second bed of marl, rested on drift gravel.

Remains of the *Castor Europæus* have been found at the depth of eight feet and a half beneath peat, resting upon a stratum of clay, with much decayed and seemingly charred wood, associated with remains of the *Megaceros*, or great Irish Deer, at Hilgay, Norfolk. I owe this information, and the opportunity of examining the specimens, to Mr. Wickham Flower, F.G.S., in whose collection they are now contained.

Mr. Patrick Neill* cites an entry in the minutes of the

* Edinburgh Philosophical Journal, vol. i. p. 183.

Society of Antiquaries of Scotland, dated 16th December, 1788, specifying that "Dr. Farquharson presented to the Society the fossil skeleton of the head, and one of the haunch-bones, of a Beaver." On comparing the specimens, which were dyed of a deep chocolate colour, Mr. Neill identified them with the genus *Castor;* and on application to Dr. Farquharson, he learnt that these specimens were the remains of a Caledonian Beaver, having been dug up in the parish of Kinloch, in Perthshire, near the foot of the Grampian Hills, out of a marl-pit on the margin of the Loch of Marlee, under a covering of peat-moss, between five and six feet thick.

In October 1818, during the progress of draining a morass, called Middlestot's Bog, in the parish of Edrom, Berwickshire, a bed of shell-marl was exposed under the peat-moss. A layer of loose whitish substance, consisting of decayed *Musci*, of the species which grow in marshy situations, was found pretty generally between the bed of compact peat-moss and the bed of marl; and the skeleton of a Beaver was found imbedded partly in this loose and spongy matter, and partly in the marl below. "Only the hard bones of the cranium and face, and the jaw-bones, retained enough of their firm texture to fit them for being removed and preserved in a dry state. Around these, however, dispersed in rather a promiscuous manner, were many bones, which, from their size and appearance, evidently belonged to the same animal. Several of the long bones and vertebræ, while they remained *in situ*, seemed perfect; but, on being touched, they were found to be nearly in a state of dissolution, and, though some were carefully taken out, they speedily mouldered down on being exposed to the air and becoming dry."* Mr. Neill

* Mr. Neill, loc. cit. p. 184.

adds in a note the following judicious observation. "The apparent dislocation of the skeleton is not to be ascribed to violence, but to the gradual separation of the parts by unequal subsidence. The appearance of the marl, in which delicate shells of the genera *Limnea* and *Succinea* can be traced, indicates a long-continued state of tranquillity."

On comparing the fossil skull of the old Berwickshire Beaver with recent ones of the North American species, the nasal bones were observed to be proportionally larger in the fossil; it is not stated whether they were proportionally longer, or had their posterior apices produced farther back between the orbits. There can be little doubt, however, that they belong to the *Castor Europæus*, like the skull of the Beaver from a peat-moss in the valley of the Jomme in Picardy, figured in the 'Ossemens Fossiles,' and with which the Scottish specimens are stated closely to agree.

The next example of the remains of the Beaver from British localities which may here be cited, is that recorded by Mr. Okes in the Cambridge Philosophical Society. The specimens consisted of two left rami of two lower jaws, which were dug up in 1818, about three miles south of Chatteris, in the bed of the old West Water, formerly a considerable branch of communication between the Ouse and river Nen, but which, according to the traditions of the fen people, has been choked up for more than two centuries.* The length of one of the lower jaws was four inches eight lines.

* Mr. Okes says, "The accuracy of this tradition respecting the old West Water, is proved by the following extract from an order of Council quoted in Dugdale's History of the Fens.

"*Anno* 1617, 9 Maii, 15 Jac.—'That the rivers of Wisbeche, and all the branches of the Nene and West Water be clensed, and made in bredth and depth as much as by antient record they have been.'" Cambridge Philosophical Transactions, vol. i. 1822, p. 175.

John Hunter had obtained, from a moss-pit in Berkshire, the upper jaw and the right ramus of the lower jaw of a Beaver. These are rather smaller, and belonged to a younger animal than the Cambridgeshire specimens; but the portion of the skull exemplifies the character of the European Beaver, in the extension of the nasal bones to beyond the middle of the orbits. This character is also well shown in the skull of a Beaver more recently disinterred from the fens of Cambridgeshire, and figured at the head of the present section. The transverse line touching the point of the nasal bones, intersects the orbits behind their middle part; in the Canadian Beaver the transverse line touching the same points of the nasal bones, usually intersects the antorbital processes. In the view of the base of the skull (*fig.* 74) the complex inflections of the enamel upon the grinding surface of the molar teeth is shown. A very characteristic part of the skull of the Beaver was, however, lost in the specimen figured. In an entire skull recovered, with a great part of the skeleton, from the Cambridgeshire fens, and now in the museum of Professor Sedgwick, the character alluded to is well shown. It is manifested in the basilar process of the occipital bone, which has a peculiar cavity on the under and outer surface, as if the bone had been pressed upwards when soft, or indented by the end of a finger. This cavity lodges a peculiar sac of the pharynx in the recent animal; some additional lubrication is perhaps requisite to facilitate the deglutition of the coarse vegetable substances which chiefly constitute the food of the Beaver. This cavity is both deeper and wider in the old British Beaver, than in the Canada species. The following are the dimensions of the skull above-cited, from the Cambridge fens:

	In.	Lines.
Length of the skull	5	8
Breadth of do.	4	2
Length of the lower jaw	4	3
Height of do. at the coronoid process . .	2	6

These remains of the Beaver were met with in nearly the same position and locality as those in which the bones of the Otter, described at pp. 119—122, were found.

Mr. Lyell cites, from the Bulletin de la Société Géologique de France, tom. ii. p. 26, M. Morren's discovery, in the peat of Flanders, of the bones of Otters and Beavers; and he observes, "but no remains have been met with belonging to those extinct quadrupeds, of which the living congeners inhabit warmer latitudes, such as the Elephant, Rhinoceros, Hippopotamus, Hyæna, and Tiger, though these are so common in superficial deposits of silt, mud, sand, or stalactite, in various districts throughout Great Britain. Their absence seems to imply that they have ceased to live before the atmosphere of this part of the world acquired that cold and humid character which favours the growth of peat."* The Ox, the Horse, the Roebuck, the Red Deer, the Wild Boar, the Brown Bear, the Wolf, and the Beaver, of which animals the bones have been found under similar circumstances in fens and peat-bogs, have doubtless all existed as wild animals in this country since the formation of the peat began, and have been either gradually domesticated or extirpated by man.

With respect to the historical records and notices of the Beaver as an indigenous quadruped of Great Britain, Mr. Neill, in an interesting Memoir on the Beavers of Scotland,† states, that no mention of such an animal occurs in any of the public records now extant. In an act, dated June

* Principles of Geology, 1837, vol. iii. p. 187.

† Edinburgh Philosophical Journal, vol. i. p. 177.

1424, c. 22., "Of the custome of furringes," he says:—"mertricks (martens,) fowmartes (polecats,) otters and tods (foxes) are specified, but not a word is said of Beavers, although these, had they existed, must have been most valuable of all, not only for their furs, but for the substance called castor, found in the inguinal (preputial) glands of the animal, which, in those days still retained some share of its ancient repute as a medicine." The Beaver might, however, have become so scarce at the beginning of the 15th century, as to be not worth the attention of the legislature. At an earlier period, towards the end of the 12th century, Giraldus de Barri, in his 'Itinerarium Cambriæ,' lib. ii. cap. 3, speaking of the river Teivi in Cardiganshire, says, "Inter universos Cambriæ seu etiam Llægriæ fluvios, solus hic castores habet;" and adds, "In Albania quippe, ut fertur, fluvio similiter unico habentur sed rari." From which it would appear that the Beaver still existed in Scotland, but had then become a scarce animal.

Hector Boethius, however, enumerates the Beavers, '*fibri*,' among the animals which abounded in and about Loch Ness, and whose furs were in request for exportation towards the end of the 15th century, when he published his Description and History of Scotland.

Dr. Walker, Professor of Natural History in the University of Edinburgh, in his 'Mammalia Scotica,'* states, on the authority of Giraldus, that Beavers formerly existed in the country; and Mr. Neill adds, that Dr. Walker in his lectures used to mention that the Scotch Highlanders still retain, by tradition, a peculiar Gaelic name for the animal; this name, he was informed by Dr. Stuart of Luss, is *Losleathan*, derived from *los*, the tail, point, or end of a

* Posthumous Essays on Natural History, &c., 8vo. Edited by Mr. Charles Stewart.

thing, and *leathan*, broad; or '*Dobran losleathan*,' the Broad-tailed Otter. Dr. Stuart adds that he recollects to have heard of a tradition among the Highlanders, that the "Beaver, or Broad-tailed Otter, once abounded in Lochaber."

The evidence of the existence of the Beaver in Wales, within the historical period, is more decisive. Pennant cites a passage from a remarkable and interesting document of the 9th century, 'Leges Wallicæ,' or the Laws of Howel the Good, (Hywel D'ha,) book iii. § 11, 12, in which the prices of furs are regulated.

The Marten's skin is valued at 24*d.*

The Otter's (Ddyfrgi, or Lutra,) at 12*d.*

The Beaver's (Llosdlydan, or Castor,) at 120*d.*

Which shows that the Beaver had become very scarce at that period, but that it was still hunted for its skin, which was held in high estimation.

Mr. Neill, who likewise cites this authority in his Memoir on the Beavers of Scotland, notices the similarity between the Welsh and Gaelic names. And then quotes the 'Itinerarium Cambriæ' of Sylvester Giraldus de Barri. "This writer," says Mr. Neill, "made his journey into Wales, towards the end of the 12th century, or about three hundred years after the date of the laws of Hywel D'ha, as the attendant of no less a personage than Baldwin, Archbishop of Canterbury, whose zeal led him personally to excite the Welshmen to join in the projected crusades. In such company, and on such an errand, Giraldus must have had ample opportunities of intercourse with the best informed people of the districts through which he passed; and that he was inclined to be an observer of nature, is proved by the single fact, that when he arrives on the confines of the river Teivi in Cardiganshire, he immediately seems to forget the object of his mission, makes a long digression on the natural history of the Beaver, and enlarges with evident satisfaction on the habits of that singular animal."—" He mentions that in the course of time the habitations of the Beavers assume the appearance of a grove of willow-trees, rude and natural without, but artfully constructed within; that the Beaver has four teeth, two above, and two below, which cut like a carpenter's axe; and that it has a broad short tail, thick like the palm of the hand, which it uses as a rudder in swimming." The passage in which Giraldus states that, in his day, the Beaver continued to exist in the river Teivi alone, of all the rivers of Wales, has been already cited.

Pennant says that " Two or three waters in the Principality still bear the name of *Llyn yr afange*, or the Beaver Lake."

Tradition refers the name and arms of the town of Beverley in Yorkshire, to the fact of Beavers having abounded in the neighbouring river Hull.

Fig. 75.

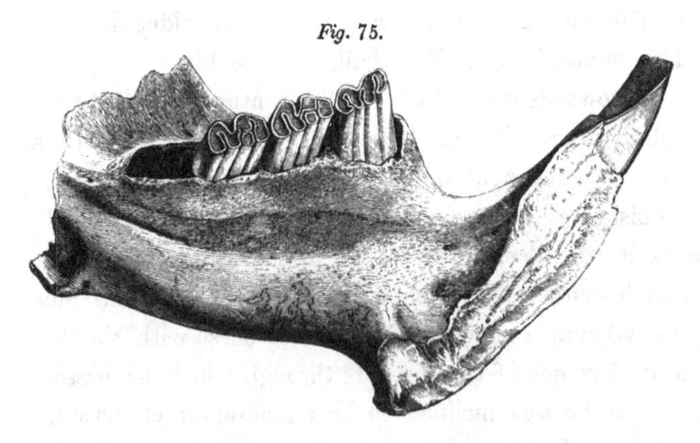

Lower jaw of Beaver, Nat. size. Peat Moss, Newbury.

RODENTIA. *CASTORIDÆ.*

Fig. 76.

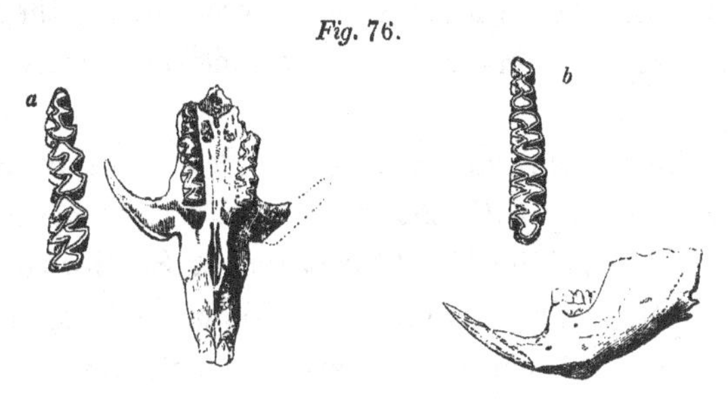

Fossil. Kent's Hole.

ARVICOLA AMPHIBIA. Water Vole.

Water-rat,	BUCKLAND, Reliquiæ Diluvianæ, p. 18, pl. 11, 12.
Campagnol des cavernes,	CUVIER, Ossem. Fossils, tom. v. pl. 1, p. 54.
Arvicola amphibia,	OWEN, Report of British Association for 1842.

MR. OKES, in his interesting Paper on the Fossil Beaver of Cambridgeshire, contrasts the recent nature of its matrix with the more ancient subjacent diluvial clay, in which remains of the Mammoth or extinct Elephant had been found; and he recurs to the authority of Cuvier, who, among the several general laws which he has laboured to establish, concerning the relations of organized remains, and the strata which contain them, has arrived at the following important conclusions: "that the bones of species which are apparently the same with those that still exist alive, are never found except in the very latest alluvial deposits, or those which are either formed on the sides of rivers, or in the bottoms of ancient lakes or marshes now dried up, or in the substance of beds of peat, or in the fissures and caverns of certain rocks, or at small depths below the present surface, in places where they may have been overwhelmed by debris, or even buried by man; and, although these bones are the most recent of all, they are almost always, owing to their superficial situation, the worst preserved." *

The fossil remains, however, of the Beaver discovered in the lacustrine clay with the submerged forest at Bacton, and those obtained by Mr. Warburton at Southwold, and by Mr. Lyell at Thorpe, from the fluvio-marine crag, carry back the date of this existing species to the pliocene tertiary period, when it was the associate of the Mammoth, Rhinoceros, and Hippopotamus.

The like antiquity of another and smaller Rodent of the Beaver family, still existing in most of our British rivers and smaller streams and ditches, is more abundantly testified by the numerous fossils of a species of Arvicola, which I have been unable satisfactorily to distinguish from the *Arvicola amphibia*, or common Water-rat.

Dr. Buckland appears to have been the first to have noticed the fossil *Arvicolæ* in British localities, observing, with regard to the Kirkdale cavern, that "the teeth which occur, perhaps in greatest abundance, are those of the Water-rat; for in almost every specimen I have collected or seen of the osseous breccia, there are teeth or broken fragments of the bones of this little animal mixed with, and adhering to, the fragments of all the larger bones. These rats may be supposed to have abounded on the edge of the lake, which I have shown probably existed at that time in this neighbourhood."†

The abundance of these small aquatic Mammals in the Kirkdale cave, at first view suggests a very common con-

* Cambridge Philosophical Transactions, vol. i. p. 177.

† Reliquiæ Diluvianæ, p. 18.

dition of limestone caves as the cause of their introduction, viz., that of being traversed by small rivers, which in some limestone countries lose themselves in cavities, and after running through a series of subterranean caverns, reappear on the surface at a distance from the spot where they first disappeared. A change in the relative levels and other physical features of a country, and a variety of other modifying circumstances, might afterwards alter the relation of a cavern to such subterraneous stream.

Dr. Buckland, however, is disposed to refer the introduction of the smaller Mammalia, as well as the larger ones, to the agency of the hyænas. He says,* "The extreme abundance of the teeth of Water-rats has also been alluded to; and though the idea of hyænas eating rats may appear ridiculous, it is consistent with the omnivorous appetite of modern hyænas, and with the fact, quoted by Johnson, that they feed on small animals, as well as carrion and bones; nor is the disproportion in size of the animal to that of its prey greater than that of wolves and foxes, which are supposed by Captain Parry to feed chiefly on mice, during the long winters of Melville Island. Hearne, in his 'Journey to the Northern Ocean,' mentions the fact 'of a hill, called Grizzly Bear Hill, being deeply furrowed and turned over like ploughed land, by bears in search of ground squirrels, and perhaps mice, which constitute a favourite part of their food.' If bears eat mice, why should not hyænas eat rats? Our largest dogs eat rats and mice; jackalls occasionally prey on mice, and dogs and foxes will eat frogs. It is probable, therefore, that neither the size nor aquatic habit of the Water-rat would secure it from the hyænas. They might occasionally, also, have eaten mice, weasels, rabbits, foxes, and

* Reliquiæ Diluvianæ, p. 33.

birds; and in masticating the bodies of these small animals with their coarse conical teeth, many bones and fragments of bone would be pressed outwards through their lips, and fall neglected to the ground."

Whatever cause may have operated on the introduction of the numerous Water-rats into the Kirkdale Cavern, a similar effect has been produced in many other caverns, both in this and other countries. Dr. Schmerling has figured characteristic remains of both large and small species of *Arvicola* from the caverns of Liège. The specimens of upper and lower jaw of the *Arvicola amphibia*, figured at the head of the present section, are amongst several specimens of this species from the cave of Kent's Hole, some of which are now in the British Museum. Remains of the *Arvicola amphibia*, (lower jaws) were found in the ossiferous cavern at Berry Head, Devon. Some of the bones from the cavernous fissures at Oreston, show marks of nibbling, which may be referred more probably to the incisors of a small Rodent, than to the canines of a weasel.*

Cuvier, to whom both specimens and drawings of the *Arvicola* from Kirkdale were transmitted, acknowledges that the jaws and teeth agree in size and other characters with the common Water-rat, but he found the other bones to be a little smaller, which led him to suspect that the species was not the same; but he adds that an entire skull of the fossil *Arvicola* could alone determine the question. So desirable a specimen has not, hitherto, been obtained from any British cavern. An os innominatum, the characteristic anchylosed tibia and fibula, and some vertebræ of the *Arvicola* from Kent's Hole, are not inferior in size to those of the existing Water-vole, with the dental

* See ante, p. 118.

and maxillary characters of which, the fossils of both Kirkdale and Kent's Hole closely agree. The upper incisors have a slightly convex and entire anterior enamelled surface, which in the fossils has lost the deep yellow colour that characterises the enamel in the recent Water-vole. The first molar consists of five triangular prisms, one anterior, two on the outer, and two on the inner side, alternately disposed; the second molar has four triangular prisms, as has also the third molar; and these teeth, (fig. 76 *a*,) progressively decrease in size from the first to the last, as in the recent species. The molars of the lower jaw, (fig. 76 *b*,) present the same close correspondence with those in the recent Water-vole.

Remains of the *Arvicola amphibia* have been found in newer pliocene deposits, associated with those of the usual extinct Mammalia, as at Erith and Stutton, at Crayford, Kent, and at Grays in Essex. They are very abundant in the lacustrine deposits, and the fluvio-marine crag along the Suffolk and Norfolk coasts. It is only from some of the older tertiary deposits in these parts, that I have noticed any well-marked indications of a species of *Arvicola* distinct from any now known to inhabit Britain. The remains to which I refer were portions of upper and lower jaws, discovered in the older pliocene crag near Norwich, from which molars of *Mastodon angustidens* have been obtained; they indicated a species of *Arvicola* intermediate in size between the Water-vole (*Arvicola amphibia*,) and the Field-vole (*Arvicola arvalis*).

RODENTIA. *CASTORIDÆ.*

Fig. 77.

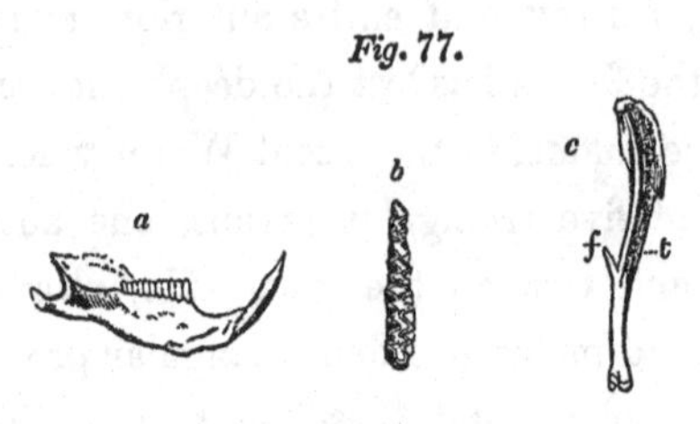

a, *c*, Nat. size. Kent's Hole.

ARVICOLA AGRESTIS. Field Vole.

Young Water-rat,	BUCKLAND, Reliquiæ Diluvianæ, p. 265, pl. xi. fig. 11. (?)
Petit Campagnol des Cavernes,	CUVIER, Ossem. Fossils, v. pl. 1, p. 54.

THE best preserved fossil specimens, from the caves at Kirkdale and Torquay, of the jaws and teeth of the species of *Arvicola* which are inferior in size to the common Water-rat, appear to me to be identical with the corresponding parts of two of our existing Voles. The jaw *a*, and leg-bone *c*, figured above, agree with those of the species with rootless molars figured by Mr. Bell in his British Quadrupeds, p. 325, as the Field-vole, *Arvicola agrestis* of Fleming, which is the *Mus arvalis* of Pallas.

Cuvier cites the jaws, teeth, and a thigh-bone, apparently of this little Rodent, from the cave at Kirkdale, which parts, he says, do not surpass in size the common Field-vole, (*Mus arvalis*, Linn.); but adds that the femur, though of the same length, is sensibly thicker, (*plus large transversalement*). The anchylosed tibia (*t*) and fibula (*f*), fig. 77, from Kent's Hole agree, like the jaws, with that of the existing Field-vole. A magnified view of the grinding surface of the three lower molar teeth is given at *b*, and the jaw *a*, is figured of the natural size. These specimens have all the characters of the fossils of the extinct Mammalia of the cave, Kent's Hole, from which they were obtained by Mr. Mac Enery; they are now in the British Museum.

A bank covering the foundations of an ancient Roman fortification near Cirencester, was pointed out to me by Mr. Brown, who has interested himself in the collection of the fossils of that neighbourhood, as being remarkable for the number of minute jaws and other bones which it contained: these were chiefly remains of Field-voles, mixed with those of Shrews; and, though they do not belong to the category of fossils, the fact seems worthy of notice on account of the extraordinary abundance in which such bones and teeth are occasionally found accumulated in similar superficial situations.

RODENTIA. *CASTORIDÆ.*

Fig. 78.

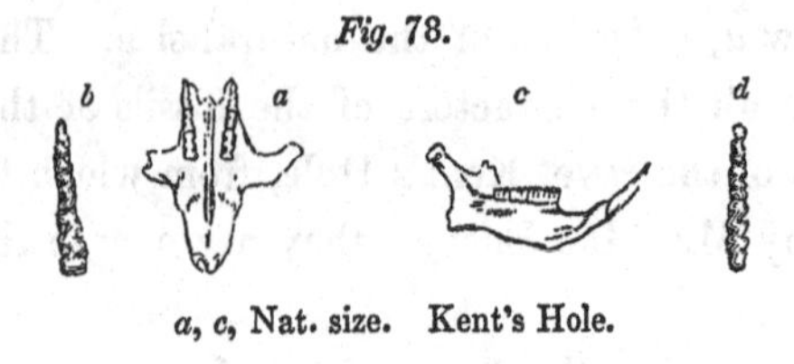

a, c, Nat. size. Kent's Hole.

ARVICOLA PRATENSIS. Bank Vole.

For the knowledge of this Vole, as a fossil of equal antiquity with the preceding, I am indebted to Mr. Waterhouse, whose special study of the osteology and dentition of the Rodent quadrupeds, particularly qualifies him for discriminating the nearly equal-sized and closely-allied species of the present genus. The Bank-vole is distinguished from the Field-vole, in addition to the characters pointed out by Mr. Bell,* by the early addition of roots to the molar teeth; the crowns of these teeth are also narrower in proportion to their antero-posterior extent, than in the *Arvicola agrestis;* both this character, and the smaller size of the jaws, are shown in the specimens figured above, where *b* is a magnified view of the grinding surface of the upper molars, and *d* that of the lower molars. They were obtained by Mr. Mac Enery from Kent's Hole, and are now in the British Museum.

These remains of the Bank Vole carry back, more unequivocally, perhaps, than those referred to the Field-vole, the date of these small and feeble Rodents to the remote antiquity of the æra of the great extinct Cave-mammalia. The existence of the little Bank-vole as a living member of the British Fauna, was discovered by Mr. Yarrell, who has described it under the name of *Arvicola riparia.*†

* British Quadrupeds, p. 331.

† Proceedings of the Zoological Society, 1832, p. 109.

RODENTIA. *MURIDÆ.*

Fig. 79.

Kirkdale, Nat. size. *a*, *b*, molars magnified.

MUS MUSCULUS. (?) Mouse.

Mouse,	Buckland, Reliquiæ Diluvianæ, pp. 19, 265, pl. 11, figs. 7, 8, 9.
Rat des Cavernes,	Cuvier, Ossemens Fossiles, tom. v., pt. i. p. 55.
Mus musculus fossilis,	Karg, Denkschriften der Vaterland: Gesellschaft der Aerzte und Naturf. 8vo. Schwabens.

Most unequivocal evidence of a species of true *Mus* has been yielded by the fossils from Kirkdale cavern, of which a lower jaw and teeth are figured in the 'Reliquiæ Diluvianæ.' Instead of the molars being rootless, and with deeply-inflected plates of enamel, a structure which approximates our so-called Water-rats and Field-mice to the Beaver, the true Rats and Mice have the crowns of the molars simply tuberculate, with the enamel bent into slight depressions on the grinding surface, and the crown is always supported by well-developed roots: this more simple form of tooth governs the mixed diet of the true *Muridæ*, the occasional carnivorous habits of which are well known. The fossil specimens of this genus differ from the common Mouse only by a slight superiority of size.

Fossil remains of species of *Mus* have been found in the tertiary beds at Œningen; in caves in the South of France,* and in Belgium.†

* Serres, Journal de Géol., iii., p. 254. † Schmerling, loc. cit.

RODENTIA. *LEPORIDÆ.*

Fig. 80.

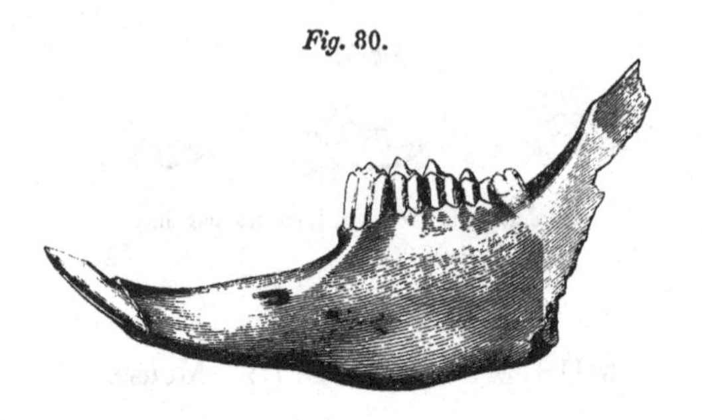

Nat. size. Kent's Hole.

LEPUS TIMIDUS. Hare.

Hare, or very large rabbit,	BUCKLAND, Reliquiæ Diluvianæ, pp. 19, 267, pl. 13, fig. 8.
Lièvre des cavernes,	CUVIER, Ossemens Fossiles, tom. v. pt. i. p. 55

THE land that could grow vegetation sufficient for the sustenance of colossal Mammoths, ponderous Rhinoceroses, and huge species of Deer and Oxen, may well be supposed to have afforded an abundant table for the smaller graminivorous quadrupeds, as the Hares and Rabbits; and it would seem that these, like the weakest species of the pliocene Carnivora, have survived and escaped those exterminating influences to which the gigantic quadrupeds have succumbed.

Dr. Buckland makes mention of "the jaw of a Hare, and a few teeth and bones of Rabbits and Mice, amongst the fossils of the Kirkdale Cave," and has given excellent figures of them in the 'Reliquiæ Diluvianæ.' Cuvier notices these illustrations of the fossil *Leporidæ*, in his great work. The heel-bone (calcaneum) figured in Pl. x. fig. 14, of the

'Reliquiæ,' has, he says, "the size and shape of that of a Hare:" the metatarsal bone, Pl. x. figs. 15 and 16, is that of the outer toe, and is nearly as long as that of a Hare, but is proportionally thicker. Cuvier adds, that he himself possesses a first phalanx of the hind foot from Kirkdale, which is also a little thicker in proportion than in the Hare; but the distal end of a tibia from the same cavern, exactly resembles the corresponding part of the Hare, and, with regard to a portion of jaw, he says, "I cannot perceive any difference that can be regarded of a specific nature;" and concludes that, "if these fragments appertain to a known species, it must be the Hare; the Rabbit would have them smaller and more slender."

The fossil lower jaws, from both Kirkdale and Kent's Hole which I have examined, have presented a somewhat shorter interspace between the molars and incisors, than in the common Hare of this country, with the same proportions of depth and other dimensions, and the same sized teeth; whereby it would appear that the Hare of the caves had a rather shorter head, and resembled in that respect the variety or species to which the name of *Lepus Hibernicus* has been given, and which has also somewhat stouter limbs than our English Hare.

I cannot detect any difference between the fossil Hare and the Irish Hare in the forms and proportions of the bones of the extremities: a very little increase of thickness being all that distinguishes the Irish from the English Hare in these parts of the skeleton.

Fossil remains of a Hare have been discovered by Croizet in the tertiary strata of the Puy de Dôme; by Serres, in the ossiferous caves of Montpelier, and by Schmerling in those of the Province of Liège.

RODENTIA. *LEPORIDÆ.*

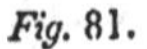

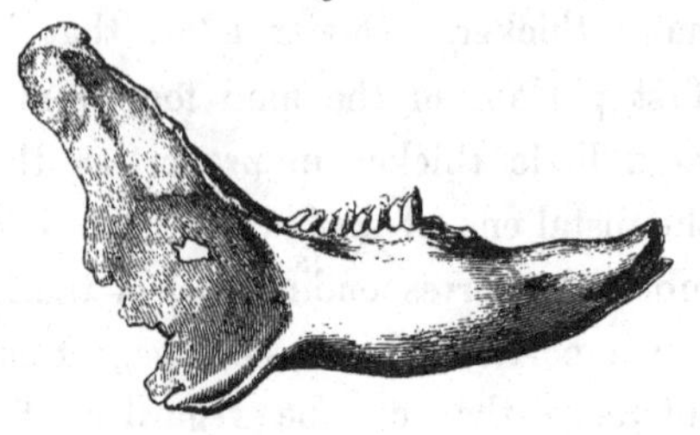

Nat. size. Kent's Hole.

LEPUS CUNICULUS. Rabbit.

Rabbit,	BUCKLAND, Reliquiæ Diluvianæ, p. 19, pl. x. xi.
Lapin des cavernes,	CUVIER, Ossemens Fossiles, v. pl. i. p. 55.

OF this smaller species of the Hare tribe portions of the jaws, teeth, and bones of the extremities, have been found fossil in the cave at Kirkdale, in Kent's Hole, and in the cave at Berry Head, Torquay; they closely accord with the corresponding parts in the existing wild Rabbit.

The specimen figured is the right ramus of the lower jaw of a young individual, from Kent's Hole; it is now in the British Museum.

Bones of the Rabbit form part of the osseous breccia of Corsica. MM. Serres, Dubreuil, and Jean-Jean,* describe and figure the fossil remains of two varieties of the *Lepus cuniculus*, which they discovered in the caverns at Lunel-Viel.

* Recherches sur les Ossemens Humatiles des Cavernes de Lunel-Viel, 4to, 1839, p. 130, pl. x.

RODENTIA. *LEPORIDÆ.*

Fig. 82. *Fig.* 83. *Fig.* 84.

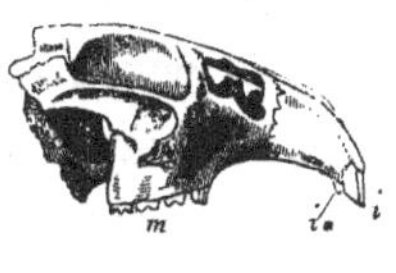

Fossil, Nat. size. Kent's Hole.

LAGOMYS SPELÆUS. Cave Pika.

THE fossil above figured possesses more than common interest. None of the circumstances attending its discovery, nor any character deducible from its colour or chemical state, indicate it to be an older fossil than the jaws and teeth of the Hares, Rabbits, Field-voles, or Water-voles already described; yet it unquestionably attests the former existence in England of a species of Rodent, whose genus not only is unrepresented at the present day in our British Fauna, but has long ceased to exist in any part of the continent of Europe.

The sole evidence of the geographical range of the Pikas, or tail-less Hares, having ever extended to Europe, has been in fact derived from fossil remains; and before natural history began to profit by the systematic study of such evidences, every other trace of the genus *Lagomys* had been so completely obliterated in Europe, that Zoologists had not the slightest knowledge of such a form in the Rodent Order, until Pallas made his journey into Siberia, when he announced the existence of three species of the tail-less Hares as the most curious little animals which he had detected in that remote and unfrequented region.

The Pikas are remarkable for their industrial instincts, which lead them in the summer season to select and dry a quantity of herbage for their winter provision. These haystacks, which are sometimes six or seven feet high, are a valuable resource for the horses of the sable-hunters. Since the time of Pallas, species of *Lagomys* have been discovered at a considerable altitude on the Himalayas, and also in North America.

The former existence of Pikas, or tail-less Hares in Europe, appears to have been first recognised by Cuvier,* who determined a species, nearly allied to the *Lagomys alpinus* of Siberia, amongst the fossils of the ossiferous breccia at Cette, in Corsica; and he was led to suspect the existence of another species of *Lagomys*, by the inspection of certain drawings of fossil jaws, and other bones from the breccias of Gibraltar, preserved in the museum of Adrien Camper.

The relations of these fossils to the Siberian genus *Lagomys* were more definitely pointed out by Wagner in Kastner's 'Archiv fur Naturgeschichte,' tom. iv.

The fossil from Kent's Hole consists of the facial or maxillary part of the skull of a full-grown individual, with the molar and incisive teeth *in situ* on one side, demonstrating the longitudinal furrow on the large anterior chisel-shaped incisor, (fig. 82,) and the small posterior supplementary incisors, (*i*, fig. 83,) which the genus *Lagomys* has in common with the ordinary Hares and Rabbits.

The dentition of the small Siberian tail-less Hares closely resembles that in the true genus *Lepus*, in the form of the teeth, and differs principally in the absence of the small molar tooth which terminates the series posteriorly

* Ossemens Fossiles, tom. iv. pp. 174, 178.

in the Hare; the number of molars is thus reduced in the *Lagomys* to five on each side of the upper jaw, instead of six, as in the Hares; and it is precisely this sub-generic distinction that the fossil from Kent's Hole demonstrates.

This fossil agrees in size with the corresponding part of the skull of the existing Siberian species, called *Lagomys pusillus*, but it resembles more in its configuration that of the *Lagomys alpinus*, which is the larger Siberian species; the fossil presents, for example, a less relative depth of the fore part of the alveolar process of the upper jaw, than in the *Lagomys pusillus*; the characteristic descending obtuse process (*a*, fig. 82) of the malar bone overhangs in a greater degree the alveolar process than in the *Lagomys pusillus*: the upper border of the zygoma is slightly convex in the *Lagomys spelæus*, not concave as in the *Lagomys pusillus*: the suborbital foramen beneath the vacuity in the nasal process of the maxillary is relatively larger than in the *Lagomys pusillus*, and is divided on both sides of the face by a slender osseous bar, which makes it double.

Pallas alludes to the idea entertained by some Naturalists of his time,* that the Cavies of South America were modified Hares or Rabbits, and he saw that the transmutation theory might be more plausibly applied to the Siberian leporine animals, which, retaining the essential character of the dentition and internal organization of the Hare, but with curtailed ears and shorter hind legs, have entirely lost the small trace of tail which that animal possesses.

The great naturalist of Asiatic Russia remarks, however, with his wonted sound judgment:—"Sed non est ea

* See Buffon's Histoire Naturelle, "Dégénération des Animaux," tom. xiv., p. 372; who does not, however, admit the application of the hypothesis of transmutation to the South American Rodentia.

Naturæ rerum paupertas, ut depravatione formarum varietatem sibi quærere velit; estque, ut omnia probant, imaginaria hæc specierum transmutatio." We can now add to the proofs of the stability of generic and specific characters which Pallas might have derived from the organization of the Siberian species of *Lagomys*, the fact that a species of that genus whose chief external characters might be explained by hypothetical degeneration of the ears, hind-legs, and tail of the Hare, claims as high an antiquity as any species of the true genus *Lepus*.

Remains of a *Lagomys* were discovered by Professor Sedgwick and Mr. Murchison in a lacustrine deposit at Œningen, associated, like the Pika of Kent's Hole, with a species of Fox, indistinguishable from that now existing, and affording another of those instructive examples which exhibit a gradual passage from an ancient Fauna, or animal population, to that which now prevails.

PACHYDERMATA. *PROBOSCIDIA.*

Fig. 85.

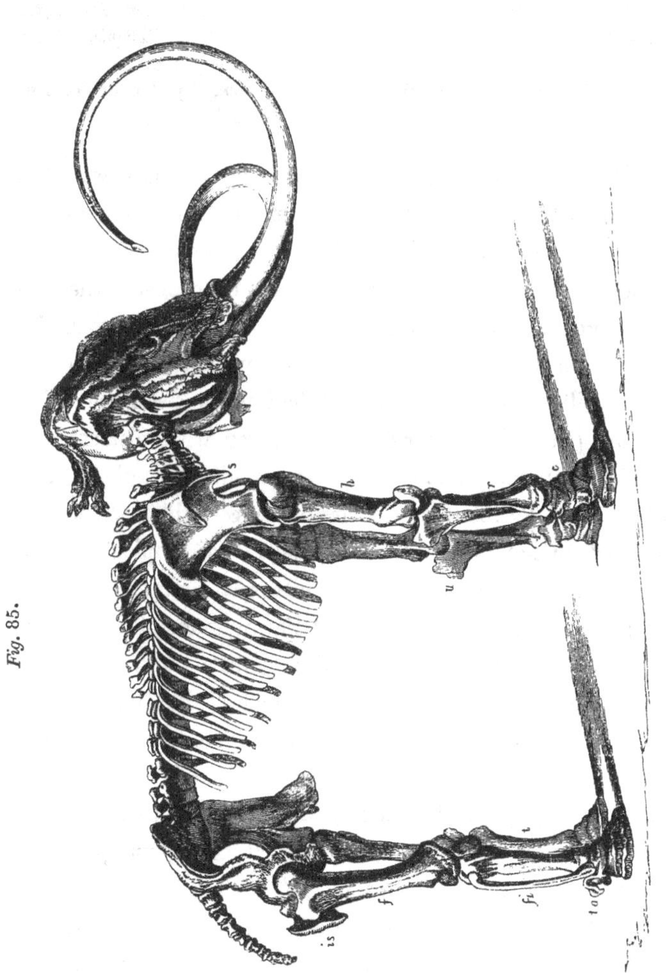

Mammoth found in Siberia, 1799.

ELEPHAS PRIMIGENIUS. Mammoth.

Elephas primigenius,	BLUMENBACH, Voigt's Magazine, Bd. v.
Éléphant fossile ou Mammouth des Russes,	CUVIER, Annales du Muséum, tom. viii. Ossem. Fossiles, t. i.
Fossil Elephant,	BUCKLAND, Reliquiæ Diluvianæ, p. 171.
Elephas primigenius or Mammoth,	OWEN, Report of British Association, 1843.

WHEN the science of fossil organic remains was less advanced than it is at present, when its facts and generalizations were new, and sounded strange not only to ears unscientific but to anatomists and naturalists, the announcement of the former existence of animals in countries where the like had not been known within the memory of man, still more of species that had never been seen alive in any part of the world, was received with distrust and doubt, and many endeavours were made to explain these phenomena by reference to circumstances which experience showed to have led to the introduction of tropical animals into temperate zones within the historical period.

When Cuvier first announced the presence of remains of Elephants, Rhinoceroses and Hippopotamuses in the superficial unstratified deposits of continental Europe, he was reminded of the Elephants that were introduced into Italy by Pyrrhus in the Roman wars, and afterwards more abundantly, and with the stranger quadrupeds of conquered tropical countries, in the Roman triumphs and games of the amphitheatre. The minute anatomical distinctions by which the great Comparative Anatomist proved the disinterred fossils to have belonged to extinct species of *Elephas*, *Hippopotamus*, *Rhinoceros*, &c., were at first hardly appreciated, and, by some of his contemporaries, were explained away or disallowed. Cuvier, therefore, appealed with peculiar satisfaction to the testimonies and records of analogous Mammalian fossils in the British Isles, to the origin of which it was obvious that the hypothesis of Roman or other foreign introduction within the historical period could not be made applicable.

"If," says the founder of palæontological science, "passing across the German Ocean, we transport ourselves into Britain, which, in ancient history, by its position, could not have received many living elephants besides that one which Cæsar brought thither according to Polyænus;* we shall, nevertheless, find there fossils in as great abundance as on the continent."

Cuvier then cites the account given by Sir Hans Sloane of an elephant's fossil tusk, disinterred in Gray's Inn Lane, out of the gravel twelve feet below the surface. Sir Hans Sloane had obtained also the molars of an elephant from the county of Northampton, which were found in blue clay beneath vegetable mould and loam, from three to six feet below the surface: these specimens were explained by Dr. Cüper as having belonged to the identical elephant brought over to England by Cæsar; but Cuvier remarks that too many similar fossils had been found in England to render that conjecture admissible. He then proceeds to quote the instances of this kind on record, at the period of the publication of the 'Ossemens Fossiles.'

Dr. Buckland adds the weighty objection, that the remains of these Elephants are usually accompanied in England, as on the continent, by the bones of the Rhinoceros and Hippopotamus, animals which could never have been attached to Roman armies; and I may add, that the natural historians of Ireland, Neville and Molineux, made known in 1715 the existence of fossil molar teeth of the

* Lib. viii. c. 23. § 5. cited in Ossem. Fossiles, 4to, 1821, tom. i. p. 134.

Elephant at Maghery, eight miles from Belturbet in the county of Cavan, and similar evidences of the Elephant have since been discovered in other localities of Ireland, where the armies of Cæsar never set foot. Some other hypothesis must therefore be resorted to in order to explain these phenomena.

Observation, the basis of all sound hypotheses, has shown in the first place that the remains of the Elephants which are scattered over Europe in the unstratified superficial deposits called 'Diluvium,' 'Drift,' 'Till,' and 'Glacio-diluvium,' as well as those from the upper tertiary strata, are specifically different from the teeth and bones of the two known existing Elephants, *Elephas indicus* and *El. africanus.* This fundamental fact, when first appreciated by Cuvier, who announced it in 1796, opened to him, he says, entirely new views of the theory of the earth, and a rapid glance, guided by the new and pregnant idea, over other fossil bones, made him anticipate all that he afterwards proved, and determined him to consecrate to this great work the future years of his life.

The differences which the skull of the fossil Elephant presents as compared with the recent species are, the more angular form and relative shortness of the zygomatic processes; the longer, more pointed and more curved form of the postorbital process; the larger and more prominent tubercle of the lachrymal bone; the greater length of the sockets of the tusks; the more parallel position of the right and left sockets of the grinders, making the anterior interspace and channels at the junction of the rami of the lower jaw proportionably wider than in the existing Elephants. Of these characteristics, I have verified the last-mentioned instance, taken from the lower jaw, by observation of English specimens; they are well displayed in the lower jaw of a young Mammoth, (fig. 86), disinterred from a Pleistocene bed near Yarmouth in the county of Norfolk,

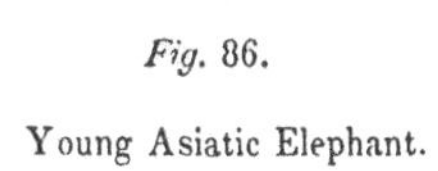

Fig. 86.

Young Asiatic Elephant.

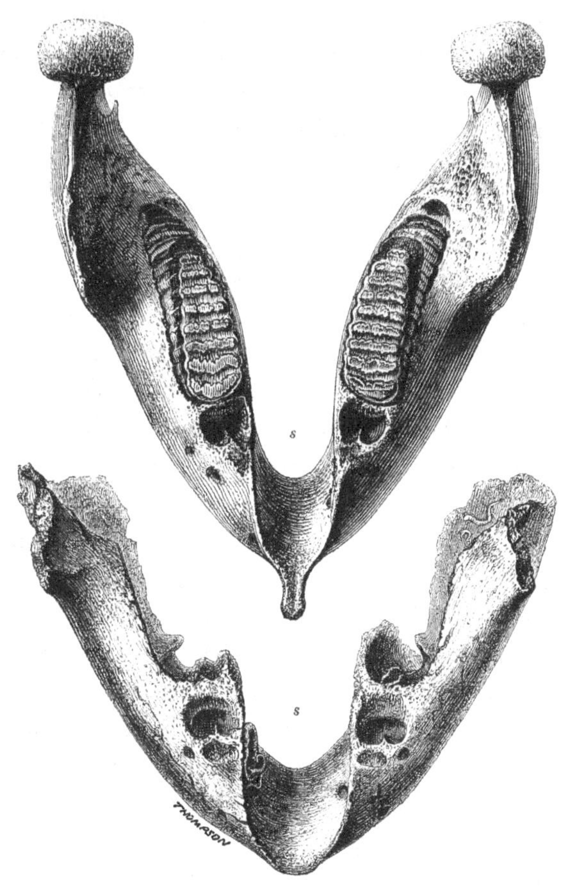

Young Mammoth.

and now in the possession of Mr. E. Stone, of Garlick Hill, London.

This lower jaw shows also that the outer contour of one ramus meets that of the other at a more open angle than in the African or Asiatic Elephant, and that the symphysis itself (*s*), though acute at this period of life, is less prolonged; in illustration of which the figure of the lower jaw of the Asiatic Elephant at a corresponding age with the fossil, is added. In the older Mammoths the symphysis becomes obtuse; were it otherwise, the prolonged alveoli of the fully developed tusks would have interfered with the motion of the lower jaw.

The difference between the extinct and existing species of Elephant, in regard to the structure of the teeth, has been more or less manifested by every specimen of fossil elephant's tooth that I have hitherto seen from British strata, and those now amount to upwards of three thousand. Very few of them could be mistaken by a comparative anatomist for the tooth of an Asiatic Elephant, and they are all obviously distinct from the peculiar molars of the African Elephant.

Cuvier, who had recognised a certain range of variety in the structure of the numerous teeth of the Mammoth from continental localities, found nevertheless, that the molars of the fossil Elephant were broader in proportion to their length or antero-posterior diameter than in the existing species; that the transverse plates were thinner and more numerous in the fossil molars than in those of the Indian Elephant; that a greater number of plates entered into the formation of the grinding surface of the tooth, and that the lines of enamel were less festooned; but to this character there are exceptions, especially in the large molars of aged individuals.

The development, progressive complication, and succession of the molar teeth, obeyed the same laws in the ancient Mammoth, as in the existing Elephant; it may, indeed, be affirmed that these most remarkable phenomena in the comparative anatomy and physiology of teeth are more fully and perfectly illustrated by the fossils which the primigenial Elephants have left in the superficial deposits of England, than by any collection of the molars of the Indian or African Elephants now existing in our metropolitan museums. John Hunter owed most of his knowledge, and his specimens illustrative of the succession and shedding of the teeth in the genus *Elephas*, to the fossil molars of the Mammoth, which, with similar remains, he had been silently collecting at a time when they attracted little if any attention, and some years before the recent Elephant's teeth brought from India by Mr. Corse, afforded the materials for Mr. Corse's and Sir Everard Home's papers on this subject in the eighty-ninth volume of the Philosophical Transactions.

In a fossil lower jaw of a Mammoth, younger than the subject of figure 86, which was obtained by the late John Gibson, Esq., of Stratford, from the pleistocene brick-earth at Ilford, the remains of the socket of the molar corresponding to the first small one in the Indian Elephant, and the crown of which is divided into four transverse plates, are still visible; it is about one inch in length.*

This tooth is succeeded by a second molar consisting of eight transverse plates, the length, or antero-posterior extent of the tooth being three inches, its breadth, one inch and a half. Dr. Buckland has figured the corresponding second molar of the upper jaw of a young Mammoth in pl. 7, fig. 1, of the 'Reliquiæ Diluvianæ:' the specimen was discovered in the Hyæna-cave at Kirkdale. The subjoined cut (fig. 87) gives a view of a second molar tooth of the

* In the Asiatic Elephant, the corresponding molar cuts the gum eight or ten days after birth, and is shed at the age of two years.

lower jaw of a young Mammoth, from the bone-cave at Kent's Hole, near Torquay: the crown of which is divided, like that from Ilford and Kirkdale, into eight transverse plates: and is supported by two fangs or roots, a small anterior, and a thick and large posterior one: the sockets of these fangs are shown in fig. 86, anterior to the empty socket of the third molar in the young Mammoth, and anterior to that molar which is in place, in the young Elephant.

Fig. 87.

⅓ Nat. size

The average size of the second lower molar tooth in the Indian Elephant,* is two inches and a half in length, and one inch in breadth, which, compared with the dimensions of the corresponding molar of the young Mammoth above given, shows that already the specific character of the *Elephas primigenius*, founded on the superior breadth of the tooth, is recognisable. I have found this character still more strongly manifested in the second molar of a young Mammoth which had perished before that tooth had come into use; it was found in the pleistocene fresh-water deposits, exposed on the sea-coast, near Cromer, Norfolk; the crown, which is divided, as in the rest, into eight plates, measures three inches in antero-posterior diameter, and two inches in breadth.† An entire second molar of the lower jaw of a young Mammoth, from the pleistocene blue-clay at Mundesley, Norfolk, had the crown, which measured three inches in length, and one inch five lines in breadth, divided into seven plates: it belongs to what will be subsequently

* This tooth is shed before the Elephant has attained its sixth year.

† The Mammoth's molar from the drift at Fouvent, figured by Cuvier in the 'Ossemens Fossiles,' vol. i. pl. vi. fig. 2, as "une vraie molaire de lait," is a much worn and naturally shed second molar: the figure is half the natural size.

described as the "thick-plated" variety of Mammoth's molar; yet, nevertheless, exhibited the characteristic superior breadth, as compared with the Indian Elephant, in a corresponding molar of which species divided into nine plates, the length of the crown was three inches, and its breadth one inch.

A third upper molar of the Mammoth from the drift at Hinton, Somersetshire, has the crown divided into twelve plates, and measures three inches, four lines in length, and one inch and a half in breadth. This would be precisely the size of the molar tooth of the young Mammoth, figured in Cuvier's 'Ossemens Fossiles,' pl. vi., *Éléphans*, fig. 4, if the figure be, as I suspect, half the size of nature. In a corresponding upper molar of an Indian Elephant of equal breadth, but greater length than the preceding, I found eleven lamellar divisions of the crown; the more common number is twelve or thirteen.*

The number of the coronal plates of the fourth grinder in the Indian Elephant is fifteen or sixteen; the greatest number in the last molar developed, the seventh or eighth in succession, is, according to Mr. Corse,† twenty-two or twenty-three. The number of the coronal plates is subject to greater variation in the Mammoth, and increases in a less regular ratio in each succeeding molar. The fourth molar of the upper jaw, with an antero-posterior extent of from seven to nine inches, varies in the number of its plates from twelve or sixteen.

The fifth molar, with an antero-posterior diameter of from ten to eleven inches, may have from sixteen to nineteen plates.

The largest upper molar of the Mammoth which I

* This tooth begins to appear above the gum at the end of the second year; and is shed during the ninth year.

† Philos. Trans. 1799, p. 224.

have yet seen, measured fifteen inches in length, and had twenty-two coronal plates: it was discovered in the drift at Wellsborne in Warwickshire. Mammoths' molars of less dimensions have come under my observation, in which the crown had been divided into twenty-five and twenty-six transverse plates.

In the lower jaw, the grinders as they succeed one another from behind forwards are also larger, and have more numerous plates than those which they displace, and the number of plates increases more gradually and with less constancy than in the Asiatic Elephant.

A lower molar of the Mammoth may always be distinguished from an upper molar, by the grinding surface being slightly concave in the direction of its longest diameter, that of the upper molar being in the same degree convex.

The largest lower molar of a Mammoth that has come under my observation, is the one represented in fig. 90: its length, or antero-posterior diameter, following the curve on the convex side is one foot seven inches: the number of the lamelliform divisions of the crown is twenty-eight. This remarkably fine molar exhibits the most complete state in which the progressive development and the actions of mastication permit so large a grinder to be seen: the anterior portion of the crown having been worn down to the common base of dentine (d'), from which the fang is continued; whilst the last, or hindmost plates, have been completed, as far as the formation of the digital divisions (f, f), which form, by their basal confluence, the transverse plate.

The complex structure, and mode of growth of the molar teeth in the genus *Elephas* is so well illustrated by this specimen, that I shall here give the brief account which is necessary for the intelligibility of subsequent references to those teeth, in the important question of the species or varieties of Mammoth that formerly inhabited England.

The crown of the molar of the Mammoth, like that of the existing species of Elephant, consists of, or is divided into, a number of transverse perpendicular plates, composed of two distinct substances, and cemented together by a third substance. The body of each plate consists of the basal constituent of a tooth called "dentine," of which ivory is a modification; it is marked *d*, in the figures of the teeth in this section. The dentine is coated by a layer of harder substance called "enamel" (*e*), and the interspaces of the plates so formed are filled by a less dense substance called "cement" (*c*), because it fastens together the several divisions of the crown, and more strikingly fulfils the office of cement when those divisions are incompletely formed and not united by mutual confluence. As the growth of each plate begins at the summit, they remain detached and like so many separate teeth or denticules, until their base is completed, when it becomes expanded and blended with the bases of contiguous plates to form the common dentinal body of the crown of the complex tooth, from which the roots are next developed.

But the composition and growth of the plates are analogous to, and almost as complex as, that of the entire tooth; each plate consists at first of a series of separate slender conical columns or digital processes, arranged transversely across the tooth. The formation of these columns begins at their summit, and descends, their bases gradually expanding until they are blended together to form a continuous transverse plate; just as the plates are subsequently blended together to form the continuous longitudinal crown of the whole grinder. The digital processes and the digitated plates of an incompletely developed tooth are held together, prior to their basal confluence, by the external cement; this substance is generally more or less decomposed in fossil grinders, and the parts of the complex tooth then become detached: a separate plate, with its digital processes, offers a rude resemblance to a hand, and such specimens have been figured by the older collectors of petrifactions, under the name of "Cheirolites," as the fossilized hand of a monkey or a child. The digital processes of the last formed plates of the large molar tooth in fig. 90, are shown at *f*, *f*, *f:* this figure well illustrates the progressive development of the Elephant's tooth from before backwards; the formation begins with the summits of the anterior plate, and the rest are completed in succession: the tooth is gradually advanced in position as the growth proceeds, and its anterior plates are brought into use before the posterior ones are formed. When it cuts the gum, the cement is first rubbed off the digital summits; then their enamelled cap is worn away and the central dentine exposed; next, the digital processes are ground down to their common uniting base and a transverse tract of dentine —now the upper margin of the plate—with its border of enamel is exposed; finally, the transverse plates themselves are abraded to their common base of dentine, and a smooth and polished tract of that substance of greater or less extent is produced. When the tooth has been thus reduced to an uniform surface it becomes useless as an instrument for grinding the coarse vegetable substances on which the Elephant subsists, and is shed.

The tooth figured (90) exhibits all the foregoing stages of attrition: the common longitudinal base of dentine is exposed at *d′;* the continuous margin of a transverse plate at *d:* and the digital summits of varying breadth, according to the degree of abrasion, appear behind, and are followed by that part of the complex grinder which was concealed in the closed recess of the socket, and which part, in the present instance, is folded upwards and laterally upon the concave side of the tooth; the sides of the digitated plates being parallel with the grinding surface of the tooth.

There are few examples of natural structures that manifest a more striking adaptation of a highly complex and beautiful structure to the exigences of the animal endowed with it, than the grinding teeth of the Elephant. Thus the jaw is not encumbered with the whole weight of the massive tooth at once, but it is formed by degrees as it is required; the subdivision of the crown into a number of successive plates, and of the plates into subcylindrical processes, presenting the conditions most favourable to progressive formation. But a more important advantage is gained by this subdivision of the grinder: each part is formed like a perfect tooth, having a body of dentine, a coat of enamel, and an outer investment of cement; a single digital process may be compared to the simple tooth of a Carnivore; a transverse row of these, therefore, when the work of mastication has commenced, presents, by virtue of the different densities of their constituent substances, a series of cylindrical ridges of enamel, with as many depressions of dentine, and deeper external valleys of cement: the more advanced and more abraded part of the crown is traversed by the transverse ridges of the enamel investing the plates, inclosing the depressed surface of the dentine, and separated by the deeper channels of the cement: the fore-part of the tooth exhibits its least efficient condition for mastication, the inequalities of the grinding surface being reduced in proportion as the enamel and cement which invested the dentinal plates have been worn away. This part of the tooth is, however, still fitted for the first coarse crushing of the branches of a tree: the transverse enamel ridges of the succeeding part of the tooth divide the food into smaller fragments, and the posterior islands and tubercles of enamel pound it to the pulp fit for deglutition. The structure and progressive development of the tooth not only give to the Elephant's grinder the advantage of the uneven surface which adapts the millstone for its office, but, at the same time, secure the constant presence of the most efficient arrangement for the finer comminution of the food, at the part of the mouth which is nearest the fauces.

One cannot contemplate the more numerous lamelliform divisions and subcylindrical subdivisions of the crown of the Mammoth's molar, and the resulting increase of the dense enamel that enters into the formation of the grinding surface, as compared with the teeth of the Indian and African Elephants, without connecting that specific difference of structure with the coarser kind of vegetable food, on which the geographical position of the Mammoth in the temperate regions of the ancient world would most probably compel it to subsist.*

VARIETIES.—QUESTION OF SPECIES.

The varieties to which the grinders of the Elephant are subject in regard to the thickness and number of their plates, increase in the ratio of the average number of the plates which characterizes the molar teeth of the different species. Thus in the African Elephant, (fig. 88,) in which

* The reader desirous of full information on the structure, growth, and succession of the teeth of Elephants, is referred to Mr. Corse's and Sir Everard Home's Papers in the 89th Volume of the Philosophical Transactions, and to the 'Ossemens Fossiles' of Cuvier, tom. i. p. 31.

the lozenge-shaped plates are always much fewer and thicker than the flattened ones in the Asiatic species, the variation which can be detected in any number of the grinders of the same size is very slight.

In the molars of the Asiatic Elephant, (fig. 89,) which, besides the difference in the shape of the plates, have always thinner and more numerous plates than those of the African species, a greater amount of variation in both these characters obtains; but it is always necessary to bear in mind the caution which Cuvier suggested to Camper, that a large molar of an old Elephant is not to be compared with a small molar of a young one, otherwise, there will appear to be a much greater discrepancy in the thickness of the plates than really exists in the species; and the like caution is still more requisite in the comparison of the molars of the Mammoth (*Elephas primigenius*), which, having normally more numerous and thinner plates than in the existing Asiatic Elephant, present a much greater range of variety.

Of the extent of this variety in the British fossils, some idea may be gained by the fact, that in one private collection, that of Miss Gurney of Cromer, of fossil Mammalian remains from a restricted locality, there are Mammoth's teeth from the drift of the adjacent coast, one of which, measuring ten inches nine lines in antero-posterior diameter, has nineteen plates, whilst another grinder, eleven inches in antero-posterior diameter, has only thirteen plates.

A greater contrast is presented by two grinders of the Mammoth from British diluvium in the collection of the late Mr. Parkinson, one of which, with a grinding surface of five inches and a half in antero-posterior extent, exhibits the abraded summits of seventeen plates, whilst the other shows only nine plates in the same extent of grinding surface.

Some palæontologists have viewed these differences as indications of distinct species of *Elephas*. But the vast number of grinders of the Mammoth from British strata which have been in my hands in the course of the last three years, have presented so many intermediate gradations, in the number of plates, between the two extremes above cited, that I have not been able to draw a well-defined line between the thick-plated and the thin-plated varieties of the molar teeth. And if these varieties actually belonged to distinct species of Mammoth, those species must have merged

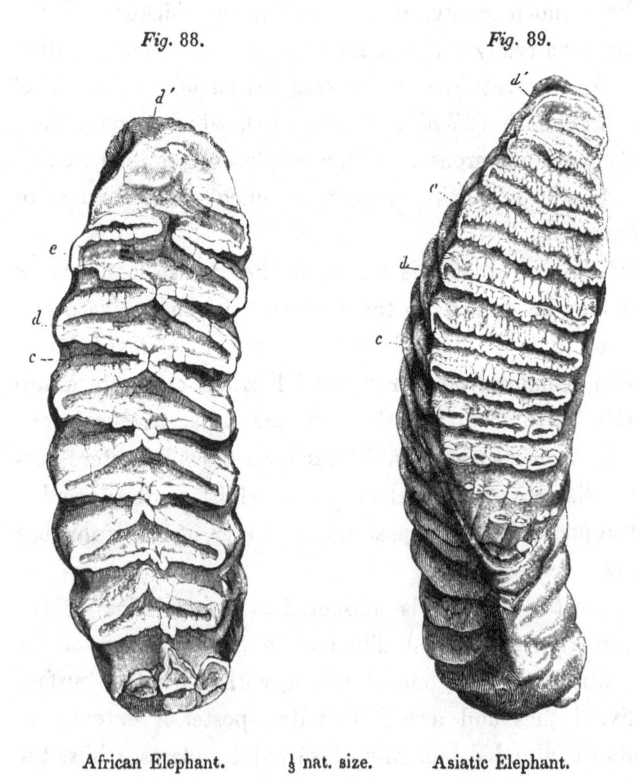

Fig. 88. African Elephant. *Fig.* 89. Asiatic Elephant. ⅓ nat. size.

Fig. 90.

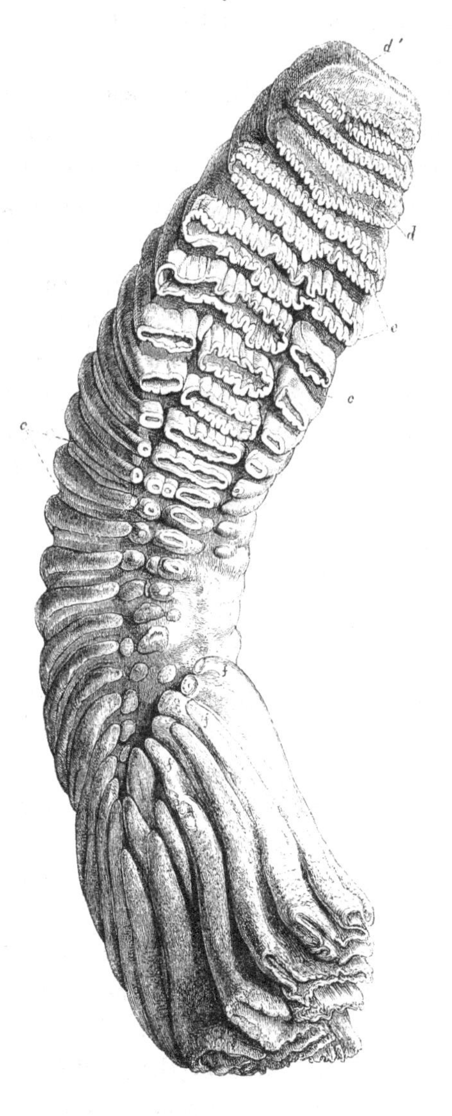

Mammoth, British drift. $\frac{1}{3}$ nat. size.

into one another, so far as the character of the grinding teeth is concerned, to a degree to which the two existing species of Elephant, the Indian and African, when compared together, offer no analogy.

Five or six molars of the Mammoth, and even a greater number, if the peculiar changes superinduced by friction on the grinding surface were not taken into account, might be selected from the series to which I have alluded as indications of so many distinct species of Mammoth: such specimens have, in fact, been so interpreted by Parkinson, and likewise by Fischer, Goldfuss, Nesti and Croizet, cited in the *Palæologica* of Hermann V. Meyer, as authorities for eight distinct species of extinct Elephant.

We must, however, enter more deeply into the consideration of these varieties, before concluding that the Mammoths which severally exemplify them in their molar teeth were distinct species. In the first place, whatever difference the molars of the Mammoth from British strata may have presented in the number of their lamellar divisions, they have corresponded in having a greater proportion of these plates on the triturating surface, and likewise, with two exceptions, in their greater proportional breadth, than are found in the molars of the Asiatic Elephant. The first exception here alluded to was from the diluvial gravel of Staffordshire, and formed part of the collection of Mr. Parkinson, the author of the 'Organic Remains'; the second exception was from the brick-earth of Essex, and is now in the collection of my friend Mr. Brown of Stanway; this molar, though it combines the thicker plates with the narrower form of the entire tooth characteristic of the Indian Elephant, differs in the greater extent of the grinding surface and the greater number of plates entering into the composition of that surface.

With regard to the first-cited exception, the following is the result of a close comparison instituted between it and a corresponding grinder of the Indian Elephant.

The fossil in question is an inferior molar of the right side of the lower jaw. It exhibits the most complete state in which so large a grinder can be met with, the anterior division of the crown not being quite worn down to the fang, and the hindmost plate being just on the point of coming into use. The whole length of the tooth is thirteen inches; the total number of lamellar divisions of the crown seventeen, of which the summits of fourteen are abraded in a grinding surface of nine inches' extent. The greatest breadth of this surface is two inches and a half. The first three fangs supporting the common dentinal base of the anterior lamellæ are well developed. The transverse ridges of enamel are festooned. Compared with the thin-plated grinders of the Mammoth, these differ not only in their more numerous, thinner, and broader plates, but likewise in the thicker coat of external cement which fills the lateral interspaces of the coronal plates, and in having the fangs developed from the whole base of the tooth, even from the posterior plate, the summit of the mammillary process of which has just begun to be abraded. But from the corresponding molar of the Indian Elephant, the present tooth of the Mammoth differs in the more equable length of the coronal plates, which in the Elephant, by their more progressive elongation, give a triangular figure to the side-view of the crown; it differs also in the greater length of the grinding surface, which includes two additional plates, although these are not thinner and are not characterized by superior breadth as in the ordinary teeth of the Mammoth.

These differences from the teeth of the Indian Elephant,

and the intermediate gradations in the fossil molars, by which such rare extreme varieties are linked to the normal type of the Mammoth's dentition, give cause for rejecting the conclusion that the *Elephas Indicus* co-existed with the Mammoth in the latitude of England during the antediluvial or anteglacial epoch: and I think it probable that such differences as have been pointed out in the molar from the Museum of Parkinson, and that of the existing Elephant, might likewise have been detected in the large molar, found at the depth of six feet in brick loam, at Hove near Brighton, and alluded to by Dr. Mantell as decidedly that of the Asiatic Elephant.* One of the molars from the Elephant bed at Brighton, now in the possession of Mr. Stone of Garlick Hill, exhibits the narrow-plated variety of the Mammoth's grinder.

The molars of the Mammoth generally contain a greater proportion of cement in the intervals of the plates than the Indian Elephant's grinders do. Those in which the plates are more numerous have the enamel less strongly plicated; but in some of the large molar teeth of old Mammoths with the thicker plates, as in fig. 90, the enamel is as strongly festooned as in the teeth of the Indian Elephant.

The bones of the Mammoth that have hitherto been disinterred, present no variations from the characteristic extinct type indicative of distinct species; and it might reasonably have been expected that the lower jaw, for example, with the broad-plated tooth should have offered as recognizable differences from that with the narrow-plated teeth, as this does from the lower jaw of the Indian Elephant, if those modifications of the teeth of the Mammoth indicated distinct species. The lower jaw, however, of the ancient British Mammoth has the same distinctive

* 'Fossils of the South Downs,' 4to, 1822, p, 283.

modification of the symphysis as that of the typical Siberian specimen figured by Cuvier, and which is equally presented by that of the Mammoth of Auvergne, figured by the Abbé Croizet,* and by that described by Nesti.†

Fig. 91.

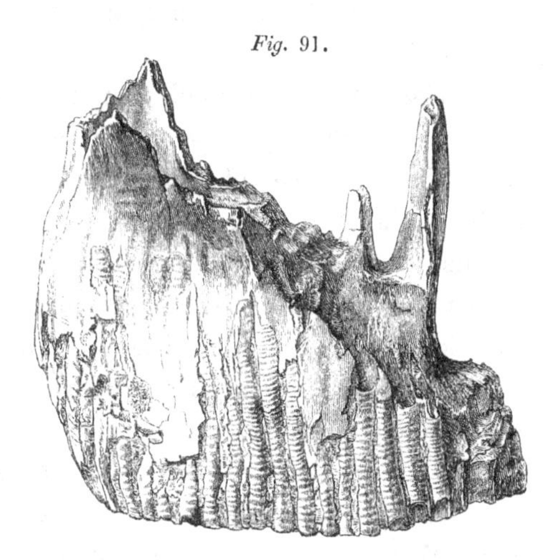

⅓ nat. size, Mammoth, Essex Till.

Fig. 92.

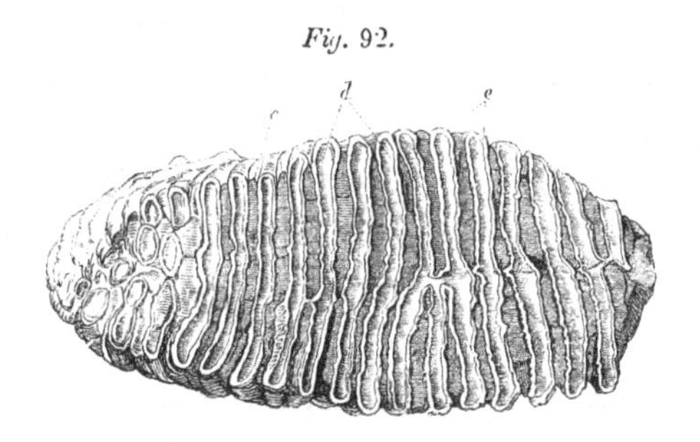

⅓ nat. size, Mammoth, Essex Till.

Both these authors being unacquainted with the intermediate varieties, incline to regard the Mammoth with the

* 'Fossiles du Puy-de-Dome,' 4to. 1828, p. 125, pl. 3, fig. 1.

† 'Nuov. Giorn. d. Letter.' 1825, p. 195.

thick-plated molars as a distinct species, which V. Meyer in his work cites as the *Elephas meridionalis*. In regard, however, to the proposed distinctive name, I may remark that the variety of molar on which this species is founded, occurs not only in England, but in Siberia, and as far north as Eschscholtz Bay.*

Most of the molars of the Mammoth from North America are characterized by thinner and more numerous plates than those of England, but the difference is not constant. The Mammoth's molar from the Norfolk coast in the collection of Miss Gurney, which shows nineteen plates in a length of ten inches, equals several of the molars from North America in the number of the plates. An upper molar of a Mammoth from the gravel of Ballingdon, with a total antero-posterior diameter of seven inches, consists of twenty plates. Mr. Parkinson describes a molar tooth, now in the Museum of the College of Surgeons, from Wellsbourne in Warwickshire, in which twenty plates exist in a length of six inches and a half; and he figures another molar from the till of Essex, which, in a length of eight inches and a half, contains twenty-four plates. The proportions of the figure in the 'Organic Remains,' not being quite accurate, I have given two original views of this remarkable molar, (figs. 91 and 92,) which appears to have been the fifth in succession in the upper jaw. On the other hand, the molars of the Mammoths from Eschscholtz Bay, North America, figured by Dr. Buckland, manifest the same kind of variety as those from the English drift; one with a grinding surface, seven inches and a half long, exhibiting nineteen plates, whilst another in the same extent of grinding surface shows only thirteen

* See Buckland in 'Beechey's Voyage of the Blossom,' 4to. On the Fossils of Eschscholtz Bay, 4to, pl. 1. (Fossils), fig. 2.

plates; both these teeth are from lower jaws, which, like the lower jaw containing the broader-plated tooth described by Professor Nesti, are precisely similar in form to the other fossil jaws of the Mammoth; they present the same specific differences from the Asiatic Elephant, and offer no modification that can be regarded as specifically distinct from the Mammoth's jaws with narrow-plated molars of Siberia or Ohio.

Mr. Parkinson* has figured a Mammoth's molar from Staffordshire, which he deemed to differ from every other

Fig. 93.

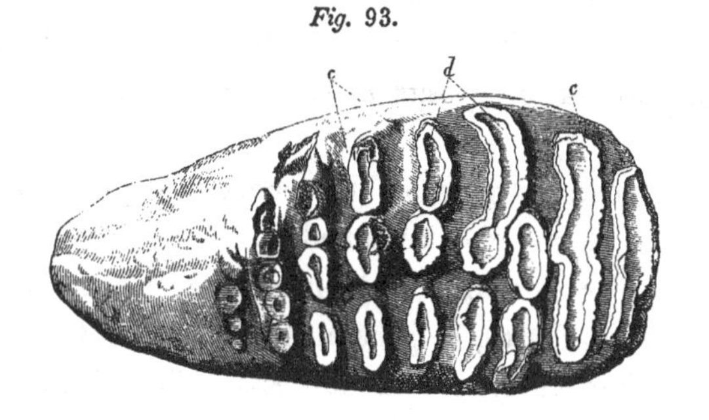

⅓ nat. size, Mammoth, Staffordshire Drift.

that had come to his knowledge in the great thickness of the plates, the smoothness of the sides of the line of enamel, and the appearance of the digitated part of the plates even in the anterior part of the tooth, and which unquestionably offers a great contrast with the preceding (fig. 92).

The specimen (fig. 93), is the posterior part of a large grinder, apparently the last of the upper jaw, of an old Mammoth. The superior thickness of the plates arises from the circumstance of the posterior plates being thicker than the anterior ones; these thick plates are more deeply cleft, or their digitated summits are longer, and advance further for-

* 'Organic Remains,' iii. p. 344.

ward upon the grinding surface of the molar before they are worn down to their common base; they appear also in the specimen to be more advanced than they really are, because of the deficiency of the fore-part of the tooth, which has been broken away. In my opinion this molar has the characters of the thick-plated variety, simply exaggerated from the accidents of age and attrition. It manifests the more constant and characteristic modifications of the *Elephas primigenius* in its relative breadth, and, notwithstanding their thickness, in the number of the plates (nine), which have been exposed by the act of mastication. I have seen a very similar molar of the Mammoth from the Norfolk freshwater deposits in the collection of Mr. Fitch of Norwich.

The abraded margins of the component plates of the Mammoth's molars most commonly present a slight expansion, often lozenge-shaped, at their centre; they are divided with more regularity, in general, than those in the Indian Elephant, into three digital processes, the middle being usually the broadest and thickest, and having its summit originally sub-divided into three smaller digitations, as is shown in the posterior plates of fig. 90. The greater thickness of the middle division of the transverse plate occasions the middle expansion of the margin of the plate, when the three digitations are worn down to their common base. Only in one small molar, from the brick-earth at Grays, Essex, in the collection of Mr. Wickham Flower, have I seen the median rhomboidal dilatation, extending, in the abraded plates, so near the end of the section as to approximate the characteristic shape of the plates of the African Elephant's molar; from which, however, the fossil was far removed by its thinner and more numerous plates. The fictitious character of the *Elephas priscus* of Goldfuss one of the eight fossil species admitted in the compilation of M. H. V. Meyer, has been left scarcely doubtful by Cuvier:* it is founded on recent molars of the *Elephas africanus;* and the great anatomist alludes to attempts that had been made to palm upon himself such teeth as fossils. I have met with no nearer approach to this nominal species among the numerous British Mammoths' grinders that I have examined, than the example just quoted from the brick-earth at Grays; I need hardly say that I regard it as another of the numerous varieties to which the molars of the Mammoth were subject.

The clefts that separate the transverse plates are deeper at the sides than at the middle of the tooth in all Mammoths' grinders; hence the ridges of enamel in a much-worn molar are confined to the outer and inner sides of the grinding surface, which is traversed along the middle by a continuous tract of dentine. The layer of enamel extends along the lateral clefts to this exposed tract, is reflected back upon the opposite side of each cleft, bends round the outer margin of the remaining base of the plate, and is continued into the next cleft, and so on. When the edge of this sinuous coat of enamel is exposed by abrasion of the masticating surface, it describes what Mr. Parkinson has called a "dædalian line," and he has figured two examples of teeth so worn down in the "Organic Remains."† An original figure of the grinding surface of one of these molars, which was dredged up from the drift-gravel forming the bed of the Thames near London, is given at fig. 94. Having noticed the structure in three specimens, Mr. Parkinson conceives it to be characteristic of a distinct species of Mammoth. But the ordinary teeth of the

* "Ossemens Fossiles," tom. v. pl. ii., Additions, p. 496.

† Pl. 20. figs. 5 and 7.

Mammoth, from the unequal vertical extent of their plates above described, must necessarily exhibit the continuous undulating lateral lines of enamel when worn down to a certain extent.

Fig. 94.

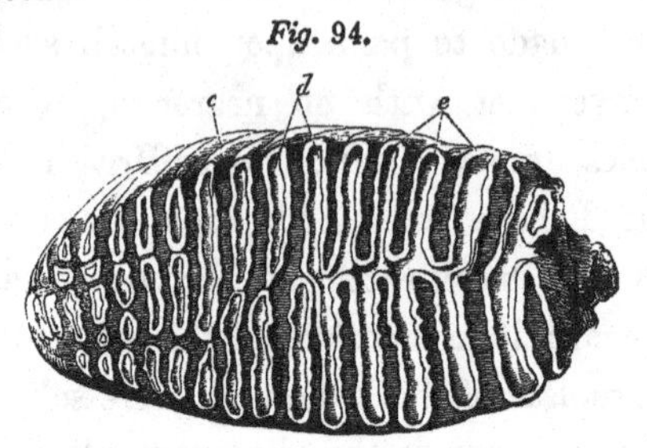

⅓ Nat. size. Mammoth, Thames gravel.

I have seen this structure in a few only amongst the numerous molars of the Mammoth examined by me, for teeth so worn down are rare. It is well shown in the remains of a very large molar, found in the beach near Happisburgh, Norfolk, which on a grinding surface of four inches nine lines in length, and four inches wide, shows seven dentinal plates worn down to their common uniting base of dentine, along the middle of the surface.

It sometimes happens that the outer and inner margins of a plate, which are always deeper than the middle part, are not on the same transverse line, but one is inclined a little in advance of the other. In this case the abraded crown of the tooth, when worn down to the common middle base of dentine, displays an alternating disposition of the folds of the outer and inner sinuous lines of enamel. This variety affords grounds of the same kind and value for a distinct species of Mammoth as for the two other new species proposed by Mr. Parkinson.

In old and much worn teeth the abrasion is sometimes so partial as to wear away the whole of the enamel on one side of the tooth, and leave a single undulating or dædalian line on the other: this is shown in the Mammoth's molar figured in pl. 20, fig. 7, of Parkinson's "Organic Remains," of which a new and more accurate figure is here given, showing the long and strong compressed fang, which is developed from the base of the crown of a tooth so worn down, (*fig.* 95.)

Fig. 95.

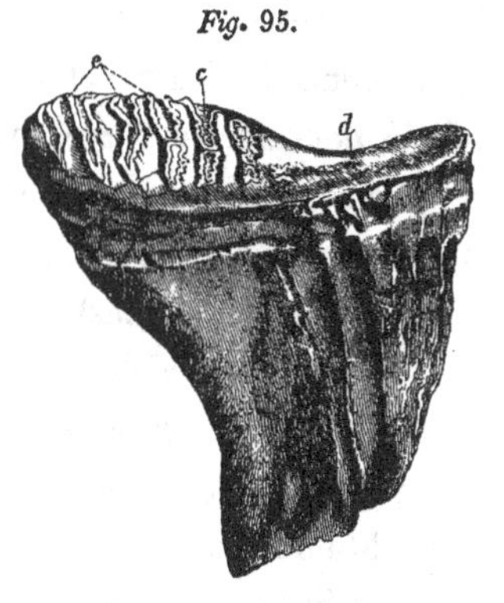

⅓ Nat. size. Mammoth.

A consideration of the anatomical structure and an extensive comparison of the teeth in question have led me to the conclusion, that whilst some of the supposed specific characters are due to effects of changes produced by age, the others depend upon the latitude of variety, to which the highly complex molars of the *Elephas primigenius* were subject.

In proof of such variety we have the analogy of existing species: that such variety is the characteristic of a particular part of the enduring remains of the Mammoth may be inferred from the absence of any corresponding differences in the bones of the Mammoth that have hitherto been found; all of which indicate but one species. And this conclusion harmonizes with the laws of the geographical distribution of the existing species of Elephant.

Throughout the whole continent of Africa but one species of Elephant has been recognized. A second species of Elephant is spread over the South of Asia and some of the adjacent islands; and extensive and accurate observations of this species, whilst they have made known some well-marked varieties, as the Mooknah, the Dauntelah, &c., founded on modifications of the tusks, have more firmly

R 2

established the unity of species to which those varieties belong.

If, on the other hand, the observed varieties in the dentition of the Mammoth are to be interpreted, as Parkinson, Nesti, Croizet, V. Meyer and others have done, as evidences of distinct species, we must be prepared to admit not merely three, but six or more distinct species of gigantic Mammoths to have roamed through the primeval swamps and forests of England.

Tusks.—All the tusks of the *Elephas primigenius* from British strata which have fallen under my observation in a sufficiently complete state for the comparison, possess the same extensive double curvature as the tusks of the great Mammoth, from the icy cliff at the mouth of the river Lena in Siberia, which is figured at the head of the present section, and as those brought to England by Captain Beechey from Eschscholtz Bay, which have been figured by Dr. Buckland, and are now in the British Museum.

They arrange themselves in two groups according to size, the larger tusks averaging about nine feet and a half, the smaller ones from five to six feet in length; the latter are readily distinguishable by their greater degree of curvature from the incompletely developed tusks of the larger kind, when of the same length as the smaller fully developed tusks. It is most probable, therefore, from the analogy of the Asiatic Elephant, that the larger tusks belonged to the male sex, and the smaller ones to the female; which, however, if this idea be correct, possessed better developed tusks than the female Asiatic Elephant, in which the tusks either do not appear beyond the lip, or project a short way straight down from the mouth, as in both sexes of the variety called 'Mooknah.'*

* See Corse in Philos. Trans. 1799, p. 208.

The finest tusk of a British Mammoth that has come under my observation, forms part of the rich collection of fossil Mammalian remains obtained from Ilford by the late John Gibson, Esq., of Stratford, Essex; this tusk measured twelve feet six inches in length, following the outward curvature. A tusk disinterred from Mr. Hobson's brick-field at Kingsland, a model of which is preserved in the Museum of the Geological Society of London, measures nine feet ten inches along the outer curve, three feet one inch in a straight line from point to base, and twenty-nine inches in its greatest circumference.

In the collection of Mr. Brown of Stanway, there is a fragment of a tusk of the Mammoth, from the freshwater formation at Clacton in Essex, which measures two feet in circumference, thus exceeding the size of the largest of the tusks brought home by Captain Beechey from Eschscholtz Bay, which measured twenty-one inches and a half at its largest circumference, nine feet two inches along the curve from the root to the tip, part of which was broken off, and five feet two inches across the chord of its curve. The tusks which were collected in this northern locality, were of two sizes: "five of them large, and weighing from one hundred to one hundred and sixty pounds each; and four small."*

A very fine tusk of the Mammoth from British strata forms part of the remarkable collection of remains of the Mammoth obtained by the Rev. J. Layton from the drift of the Norfolk coast, near the village of Happisburgh; it was dredged up in 1826, measured nine feet six inches in length, and weighed ninety-seven pounds.

At Knole-sand, near Axminster, about twenty miles

* Dr. Buckland in the Appendix to "Beechey's Voyage," "Fossils," p. 2.

from the coast, Sir H. De la Beche obtained a tusk nine feet eight inches in length.

A tusk of a young male Mammoth, whose double oblique curve describes a semicircle, from a bed of drift at Newnham in Warwickshire, and now in Dr. Buckland's collection at Oxford, measures seven feet in length.* Other tusks from the same locality, present the same considerable curvature outward towards the point. A Mammoth's skull containing two tusks of enormous length, as well as grinding teeth, was discovered in 1806, at Kingsland near Hoxton, Middlesex. The large tusk, mentioned by Parkinson,† which was discovered in the brick-fields at Kingsland between the drift-gravel and a bed of clay, described nearly four-fifths of a circle.

Most of the largest and best preserved tusks of the British Mammoth, have been dredged up from submerged drift, near the coasts. In 1827, an enormous tusk was landed at Ramsgate: although the hollow implanted base was wanting, it still measured nine feet in length, and its greatest diameter was eight inches; the outer crust was decomposed into thin layers, and the interior portion had been reduced to a soft substance resembling putty. A tusk, likewise much decayed, which was dredged up off Dungeness, measured eleven feet in length: and yielded some pieces of ivory fit for manufacture. Captain Byam Martin who has recorded this and other discoveries of remains of the Mammoth in the British Channel in the Geological Transactions,‡ procured a section of ivory near the alveolar cavity of the Dungeness tusk, of an oval form, measuring nineteen inches in circumference. A tusk dredged up from the Goodwin Sands,

* Buckland, "Reliquiæ Diluvianæ," p. 177.

† "Organic Remains," p. 350.

‡ Second Series, vol. vi. p. 161.

which measured six feet six inches in length, and twelve inches in greatest circumference, probably belonged to a female Mammoth: Captain Martin describes its curvature as being equal to a semicircle turning outwards on its line of projection. This tusk was sent to a cutler at Canterbury, by whom it was sawn into five sections, but the interior was found to be fossilized and unfit for use: it is now in Captain Martin's possession. The tusks of the extinct Elephant which have thus reposed for thousands of years in the bed of the ocean which washes the shore of Britain, are not always so altered by time and the action of surrounding influences as to be unfit for the purposes to which recent ivory is applied. Mr. Robert Fitch of Norwich possesses a segment of a Mammoth's tusk, which was dredged up by some Yarmouth fishermen off Scarborough, and which was so slightly altered in texture, that it was sawn up into as many portions as there were men in the boat, and each claimed his share of the valuable product.

Of the tusks referable by their size to the female Mammoth which have been disinterred on dry land, I may cite the following instances.—A tusk in the Museum of the Geological Society, from the lacustrine pleistocene bed exposed to the action of the sea on the coast of Essex at Walton, which measures five feet and a half in length; and another from the same locality, in the possession of John Brown Esq., of Stanway, Essex, which measures four feet in length. A tusk recently discovered near Barnstaple, on a bed of gravel, beneath a stratum of blue clay five feet deep, and one of yellow clay about six feet deep, with several feet of coarse gravel and soil above. This tusk was broken by the pickaxes of the men, but must have been about six feet in length: it had the grain and markings of ivory, but was reduced to the colour and consistency of horn, and retained a considerable degree of elasticity.

A very perfect specimen was dug up entire in 1842, twelve feet below the surface, out of the drift gravel of Cambridge; it measured five feet in length and two feet four inches across the chord of its curve, and eleven inches in circumference at the thickest part of its base: this tusk was purchased by the Royal College of Surgeons. The smallest Mammoth's tusk which I have seen is in the museum of Mr. Wickham Flower; it is from the drift or till at Ilford, Essex, and has belonged to a very young Mammoth; its length measured along the outer curve is twelve inches and a half, and the circumference of its base four inches. It has nevertheless been evidently put to use by the young animal, the tip having been obliquely worn.

Mr. Robert Bald * has described a portion of a Mammoth's tusk, thirty-nine inches long and thirteen inches in circumference, which was found imbedded in diluvial clay at Clifton Hall, between Edinburgh and Falkirk, fifteen or twenty feet from the present surface. Two other tusks of nearly the same size have been discovered at Kilmaurs in Ayrshire, at the depth of seventeen feet and a half from the surface, in diluvial clay. The state of preservation of these tusks was nearly equal to that of the fossil ivory of Siberia; that described by Mr. Bald was sold by the workmen who found it to an ivory-turner in Edinburgh for two pounds: it was sawn asunder to be made into chessmen. The tusks of the Mammoth found in England are usually more decayed: but Dr. Buckland alludes to a tusk from argillaceous diluvium on the Yorkshire coast, which was hard enough to be used by the ivory-turners. A por-

* Wernerian Trans. vol. iv. p. 58.

tion of this tusk is now preserved in the museum at Bridlington.

The tusks of the Mammoth are so well preserved in the frozen drift of Siberia, that they have long been collected in great numbers for the purposes of commerce. In the account of the Mammoth's bones and teeth of Siberia, published more than a century ago in the Philosophical Transactions,* tusks are cited which weighed two hundred pounds each, and "are used as ivory, to make combs, boxes, and such other things; being but a little more brittle, and easily turning yellow by weather or heat." From that time to the present there has been no intermission in the supply of ivory furnished by the extinct Elephants of a former world; and I am informed by Mr. Warburton, M.P., President of the Geological Society, that Mammoths' tusks are now imported from Russia to Liverpool, and find a ready sale to comb-makers and other workers in ivory.

Bones.—There is reason to believe that instances have occurred in which a more or less entire skeleton of the Mammoth might have been recovered from British strata, if due care and attention had been devoted to the task. About three years ago, the workmen in a brick-ground, near the village of Grays in Essex, disinterred a quantity of bones of an enormous Mammoth, which they broke up as they were discovered, and sold the fragments for three-halfpence a pound to a dealer in old bones. This traffic went on weekly for more than half a year, and accidentally came to the knowledge of Mr. R. Ball, F.G.S., a sedulous collector of fossil remains, who recovered from the workmen some magnificent bones of the fore foot, with portions of the scapula and ribs. I had the account from Mr. Ball, to whom I am indebted for casts of the bones which he was

* No 446, 4to, 1737, p. 128.

so fortunate as to rescue from the destruction that awaited them.

Of the numerous detached bones of the trunk and extremities of the Mammoth which have been obtained from various British localities, I shall limit myself to a notice of a few of the most entire and remarkable examples. Of two specimens of the atlas of the Mammoth from the newer pliocene deposits near Cromer, in the collection of Miss Gurney, the most perfect measures

	In.	Lines.
In breadth	16	6
Breadth of the anterior condyles	7	10
Breadth of the posterior ditto	9	8
In vertical diameter	10	0

A vertebra dentata from the freshwater deposits at Clacton, Essex, twenty feet above high-water mark, in the collection of Mr. Brown of Stanway, measures six inches nine lines in transverse diameter, five inches in vertical diameter, and has a spinal canal three inches in transverse diameter.

A dorsal vertebra, in the same collection, measures in height one foot ten inches, the spinous process being nine inches high. The transverse diameter of the vertebra is eight inches six lines, that of the spinal canal being three inches.

In Mr. Brown's collection is also preserved the os sacrum of a Mammoth from the freshwater formations of Essex. It is of a triangular form; the transverse diameter of the fore part of the body of the first sacral vertebra is six inches six lines; the diameter of the largest nervous foramen was two inches four lines.

A scapula, with the spine, the supra-spinal plate and base broken away, from the same formation, shows the characteristic superior breadth of the glenoid articular cavity at its inferior part, and the shortness of the neck supporting it, which Cuvier has recognized in the scapula of the Siberian Mammoth (*fig.* 85, *s*).

The scapula of the Essex Mammoth gave the following dimensions:—

	Ft.	In.
From the glenoid cavity to the inferior angle . .	1	10
From ditto to the spine	0	4
From the middle of the spine to the lower costa of the scapula	0	8

In a fragment of a Mammoth's scapula from Happisburgh, in the collection of Mr. Fitch of Norwich, the long diameter of the glenoid articulation was ten inches, its short diameter four inches and a half. The head of the humerus, in the state of an epiphysis, found with the above fragment, measures ten inches and a half in its longest diameter. These parts, notwithstanding their dimensions, have belonged to an immature specimen of the Mammoth.

Of the stupendous magnitude to which some individuals, doubtless the old males, of the *Elephas primigenius* arrived, several fossils from the British drift afford striking evidence. In the skeleton of the Mammoth now at St. Petersburg, which was found entire in the frozen soil of the banks of the Lena, the humerus (*fig.* 85, *h*) is three feet four inches in length; that of the skeleton of the large Indian Elephant (Chuny) which was killed at Exeter Change in 1826, is two feet eleven inches in length. In the rich collection of Mammalian remains from the Norfolk coast, belonging to Miss Gurney of North-repps Cottage, near Cromer, there is an entire humerus of the Mammoth which measures four feet five inches in length.

Subjoined are a few of the dimensions of this enormous bone and of its analogue in the above-mentioned skeleton of the Indian Elephant in the Museum of the College of Surgeons:—

	El. primigenius.			El. Indicus.		
	Ft.	In.	Lin.	Ft.	In.	Lin.
Humerus, entire length	4	5	0	2	11	0
Circumference at the middle . . .	2	2	6	1	1	6
Ditto at proximal end . . .	3	5	0	2	8	0
Breadth of distal end	1	2	0	0	10	6
From summit of supinator ridge to end of outer condyle	1	7	0	1	0	6

The above gigantic fossil bone was found in 1836, after a very high tide, partially exposed in the cliff, composed of interblended blue clay and red gravel, near the village of Bacton in Norfolk. The outer crust of the bone is much shattered; it manifests the specific distinction of the humerus of the Mammoth in the relatively shorter proportions of the great supinator ridge, as is shown by the last admeasurement, and the bicipital canal is also relatively narrower.

A portion of a large tibia was obtained from the same bed in 1841; this bone likewise is in Miss Gurney's collection.

A Mammoth's humerus, of dimensions very nearly equal to those of the great specimen in Miss Gurney's collection, was obtained under the following circumstances described by Captain Byam Martin. "In 1837, while trawling in mid-channel between Dover and Calais, between the two shoals, called the Varn and Ridge, covered at low tide with twenty fathoms water, a fisherman suddenly encountered a heavy mass, which proved to consist of enormous bones; the net broke, but a humerus, which I purchased, was secured."* Such occurrences recall to mind the adventures of the fishermen narrated in the Arabian Nights; but the fancy of the eastern romancer falls short of the reality of this hawling up, in British seas, of Elephants more stupendous than those of Africa or Ceylon.

* Loc. cit. p. 162.

A humerus of the Mammoth, wanting the proximal end, from Clacton, Essex, in the collection of Mr. Brown of Stanway, measures two feet ten inches in length, and fifteen inches six lines in median circumference, showing the thicker proportions as compared with the existing Elephant.

The bones of the fore leg of the Mammoth from British localities have not offered any characters worthy of notice. In the figure of the Siberian Mammoth, (fig. 85,) *r* is the radius; *u* the ulna.

Of the bones of the fore foot, the specimens obtained by Mr. Ball from the brick-loam near Grays, Essex, must have belonged to a Mammoth as large as that which furnished the great humerus from Cromer above described. The following are the comparative dimensions of some of those bones and of their analogues in the skeleton of Chuny, the great Asiatic Elephant of Exeter Change: —

	El. primigenius.		*El. Asiaticus.*	
	In.	Lin.	In.	Lin.
Os magnum, vertical diameter . .	4	3	3	0
Middle metacarpal, length . . .	10	0	7	0
Middle breadth of distal end . .	4	9	3	4

Mr. J. Wickham Flower possesses a fine and perfect specimen of the femur of the Mammoth from the Essex till, which offers the usual characteristic of the extinct species in the relatively narrower posterior interspace between the two condyles and in the thicker shaft. The outer ridge of the femur extends about two-thirds down the bone. The following are some of its dimensions, compared with that of the Indian Elephant: —

	El. primigenius.			*El. Indicus.*		
	Ft.	In.	Lin.	Ft.	In.	Lin.
Length	3	4	0	3	6	0
Breadth across proximal end . .	1	1	6	1	1	0
Breadth across back part of condyles .	0	7	6	0	7	0
Circumference of shaft . . .	1	2	6	1	0	0

A femur of the Mammoth, from the drift gravel at Abingdon, is preserved in the Ashmolean Museum. It is remarkable for its fine state of preservation, and exhibits the same character of the extinct species as the foregoing specimen.

Captain Byam Martin has recorded the following dimensions of the femur of a Mammoth, which was trawled up in twenty to twenty-five fathoms water about midway between Yarmouth and the coast of Holland.

	Inches.
Entire length	49
Circumference of the head of the bone	24
" of the middle of the shaft	18
" above the condyles	29
Width across the head and great trochanter	18

The femur of the Mammoth, described by the notable French surgeon Habicot, in his "Gigantosteologie, 1613," as the thigh-bone of Theutobochus, king of the Cimbrians, which was said to be five feet in length, indicates a specimen larger than that to which the humerus from Cromer belonged. M. de Blainville is, however, of opinion that the femur in question belonged to a Mastodon.

In the skeleton (*fig.* 85), *i* is the iliac bone, *is* the ischium, *f* the femur or thigh-bone, *t* the tibia or leg-bone, *fi* the fibula or small bone of the leg, *ta* the tarsus or ankle-bones.

Strata and Localities.—Of all the extinct Mammalia which have left their fossil remains in British strata, no species was more abundant or more widely distributed than the Mammoth or *Elephas primigenius.*

Wherever the last general geological force has left traces of its operations upon the present surface, in the form of drift or unstratified transported fragments of rock and gravel, and wherever the contemporary or immediately antecedent, more tranquil and gradual operations of the sea or fresh waters have formed beds of marl, of brick-earth or loam, there, with few exceptions, have bones or teeth of the Mammoth been discovered.

It would be tedious to specify all the particular localities which have been recorded, in collecting the materials for the present Work, as yielding fossil remains of this gigantic quadruped. They are most remarkable for their abundance in the drift along the east coast of England, as at Robin Hood's Bay near Whitby; at Scarborough, at Bridlington, and various places along the shore of Holderness.

Mr. Woodward, in his "Geology of Norfolk," supposes that upwards of two thousand grinders of the Mammoth have been dredged up by the fishermen off the little village of Happisburgh in the space of thirteen years. The oyster-bed was discovered here in 1820, and during the first twelve months hundreds of the molar teeth of Mammoths were landed in strange association with the edible mollusca. Great quantities of the bones and tusks of the Mammoth are doubtless annually destroyed by the action of the waves of the sea. Remains of the Mammoth are hardly less numerous in Suffolk, especially in the pleistocene beds along the coast, and at Stutton; they become more rare in the fluvio-marine crag at Southwold and Thorp. The village of Walton near Harwich is famous for the abundance of these fossils, which lie along the base of the sea-cliffs, mixed with bones of species of Horse, Ox and Deer.*

* The more bulky fossils of this locality appear to have early attracted the notice of the curious. Lambard in his Dictionary, says, that, "In Queen Elizabeth's time bones were found, at Walton, of a man whose skull would contain five pecks, and one of his teeth as big as a man's fist, and weighed ten ounces. These bones had sometimes bodies, not of beasts, but of men, for the difference is

Reference has already been made to other localities in Essex, as Clacton, Grays, Ilford, Copford and Kingsland, where, in the freshwater deposits, the remains of the extinct Elephant occur, associated with the above-mentioned Herbivora, and with more scanty remains of Rhinoceros.

Abundant Mammalian fossils, which once lay in the drift that capped the cliffs of the coast of Herne Bay, have fallen by the undermining action of the tide and waves, and are dredged up from outlying oyster-beds. Amongst these Dr. Richardson has noticed bones and teeth of the Mammoth associated with remains of Rhinoceros, Horse, Ox, Deer, Bear, and Wolf; all the bones being characterised by the total absence of albuminous matter.

In the valley of the Thames remains of the Mammoth have been discovered at Sheppy, Lewisham, Woolwich, and the Isle of Dogs; in the drift gravel beneath the streets of the metropolis, as in Gray's Inn Lane, twelve feet deep; in Charles Street, near Waterloo Place, thirty feet deep.

Proceeding westward we encounter Mammoths' remains at Kensington, at Brentford, at Kew, at Hurley-bottom, Wallingford, and Dorchester; in the gravel-pits at Abingdon and Oxford, and at Witham Hill and Bagley Wood.* Bones of the great extinct Elephant again occur in the valley of the Medway, at the Nore, at Chatham, and at Canterbury; at Betchworth in Surrey. On the south coast of England, they have been discovered at Brighton, Hove and Worthing; at Lyme Regis and Charmouth;

manifest." The remains of Mammoths have everywhere been the prolific source of the traditions and histories of giants, and sometimes of saints: Ludovicus Vives relates that a molar tooth, bigger than a fist (dens molaris pugnâ major), was shown to him for one of St. Christopher's teeth, and was kept in a church that bare his name.

* Dr. Kidd's Geological Essays, ch. xvii., and Dr. Buckland's "Reliquiæ Diluvianæ," p. 174; where numerous other localities of the Mammoth are recorded.

also at Peppering near Arundel, about eighty feet above the present level of the Arun.

Passing inland from the south coast, we find remains of the Mammoth at Burton and Loders, near Bridport, and near Yeovil in Somerset. At Whitchurch, near Dorchester, Dr. Buckland observes that the remains of the Mammoth lie in gravel above the chalk, and are found in a similar position on Salisbury Plain; they again occur at Box and Newton near Bath, and at Rodborough in Gloucestershire.

Mr. Randall of Stroud has lately acquainted me, that in some recent railway excavations in the neighbourhood of that town, tusks and molar teeth of a Mammoth have been discovered in drift gravel from fourteen to twenty feet below the surface: one of the tusks was recovered in a tolerably perfect state, and measured nine feet in length; it is in the possession of — Carpenter, Esq., of Gannicox House near Stroud.

In Worcestershire, on the borders of the Principality, remains of the Mammoth are noticed by Mr. Murchison as occurring in a gravel-pit south of Eastnor Castle. This pit is in the midst of a group of Silurian rocks, and the fragments consist exclusively of those rocks and of the sienite of the adjacent hills, whence Mr. Murchison rightly infers that this extinct species of Elephant formerly ranged over that country.* In North Wales, Pennant mentions two molar teeth and a tusk found at Holkur, near the mouth of the Vale of Clwyd, in Flintshire, and near Dyserth; they occurred in a bed of drift gravel containing pebbles of lead-ore, which are worked like the analogous stream-works in Cornwall, which contain pebbles and sand of tin-ore.

Bones of the Mammoth, with those of the Rhinoceros

* Silurian System, p. 554.

and Hippopotamus, have been found in coarse gravelly drift with overlying marl and clay in the valley of the Severn, at Fleet's bank near Sandlin. Marine shells occur in the coarse drift, and freshwater shells in the superficial fluviatile deposits.

Mr. Strickland found remains of the Mammoth associated with Hippopotamus, Urus, &c., in the valley of the Avon, in apparently a local fluviatile drift, containing land and freshwater shells: this geologist supposes that after those parts of Worcestershire and Warwickshire had been long under the sea, an elevation of some hundred feet converted them into dry land, and that a river or chain of lakes then descending from the north-east, re-arranged much of the gravel of the great northern glacial drift, disposing it in thin strata and imbedding in it the shells of mollusks and the bones of the extinct quadrupeds.*

In the centre of England, Dr. Buckland notices the occurrence of the Mammoth at Trentham in Staffordshire, in different parts of Northamptonshire, and at Newnham and Lawford, near Rugby in Warwickshire; there the Mammoth's bones lay by the side of those of the Rhinoceros and Hyæna.

Mammoth-fossils occur at Middleton in the Yorkshire Wolds, in Brandsburton gravel-hills, and at Overton near York. Remains of the Mammoth are noticed by the Rev. Vernon Harcourt, F.R.S. and Professor Phillips, as having been found associated with the great Cave Tiger, Rhinoceros, Aurochs, Deer, &c., in blue marl, beneath strata of gravel and sand at Bielbecks, near North Cliff, Yorkshire. Tusks of the Mammoth, valuable from the condition of the ivory, have been discovered at Atwick, near Hornsea, in the same county.

* Proceedings of the Geological Society, vol. ii. p. 111.

In Scotland remains of the Mammoth have been found in the drift clay between Edinburgh and Falkirk, at Kilmuir in Ayreshire. In Ireland they have been found at Maghery in the county of Cavan, and in the drift near Tully-doly, county of Tyrone.

The celebrated cave at Kirkdale concèaled remains of Mammoths: the molars here detected were all of small size; very few of them exceed three inches in their longest diameter, and they must have belonged to extremely young animals, which had been dragged in by the Hyænas for food with Rhinoceroses, Hippopotamuses, and large Ruminantia.

The molars of the Mammoth which I have hitherto seen from the cave called Kent's Hole near Torquay are of similar young specimens; here they are associated with the Hyæna, the great Cave Tiger, the Cave Bear, &c.: and I entirely accede to Dr. Buckland's explanation, that the bones or bodies of these young Mammoths were introduced into the cave by the Carnivora which co-existed with them.

Quitting the dry land and caves of Great Britain, we find the bed of the German Ocean a most fertile depositary of the remains of the *Elephas primigenius*, and they are generally remarkable for their fine state of preservation.

Captain Byam Martin, the harbour-master at Ramsgate, possesses several well-preserved specimens which have been from time to time brought up by the deep-sea nets of the fishermen. A fine lower jaw of a young Mammoth, in the possession of Mr. G. B. Sowerby, was thus dredged up off the Dogger Bank; and a femur and portion of a large tusk, before described, were raised from twenty-five fathoms at low water, midway between Yarmouth and the Dutch coast. Remains of the Mammoth have also been raised in the British Channel from the shoals called Varn and Ridge, which lie midway between Dover and Calais.

The evidences of an enormous crushing and breaking power are very remarkably exemplified in some of the Mammalian fossils from the "till," or drift, at Walton in Essex. Mr. Brown, of Stanway, possesses molars of the Mammoth from this locality which have been split vertically and lengthwise, across all the component plates of dentine and enamel; other molars have been so crushed and squeezed that the enamel-plates are shivered in pieces, which are driven into the conglomerate of the different substances, and the fragments of enamel stick out like the bits of glass from the plaster which caps a garden wall.

The ramus of a lower jaw of a Rhinoceros from the drift near the sea-coast of Essex, has been split vertically and lengthwise through all the molars.

A similar condition of some of the Mammalian fossil remains, including parts of the Mammoth, discovered by Mr. Stutchbury in a cavernous fissure at Durdham Down near Bristol, has been explained on the hypothesis of considerable relative movement having taken place in the walls of the fissure of the cavern since the deposit of the organic remains; and Mr. Stutchbury adduces, in confirmation of this view, the fact, that a calcareous spar-vein in the vicinity bears undoubted evidence of having been moved and reconstructed.

Other forces than the concussion of rocks by earthquakes seem, however, to have operated in producing the fractures of the teeth and bones in the beds of Essex gravel or drift above adverted to; and I cannot suggest any more probable dynamic than the action of masses of ice, on the supposition of such being chiefly concerned in the deposition and dispersion of the superficial drift itself.

It is remarkable that the bones and teeth of the Elephant are very rarely rolled or water-worn; the fractured surfaces are generally entire, and sometimes the bones are found, like that in the Ashmolean Museum, in a remarkable state of integrity.

General Geographical Distribution, and probable Food and Climate of the Mammoth.

The remains of the Mammoth occur on the Continent, as in England, in the superficial deposits of sand, gravel, and loam, which are strewed over all parts of Europe; and they are found in still greater abundance in the same formations of Asia, especially in the higher latitudes, where the soil which forms their matrix is perennially frozen.* Remains of the Mammoth have been found in great abundance in the cliffs of frozen mud on the east side of Behring's Straits, in Eschscholtz's Bay, in Russian America, lat. 66° N. lat.; and they have been traced, but in scantier quantities, as far south as the States of Ohio, Kentucky, Missouri, and South Carolina. But no authentic relics of the *Elephas primigenius* have yet been discovered in tropical latitudes,† or in any part of the southern hemisphere. It would thus appear that the primeval Elephants formerly ranged over the whole northern hemisphere of the globe, from the 40th to the 60th, and possibly to near the 70th degree of latitude. Here, at least, at the mouth of the River Lena, the carcass of a Mammoth has been discovered, preserved entire, in the icy cliffs and frozen soil of

* Hedenström, in his 'Survey of the Laechow Islands' on the north-eastern coast of Siberia, remarks that the first of these islands is little more than one mass of these bones; and that although the Siberian traders have been in the habit of bringing over large cargoes of them (tusks) for upwards of sixty years, yet there appears to be no sensible diminution.

† The fossil elephantine remains discovered in India, belong to a species more nearly allied to the *Elephas Indicus*.

that coast. To account for this extraordinary phenomenon, geologists and naturalists, biassed more or less by the analogy of the existing Elephants, which are restricted to climes where the trees flourish with perennial foliage, have had recourse to the hypothesis of a change of climate in the northern hemisphere, either sudden, and due to a great geological cataclysm,* or gradual, and brought about by progressive alternations of land and sea.†

I am far from believing that such changes in the external world were the cause of the ultimate extinction of the *Elephas primigenius;* but I am convinced that the peculiarities in its ascertained organization are such as to render it quite possible for the animal to have existed as near the pole as is compatible with the growth of hardy trees or shrubs. The fact seems to have been generally overlooked, that an animal organized to gain its subsistence from the branches or woody fibre of trees, is thereby rendered independent of the seasons which regulate the development of leaves and fruit; the forest-food of such a species becomes as perennial as the lichens that flourish beneath the

* Cuvier, 'Discours sur les Révolutions de la Surface du Globe.' It is obvious that the frozen Mammoth at the mouth of the Lena forms one of the strongest, as well as the most striking, of the celebrated anatomist's assumed "proofs that the revolutions on the earth's surface had been sudden." Cuvier affirms that the Mammoth could not have maintained its existence in the low temperature of the region where its carcass was arrested, and that at the moment when the beast was destroyed, the land which it trod became glacial. "Cette gelee éternelle n'occupait pas auparavant les lieux où ils ont été saisis; car ils n'auraient pas pu vivre sous une pareille température. C'est donc le même instant qui a fait périr les animaux, et qui a rendu glacial le pays qu'ils habitaient. Cet événement a été subit, instantané, sans aucune gradation, &c."—Ossemens Fossiles, 8vo. ed. 1834, tom. i. p. 108.

† Lyell, 'Principles of Geology,' in which the phenomena that had been supposed "to have banished for ever all idea of a slow and gradual revolution,"‡ were first attempted to be accounted for by the gradual operation of ordinary and existing causes.

‡ Jameson's 'Cuvier's Theory of the Earth,' 8vo. 1813, p. 16.

winter snows of Lapland; and, were such a quadruped to be clothed, like the Reindeer, with a natural garment capable of resisting the rigours of an arctic winter, its adaptation for such a climate would be complete. Had our knowledge of the Mammoth, indeed, been restricted, as in the case of almost every other extinct animal, to its bones and teeth, it would have been deemed a hazardous speculation to have conceived, *a priori*, that the extinct ancient Elephant, whose remains were so abundant in the frozen soil of Siberia, had been clad, like most existing quadrupeds adapted for such a climate, with a double garment of close fur and coarse hair; seeing that both the existing species of Elephant are almost naked, or, at best, scantily provided when young with scattered coarse hairs of one kind only.

The wonderful and unlooked for discovery of an entire Mammoth, demonstrating the arctic character of its natural clothing, has, however, confirmed the deductions which might have been legitimately founded upon the localities of its most abundant remains, as well as upon the structure of its teeth, viz., that, like the Reindeer and Musk Ox of the present day, it was capable of existing in high northern latitudes.

The circumstances of this discovery have been recorded by Mr. Adams, in the 'Journal du Nord,' printed at Petersburg in 1807, and in the 5th volume of the 'Memoirs of the Imperial Academy of Sciences at St. Petersburg,' of which an excellent English translation was published in 1819.

Schumachoff, a Tungusian hunter and collector of fossil ivory, who had migrated in 1799 to the peninsula of Tamut, at the mouth of the river Lena, one day perceived amongst the blocks of ice a shapeless mass, not at all resembling the large pieces of floating wood which are commonly found there. To observe it nearer, he landed, climbed up a rock, and examined this new object on all sides, but without being able to discover what it was. The following year he perceived that the mass was more disengaged from the blocks of ice, and had two projecting parts. Towards the end of the next year (1801), the entire side of the animal, and one of its tusks, were quite free from the ice. On his return to the borders of the Lake Oncoul, he communicated this extraordinary discovery to his wife and some of his friends, but their reception of the news filled him with grief. The old men related how they had heard their fathers say, that a similar monster had been formerly discovered on the same peninsula, and that all the family of the person who discovered it had died soon afterwards. The Mammoth was consequently regarded as an augury of future calamity, and the Tungusian was so much alarmed that he fell seriously ill; but becoming convalescent, his first idea was the profit he might obtain by selling the tusks of the animal, which were of extraordinary size and beauty. The summer of 1802 was less warm and more stormy than usual, and the icy shroud of the Mammoth had scarcely melted at all. At length, towards the end of the fifth year (1803), the desires of the Tungusian were fulfilled; for, the part of the ice between the earth and the Mammoth having melted more rapidly than the rest, the plane of its support became inclined, and the enormous mass fell by its own weight on a bank of sand. Of this, two Tungusians, who accompanied Mr. Adams, were witnesses. In the month of March, 1804, Schumachoff came to his Mammoth, and having cut off the tusks, exchanged them with a merchant, called Bultunoff, for goods of the value of fifty rubles.

Two years afterwards, or the seventh after the discovery

of the Mammoth, Mr. Adams visited the spot, and "found the Mammoth still in the same place, but altogether mutilated. The prejudices being dissipated because the Tungusian chief had recovered his health, there was no obstacle to prevent approach to the carcass of the Mammoth; the proprietor was content with his profit from the tusks; and the Jakutski of the neighbourhood had cut off the flesh, with which they fed their dogs during the scarcity. Wild beasts, such as white bears, wolves, wolverenes, and foxes, also fed upon it, and the traces of their footsteps were seen around." The skeleton, almost entirely cleared of its flesh, remained whole, with the exception of one fore-leg, (probably dragged off by the bears). The spine, from the skull to the os coccygis, one scapula, the pelvis, and the three remaining extremities, were still held together by the ligaments, and by parts of the skin. The head was covered with a dry skin; one of the ears, well preserved, was furnished with a tuft of hair. The point of the lower lip had been gnawed; and the upper one, with the proboscis, having been devoured, the molar teeth could be perceived. The brain was still in the cranium, but appeared dried up. The parts least injured were one fore-foot and one hind-foot: they were covered with skin, and had still the sole attached. According to the assertion of the Tungusian discoverer, the animal was so fat, that its belly hung down below the joints of the knees. This Mammoth was a male, with a long mane on the neck; the tail was much mutilated, only eight, out of twenty-eight or thirty caudal vertebræ, remaining; the proboscis was gone, but the places of the insertion of its muscles were visible on the skull. The skin, of which about three-fourths were saved, was of a dark grey colour, covered with a reddish wool, and coarse long black hairs.

The dampness of the spot where the animal had lain so long, had in some degree destroyed the hair. The entire skeleton, from the fore part of the skull to the end of the mutilated tail, measured sixteen feet four inches; its height was nine feet four inches. The tusks measured along the curve nine feet six inches, and in a straight line from the base to the point three feet seven inches.

Mr. Adams collected the bones, and had the satisfaction to find the other scapula, which had remained, not far off. He next detached the skin on the side on which the animal had lain, which was well preserved; the weight of the skin was such that ten persons found great difficulty in transporting it to the shore. After this, the ground was dug in different places to ascertain whether any of its bones were buried, but principally to collect all the hairs which the white bears had trod into the ground while devouring the flesh, and more than thirty-six pounds' weight of hair were thus recovered. The tusks were repurchased at Jatusk, and the whole expedited thence to St. Petersburg; the skeleton is now mounted in the museum of the Petropolitan Academy, as it is represented at p. 217.*

It might have been expected that the physiological consequences deducible from the organization of the extinct

* A part of the skin, and some of the hair of this animal, were sent by Mr. Adams to Sir Joseph Banks, who presented them to the Museum of the Royal College of Surgeons. The hair is entirely separated from the skin, excepting in one small part, where it still remains firmly attached. It consists of two sorts, common hair and bristles, and of each there are several varieties, differing in length and thickness. That remaining fixed on the skin is thick-set and crisply curled; it is interspersed with a few bristles, about three inches long, of a dark reddish colour. Among the separate parcels of hair are some rather redder than the short hair just mentioned, about four inches long, and some bristles nearly black, much thicker than horse-hair, and from twelve to eighteen inches long. The skin, when first brought to the Museum, was offensive to the smell. It is now quite dry and hard, and where most compact is half an inch thick. Its colour is the dull black of the living Elephants.

species which was thus, in so unusual a degree, brought to light, would have been at once pursued to their utmost legitimate boundary, in proof of the adaptation of the Mammoth to a Siberian climate; but, save the remark that the hairy covering of the Mammoth must have adapted it for a more temperate zone than that assigned to existing elephants,* no further investigations of the relation of its organization to its habits, climate, and mode of life, appear to have been instituted; they have in some instances, indeed, been rather checked than promoted.

Dr. Fleming has observed that "no one acquainted with the gramineous character of the food of our Fallow-deer, Stag, or Roe, would have assigned a lichen to the Reindeer." But we may readily believe that any one cognizant of the food of the Elk, might be likely to have suspected cryptogamic vegetation to have entered more largely into the food of a still more northern species of the deer tribe. And I can by no means subscribe to another proposition by the same eminent naturalist, that "the kind of food which the existing species of Elephant prefers, will not enable us to determine, or even to offer a probable conjecture concerning that of the extinct species." The molar teeth of the Elephant possess, as we have seen, a highly complicated, and a very peculiar structure, and there are no other quadrupeds that derive so great a proportion of their food from the woody fibre of the branches of trees. Many mammals browse the leaves; some small rodents gnaw the bark; the Elephants alone tear down and crunch the branches, the vertical enamel-plates of their

* "La longue toison dont cet animal était couvert semblerait même démontrer, qu'il était organisé pour supporter un degré de froid plus grand que celui qui convient à l'éléphant de l'Inde." Pictet, Paleontologie, 8vo. tom. i. 1844, p 71.

huge grinders enabling them to pound the tough vegetable tissue and fit it for deglutition. No doubt the foliage is the most tempting, as it is the most succulent part of the boughs devoured; but the relation of the complex molars to the comminution of the coarser vegetable substance is unmistakeable. Now, if we find in an extinct Elephant the same peculiar principle of construction in the molar teeth, but with augmented complexity, arising from a greater number of the triturating plates and a greater proportion of the dense enamel, the inference is plain that the ligneous fibre must have entered in a larger proportion into the food of such extinct species. Forests of hardy trees and shrubs still grow upon the frozen soil of Siberia, and skirt the banks of the Lena as far north as latitude 60°. In Europe arboreal vegetation extends ten degrees nearer the pole, and the dental organization of the Mammoth proves that it might have derived subsistence from the leafless branches of trees, in regions covered during a great part of the year with snow.

We may therefore safely infer, from physiological grounds, that the Mammoth would have found the requisite means of subsistence at the present day, and at all seasons, in the sixtieth parallel of latitude; and, relying on the body of evidence adduced by Mr. Lyell, in proof of increased severity in the climate of the northern hemisphere, we may assume that the Mammoth habitually frequented still higher latitudes at the period of its actual existence. "It has been suggested," observes the same philosophic writer, "that, as in our own times, the northern animals migrate, so the Siberian Elephant and Rhinoceros may have wandered towards the north in summer." In making such excursions during the heat of that brief season, the Mammoths would be arrested in their northern progress by a condition to which the Rein-deer and Musk Ox are not subject, viz., the limits of arboreal vegetation, which, however, as represented by the dominating shrubs of Polar lands, would allow them to reach the seventieth degree of latitude.* But with this limitation, if the physiological inferences regarding the food of the Mammoth from the structure of its teeth be adequately appreciated and connected with those which may be legitimately deduced from the ascertained nature of its integument, the necessity of recurring to the forces of mighty rivers, hurrying along a carcass through a devious course, extending through an entire degree of latitude, in order to account for its ultimate entombment in ice, whilst so little decomposed as to have retained the cuticle and hair, will disappear. And it can no longer be regarded as impossible for herds of Mammoths to have obtained subsistence in a country like the southern part of Siberia where trees abound, notwithstanding it is covered during a great part of the year with snow, seeing that the leafless state of such trees during even a long and severe Siberian winter, would not necessarily unfit their branches for yielding sustenance to the well-clothed Mammoth.

With regard to the extension of the geographical range of the *Elephas primigenius* into temperate latitudes the distribution of its fossil remains teaches that it reached the fortieth degree north of the equator. History, in like manner, records that the Rein-deer had formerly a more extensive distribution in the temperate latitudes of Europe than it now enjoys. The hairy covering of the Mammoth

* In the extreme points of Lapland, in 70° north latitude, the pines attain the height of sixty feet; and at Enontekessi, in Lapland in 68° 30′ north latitude, Von Buck found corn, orchards, and a rich vegetation, at an elevation of 1356 feet above the sea. Lindley, Intr. to Botany, pp. 485, 490.

concurs, however, with the localities of its most abundant remains, in showing that, like the Rein-deer, the northern extreme of the temperate zone was its metropolis.

Attempts have been made to account for the extinction of the race of northern Elephants by alterations in the climate of their hemisphere, or by violent geological catastrophes, and the like extraneous physical causes. When we seek to apply the same hypothesis to explain the apparently contemporaneous extinction of the gigantic leaf-eating Megatheria of South America, the geological phenomena of that continent appear to negative the occurrence of such destructive changes. Our comparatively brief experience of the progress and duration of species within the historical period, is surely insufficient to justify, in every case of extinction, the verdict of violent death. With regard to many of the larger Mammalia, especially those which have passed away from the American and Australian continents, the absence of sufficient signs of extrinsic extirpating change or convulsion, makes it almost as reasonable to speculate with Brocchi,* on the possibility that species like individuals may have had the cause of their death inherent in their original constitution, independently of changes in the external world, and that the term of their existence, or the period of exhaustion of the prolific force, may have been ordained from the commencement of each species.

* Cited by Lyell, 'Principles of Geology,' (1835,) vol. iii. p. 104.

PACHYDERMATA. *PROBOSCIDIA.*

Fig. 96.

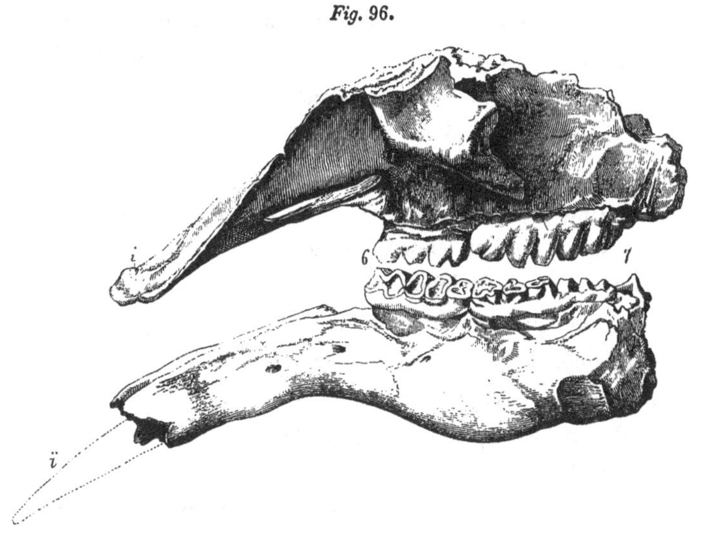

Upper and lower jaws of the *Mastodon angustidens*, from Epplesheim, after Kaup.

MASTODON ANGUSTIDENS. Narrow-toothed Mastodon.

Mastodonte à dents étroites,	CUVIER, Annales du Muséum, tom. viii., Ossemens Fossils, tom. i. 4to, 1821.
Mastodon Avernensis,	CROIZET and JOBERT, Ossem. Foss. du Puy de Dome, 4to. 1828.
" *longirostris,*	KAUP, Ossem. Foss. de Mammifères de Darmstadt. 1836.
" *angustidens,*	OWEN, Report of British Association, 1843.

NATURALISTS are most familiar with that gigantic type of quadrupeds called, from the peculiar prehensile development of the nose and upper lip, "proboscidian," as it is manifested by the existing species of Elephants which have been at different times introduced into Europe from the tropical regions of Asia and Africa; and we have seen in the preceding section, that an extinct species of this genus once ranged over the whole of the temperate, and part of the arctic zones of the northern hemisphere of the globe, and has left abundant evidence of its former existence in our own island.

In like manner we learn from the study of fossil remains, that other quadrupeds, as gigantic as Elephants, armed with two as enormous tusks projecting from the upper jaw, and provided with a proboscis, once trod the earth; the presence of the latter flexible organ being inferred, not only by its necessary coexistence with long tusks, which must have prevented the mouth reaching the ground, but also by the configuration of the skull, by the holes which gave passage to large nerves, and by depressions for the attachment of particular muscles, analogous to those which relate exclusively to the organization of the trunk in the Elephant. Like the Elephants, also, these other huge proboscidian quadrupeds were destitute of canine teeth, and provided with a small number of large and complex molar teeth, successively developed from before backwards in the jaws, with a progressive increase of size and complexity, from the first to the last. The broad crowns of the molar teeth were also cleft by transverse fissures; but these clefts were fewer in number, of less depth, and greater width than in the Elephants: the transverse ridges were more or less deeply bisected, and the divisions more or less produced in the form of udder-shaped cones, whence the name *Mastodon*,* assigned by Cuvier to the great proboscidian quadrupeds with teeth of this kind.

A more important difference presents itself when the

* Etym. Gr. *mastos*, udder, *odos*, a tooth.

teeth of the typical species of Mastodon are compared with those of the Elephants, in reference to their structure. The dentine, or principal substance of the crown of the tooth, is covered by a very thick coat of dense and brittle enamel; a thin coat of cement is continued from the fangs upon the crown of the tooth, but this third substance does not fill up the interspaces of the divisions of the crown, as in the Elephants. Such, at least, is the character of the molar teeth of the first discovered species of Mastodon, which Cuvier has termed *Mastodon giganteus*, and *Mastodon angustidens*. Fossil remains of proboscidians have subsequently been discovered, principally in the tertiary deposits of Asia, in which the number and depth of the clefts of the crown of the molar teeth, and the thickness of the intervening cement, are so much increased as to establish transitional characters between the lamello-tuberculate teeth of Elephants, and the mammillated molars of the typical Mastodons;* showing that the characters deducible from the molar teeth are rather the distinguishing marks of species than of genera, in the gigantic proboscidian family of mammalian quadrupeds.

* Mr. Clift had foreseen the possibility of the discovery of such a link, since supplied by the praiseworthy exertions of Captain Cautley, and Dr. Falconer; and in his description of the Fossil Remains from Ava, in the Geological Transactions, second series, vol. ii., he says, "It is not impossible that there may yet be a link wanting, which might be supplied by an animal having a tooth composed of a greater number of denticles, increasing in depth, and having the rudiments of *crusta petrosa*, that necessary ingredient in the tooth of the Elephant: the entire absence of which distinguishes the tooth of the *Mastodon*." Cuvier had previously enunciated the same supposed distinctive character between the structure of the teeth of the Elephant and Mastodon. "Dans l'éléphant ces vallons sont entièrement comblés par *le cortical*, tandis que dans le Mastodonte ils ne sont remplis de rien." Ossemens Fossiles, tom. i. 4to, 1821, p. 225. Mr. S. Woodward put forth a more remarkable, but not less erroneous opinion on this subject. He says, "The distinctive characters of the grinders of the Elephant and Mastodon are so decided, that it is scarcely possible to mistake the one for the other. The enamel of the former is disposed in pairs transversely, to the num-

Two dental characters, however, exist, though hitherto I believe unnoticed as such, which distinguish in a well-marked and unequivocal manner, the genus *Mastodon* from the genus *Elephas*. The first is the presence of two tusks in the lower jaw of both sexes of the *Mastodon*, one or both of which are retained in the male, and acquire a sufficiently conspicuous size, though small in proportion to the upper tusks; while both are early shed in the female. The second character is equally decisive; it is the displacement of the first and second molars in the vertical direction, by a tooth of simpler form than the second, a true *dent de remplacement*, developed above the deciduous teeth in the upper, and below them in the under jaw.

These two dental characters, which are of greater importance than many accepted by modern zoologists as sufficient demarcations of existing generic groups of Mammalia, have been recognised in the species called *Mastodon giganteus*, most common in North America, and in the *Mastodon angustidens*, which is the prevailing species of Europe.

To the last-named species I refer the comparatively few remains of the *Mastodon* that have been discovered in Eng-

ber of about ten, surrounded and held together by what Parkinson terms the *crusta petrosa*. Now the enamel of the grinder of the Mastodon is all external, whilst the crusta petrosa, or a substance resembling it, is internal." In fully-formed and worn teeth of Mastodons, the dentine or substance which supports the enamel degenerates, near the pulp cavity, into a kind of coarse bone-like tissue approaching in structure to crusta petrosa, or cement; but the same tissue is found in the internal part of the dentine of the old grinders of the Elephant. The truth is, that the exterior of the fangs in all Mastodons is covered by a moderately thick coat of cement (*cortical* of Cuvier, *crusta petrosa* of Clift); and that this substance extends upon the enamel of the crown, in a very thin layer, requiring microscopical sections and examination for its detection in the typical Mastodons; but augmenting in thickness in the elephantoid and other Mastodons, with thinner and more numerous transverse divisions of the crown of the grinders.

land, and hitherto exclusively in these deposits, consisting of sand, shingle, loam, and laminated clay, containing an intermixture of the shells of terrestrial, fresh-water, and marine mollusca, which extend along the coast of Norfolk and Suffolk, and have been so admirably described by Mr. Lyell under the name of the "fluvio-marine crag," and referred to the "older pliocene" division of his tertiary system.

The first representation of any fossil relic of a *Mastodon* from English strata, was given by the father of English Geology, William Smith; it forms the frontispiece of his famous 4to work, 'Strata identified by Organised Fossils,' 1816, and is a coloured engraving of the natural size of the last molar tooth of the upper jaw of the *Mastodon angustidens.*

Cuvier has given figures of the corresponding tooth of the *Mastodon angustidens*, from three individuals of different ages, and from three different localities, in his 'Ossemens Fossiles,' 4to. 1821, vol. i. The first 'Divers Mastodontes,' (pl. i. fig. 5,) is a young tooth, the udder-shaped processes of the crown being unworn, and the fangs not developed; from the tertiary deposits at Trevoux: the second specimen, (pl. i. fig. 6,) with the two anterior pairs of mammillæ worn down, is from Peru: the third, (pl. ii. fig. 10,) having the summits of all the five pairs of mammillæ abraded, and the roots of the crown fully developed, is stated to have been from the collection of M. Hammer, and was most probably a German specimen; each of these molar teeth is referred by Cuvier to his narrow-toothed species, "Mastodonte à dents étroites."

If the subjoined cut (fig. 97,) of the tooth figured by Mr. Smith, be compared with the Cuvierian figures above cited, the specific identity will be readily recognised. The figures in the 'Ossemens Fossiles' are reduced one half, and, like Mr. Smith's figure, are drawn in a position the reverse of the natural one. Until very recently, I knew the present early and striking evidence of a British Mastodon only by Mr. Smith's beautiful engraving of it; he,

Fig. 97.

Last upper molar, *Mastodon angustidens*, Fluvio-marine Crag, Norfolk. ⅓ nat. size.

however, makes no mention of the specimen in his work, nor gives any reference to the locality from which it had been obtained. Indeed, it seems to have produced little impression upon the contemporary labourers in his domain of science, and to have been regarded as apocryphal by some sound geologists of that period.

Mr. Bakewell, in his Memoir 'On the Fossil Remains of large Mammalia, found in Norfolk,'* under the head *Mastodon*, says, "The remains of this animal have not hitherto been discovered in any part of England, except in the county of Norfolk, and even there I think their occurrence at present problematical;" adding, "The tooth of the supposed Mastodon, described by Mr. William Smith, I have never seen." No reference to such description is given by Mr. Bakewell, and I presume the remark refers to the figure above cited. Mr. S. Woodward, however, affirms that "The large grinder figured by

* 'Loudon's Magazine of Natural History,' vol. ix (1836).

Smith, in his 'Strata Identified,' was reported to have been found at Whitlingham; and, when at Scarborough last summer, I put the question to him, and he assured me that it was so found."* Whitlingham is a village on the right bank of the Yare, within five miles of Norwich, where the fluvio-marine crag is well developed.

Mr. Morris, in his valuable 'Catalogue of British Fossils,'† refers the tooth in question, I know not on what authority, to "Horstead, Norfolk." I have only recently ascertained that the tooth itself forms part of a collection of the late Mr. Smith's fossils, purchased by the British Museum, but not yet arranged, or brought into public view.

Mr. König kindly favoured me with the opportunity of examining the tooth, which in the manuscript catalogue of Mr. Smith's collection, is thus noticed:—"Middle-sized grinder of *Mastodon*, with numerous subdivided irregularly shaped mammillæ, one half of which only is worn; ivory converted into a brown semiopal-like mass. Found in Norfolk." Mr. König at the same time informed me, that when he exhibited this tooth to Cuvier, during his visit to the British Museum, the great anatomist warned him against placing implicit reliance on the statement of its British origin; and referred to the molar tooth of the *Mastodon angustidens* from Peru, which Mr. Smith's specimen closely resembles. The similarity is not greater, however, than that which the same specimen presents to the continental Mastodon's grinders, figures of which have been already cited from Cuvier's great work; and the same resemblance may be affirmed in regard to the smaller varieties of the last upper molar tooth of the *Mastodon*

* Ib. p. 152.

† 8vo. Van Voorst, 1843, p. 213.

angustidens from Eppelsheim, figured in tab. xviii. of the great work by Dr. Kaup on the Mammalian Fossils of that locality. To place the source and matrix of Mr. Smith's fossil beyond doubt, I applied to Mr. R. Fitch, of Norwich, for the loan of his specimens of the molars of a *Mastodon angustidens* from the crag-pits near that city, and took the largest specimen to the British Museum for comparison with the tooth in question. The identity of structure and colour between the two fossils was complete; the dentine in both had the same rich brown tint, brittle texture, and superficial ferruginous stain; both belonged to the same species of Mastodon, and alike manifested the well-known characters of crag fossils.

The crown of the molar from Mr. Smith's collection measures seven inches in length, three inches in breadth across its base, and the height of the highest unworn mammilla is two inches and a half; the crown is divided into five pairs of mastoid eminences and a strong tuberculated posterior ridge, or talon; resembling, in this respect, that of the molars above cited from Cuvier, and the figures 6, 7, and 8* of tab. xviii. of Dr. Kaup; and showing a greater complexity than do the more simple varieties of the last molar of the *Mastodon angustidens*, represented in figures 1, 2, and 3 of the same plate. The mastoid eminences have a subalternate disposition, and the smaller connecting eminence, which rises from the middle of each transverse valley, is well developed: the summits of the larger processes are more or less subdivided; but this character is best seen in the unworn teeth. The summits of the first, second, and third pairs of mammillæ have

* These are all referred by Dr. Kaup to his *Mastodon longirostris*, who, nevertheless, distinguishes that nominal species from the *Mastodon angustidens* of Cuvier by a more complex last molar tooth.

been abraded by mastication, but more gradually than those of the Peruvian Mastodon's molar figured by Cuvier. The fangs and a portion of the anterior part of the crown have been broken away in the specimen.

In Mr. Smith's figure, the mirror has not been used by the engraver, and it consequently, like the woodcut fig. 97, represents the molar as having belonged to the left side of the jaw, but the specimen is from the right side.

A fine example of the last molar tooth of the left side of the upper jaw, obtained by Captain Alexander from the sea-shore at Sizewell Gap, Suffolk, so closely corresponds in size and configuration with the molar in fig. 97, that, but for the greater extent of abrasion, it might pass for the opposite grinder of the same individual *Mastodon*. A cast of this tooth was presented by Captain Alexander to the Geological Society of London, and the following notice of it is recorded in the third volume of the 'Proceedings' of the Society:

"The larger cast was taken from a Mastodon's tooth, found on the shore at Sizewell Gap, about seven miles from Southwold. When the original came into Captain Alexander's possession, crag adhered to it in considerable quantity, and he has no doubt that it had been washed from Easton, about a mile and a half north of Southwold. The weight of the tooth is two pounds, five ounces and a half; its length is about six inches, and its breadth three inches and a half; and, although it had been washed eight miles, only three of the crowns had been injured."

From an inspection of the cast, it appears that the first and fifth pairs of tubercles, and the posterior tuberculate talon, have suffered fracture. The effects of abrasion from the acts of mastication have extended to the fourth pair of mastoid eminences: besides the injury to the crown from accidental violence, all the fangs of the tooth have been broken away.

The molar tooth (fig. 98) of the *Mastodon angustidens*, which was obtained by Mr. Robert Fitch, F.G.S. of Norwich, from the fluvio-marine crag at Thorpe, near that city, strikingly demonstrates the generic differential characters

Fig. 98.

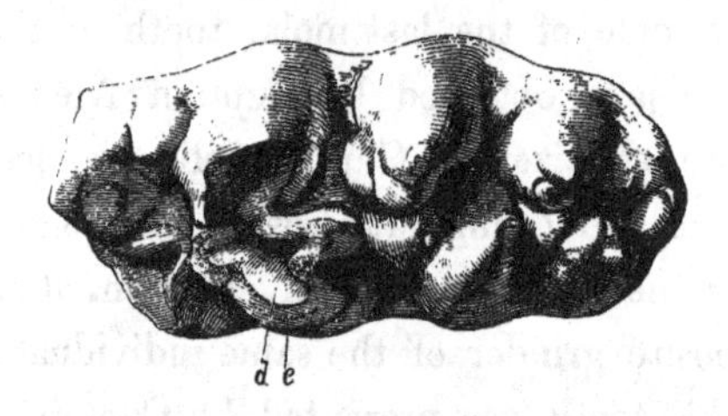

Penultimate upper Molar, *Mastodon angustidens*, Fluvio-marine Crag, Thorpe, Norfolk. ⅓ nat. size.

between the molars of the Mastodon and those of the Mammoth. The coat of enamel (*e*), which invests the dentinal eminences (*d*) of the crown, is three times as thick as that in the Mammoth's molar of thrice the size, which is figured at p. 231 (fig. 90); and it is almost twice as thick as the enamel of the molar tooth (fig. 88) of the African Elephant,—the existing species which makes the nearest approach to the Mastodon in the structure of its teeth and in its general proportions. The cement of the Mastodon's tooth, on the other hand, forms so thin a layer, that it can only be detected by the naked eye at the bottom of the clefts between the mastoid eminences. These are arranged subalternately in four pairs, and a tuberculate eminence terminates the base of the crown. This number of the chief divisions of the grinding surface, together with the size of the tooth, which is six inches in length, shows it to be the last but one in the molar series; and the convex bend of the grinding surface in the longitudinal direction, proves it to have come from the upper jaw.

The work of mastication first impresses the fore part of the grinding surface, and the inner tubercles in the upper molars are always worn lower than the outer ones. By these marks, a Mastodon's grinder may be readily referred to the jaw and the side of the jaw from which it originally came,—the tooth in question being the penultimate grinder of the right side of the upper jaw. It must have belonged to a Mastodon that perished in the vigour of youth, before the attainment of full maturity; for the two hinder pairs of tubercles had not been used in mastication: and it could not, therefore, have been shed in the ordinary course of dental change, since the last molar tooth must still have been concealed in its alveolar nidus of growth.

The mutilated molar tooth of the *Mastodon angustidens*, represented in figure 99, was likewise obtained by Mr.

Fig. 99.

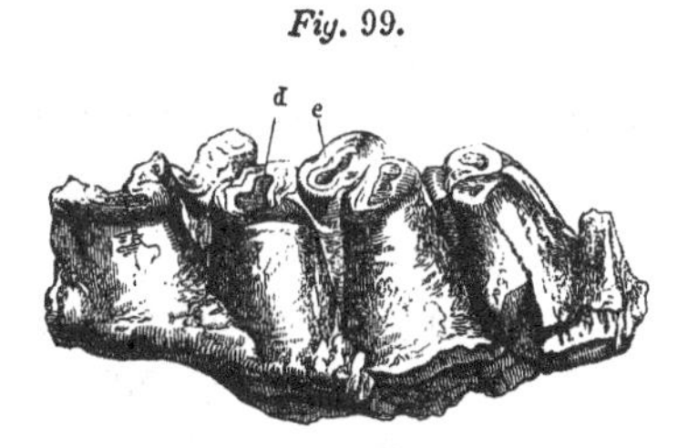

Penultimate lower molar, *Mastodon angustidens*, Fluvio-marine Crag, Norwich, $\frac{1}{3}$ nat. size.

Robert Fitch, F.G.S., "from a crag-pit in the immediate vicinity of Norwich." It is the penultimate molar of the left side of the lower jaw, and had done good service to the old Mastodon, the summits of all the mastoid eminences having been abraded by mastication; but they are not so much worn as in a naturally shed tooth. There are eight principal tubercles, with a small anterior basal ridge, and a larger posterior talon. The intermediate connecting eminences in the first and second valleys are worn down to their basal confluence with the larger mastoid tubercles, and thereby occasion a more complete alternate arrangement of these principal divisions of the crown. The wavy fibrous texture of the enamel is remarkably well shown in the fractured surfaces of the very thick layer of that substance which invests the crown. The dentine is reduced to a very brittle friable condition, and the fangs are entirely broken away.

Two views of a large portion of a corresponding tooth of the *Mastodon angustidens* are given by Mr. Samuel Woodward in a Paper 'On Remains of *Mastodon giganteus* and *Mastodon latidens*, found in the Tertiary Beds of Norfolk.'* I have examined the casts of the original specimen figured, which are now in the Geological Society's Museum, and can affirm that this tooth belonged neither to the American nor to the Indian species of Mastodon, cited by Mr. Woodward, but to the *Mastodon angustidens*. Mr. Layton thus recounts some of the circumstances attending the discovery of the molar tooth figured by Mr. Woodward, in a communication printed by Mr. Fairholme, in his 'Geology of Scripture,' p. 281:—

"In 1820, an entire skeleton of the great Mastodon was found at Horstead, near Norwich, lying on its side, stretched out between the chalk and the gravel. A grinder was brought to me; but, so long after it was discovered, that scarcely any other part of the animal could be preserved."

* 'Loudon's Magazine of Natural History,' vol. ix. (1836,) p. 131.

Afterwards, in reply to some misgivings of Mr. Fairholme respecting this discovery, Mr. Layton says, "Your doubt, as to the great Mastodon being found in Norfolk, came not at all unexpected. I should have doubted it myself, under almost any other circumstances; as it is, I feel sure and certain of the fact. I lived at Catfield, in Norfolk, six miles from Hasborough, and about as far from Horstead. From this latter place, marl is carried to all the villages in the neighbourhood, to be spread upon the ground. A boatman, who was in the habit of bringing me fossils, brought a grinder of this *Mastodon* as a curiosity, saying it had been found in the marl, and given him by the head pitman. It was the posterior portion of the grinder of the great *Mastodon* (I am certain of the fact); containing, as far as I recollect, eight points, none of which had been cut or brought into use. On the first opportunity, I went to make inquiry about it at the chalk-pit. The pitman pointed out to me the place where it was found; and said that the whole animal was, as it were, lying on its side, stretched out on the surface of the marl. He described it as being very soft, and that a great part of it would at first spread like butter; the whole, however, had been thrown down along with the marl, and carried away. He said he had looked upon it as very curious indeed, but of no use, and he had kept that piece of tooth merely by accident. He afterwards found another fragment or two of the bones in his garden, where he had thrown them, and he sent them to me. They are now in my possession; but I am not able to identify them with the Mastodon, as distinguished from the Mammoth or Elephant. The grinder I sent to Dawson Turner, Esq., of Great Yarmouth, who probably has it now."

How far the testimony of the workmen may be relied upon as indicative of arrested decomposition and change to adipocere of the soft parts of an entire Mastodon, may be questioned; but the tooth is, without doubt, such as Mr. Layton has described, showing eight alternating unworn mastoid eminences; but belonging, not to the *Mastodon giganteus*, as might be inferred from the term Great Mastodon, but to the *Mastodon angustidens*.

Fig. 100.

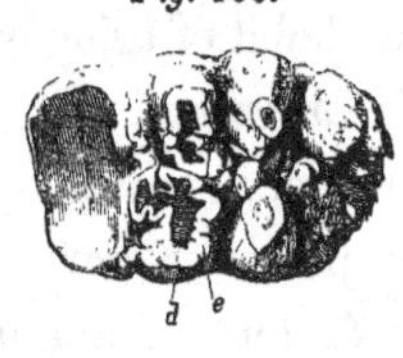

Fourth upper Molar, *Mastodon angustidens*, Fluvio-marine Crag, Norwich, ⅓ nat. size.

The molar tooth, the grinding surface of which is represented at fig. 100, is the fourth in the order of size, and the third in the order of position, counting backwards in the upper jaw, before any of the teeth are shed; and it belonged to the left side of the mouth. This beautiful specimen also forms part of the collection of Mr. Fitch, and was obtained by that zealous collector of the organic fossils of Norfolk from a crag-pit at Postwick, in the vicinity of Norwich: it was imbedded in the fluvio-marine crag, with the characteristic shells of that formation, immediately above the chalk.

This tooth corresponds with the larger molar in the portion of the upper jaw of the *Mastodon angustidens*, from the tertiary deposits at Dax, figured by Cuvier in his 'Divers Mastodontes,' pl. iii., fig. 2; and with the largest molar in a similar portion of the upper jaw of the same species of Mastodon from Eppelsheim, figured by Dr. Kaup in tab. xvi., figs. 1 and 1 *a* of his work on the Mammalian Fossils of that locality. In Dr. Kaup's figure, the tooth in question is associated with the first and second molars of the *Mastodon angustidens*, which are much worn, and are true deciduous teeth, the only ones, in fact, which strictly correspond with the deciduous teeth of ordinary Pachyderms. Cuvier's specimen shows the first of the series of permanent teeth just coming into place, with its mastoid eminences fresh and unworn; this permanent tooth, the only one corresponding to the teeth called premolars, false molars, and bicuspides in other Mammalia, is developed above the deciduous molars in the upper jaw, beneath them in the lower jaw, and it succeeds and displaces them vertically in both jaws. Its crown is divided into four tubercles, and it is consequently more simple than the second deciduous tooth which it displaces, agreeing in this respect with the premolar teeth, or *dents de remplacement*, in other Mammalia.

When the milk-teeth are shed, and their quadri-tuberculate successor is in place, the molar tooth (fig. 100) is the second of the molar series, but presents a character which must seem strange to one unacquainted with the law of the succession of teeth in the Mastodons, viz., a much more abraded crown than the smaller tooth which precedes it. The smaller tooth is therefore the first of the permanent series of molars, and the one figured in cut 100 is the second of that series; but, although they are termed permanent molars, agreeably with the general analogies of the Mammalian dentition, their duration is brief in comparison with the life-time of the animal; and they are successively shed, as in the Elephant, the Hog, and the Kangaroo, from before backwards; the dentition in the Mastodon being ultimately reduced to the last great molar tooth, which is the seventh in the order of development.

To facilitate the determination of the teeth of the *Mastodon angustidens*, I shall briefly denote their general characters, as they succeed each other in the order of their development.

The *first* molar tooth has a square-shaped crown, broadest behind, divided into four mastoid tubercles, and averages an inch in length, (antero-posterior diameter,) and three-fourths of an inch in breadth. The *second* has an oblong crown, supporting three pairs of mastoid tubercles, and averages two inches in length, and one inch one-third in breadth. The *third* tooth, which takes the place of the above when shed, has a square crown, with two pairs of mastoid tubercles, and an anterior and posterior basal ridge; it averages a length of two inches, with nearly the same breadth, ranging from one inch and a half, to two inches and a quarter. The *fourth* tooth has an oblong crown, and supports three pairs of mastoid tubercles, with usually a large posterior tuberculate talon; its average length is two inches three-fourths; its breadth one inch two-thirds. The *fifth* tooth resembles the preceding, but averages four inches one-third in length, and two inches three-fourths in breadth. The *sixth* tooth supports four pairs of mastoid tubercles, and a posterior talon, usually of small size; its average length is six inches; its breadth three inches and a quarter. The *seventh*, and last molar tooth, has generally five pairs of mastoid tubercles, and a posterior tubercular talon; its average length is seven inches and a quarter, its breadth three inches and a third. The observed extremes of size of this complex tooth, which is subject to more varieties than the preceding teeth, is five inches and a half, and nine inches; in general the pairs of tubercles gradually and slightly decrease in size from the first to the last: in the small-sized specimens of the seventh molar the decrease is more rapid, and the fifth pair is reduced almost to a tubercular talon, which is succeeded by a small basal ridge;* in the middle-sized teeth the fifth

* The tooth in the lower jaw of the *Mastodon angustidens*, figured by Cuvier

pair of mastoid tubercles acquire their normal proportions, and the posterior ridge is developed into a group of tubercles; in the large-sized variety, like that exemplified in the old male *Mastodon angustidens*, figured by Dr. Kaup in his Pl. xvi. fig. 5, and Pl. xviii. fig. 9, the tubercular talon assumes the character of a sixth pair of mastoid eminences, succeeded by a small tubercular ridge. (See figs. 96 and 97.)

Analogous varieties of form and size are manifested by the last molar tooth of the *Mastodon giganteus*. In the *Mastodon angustidens* the larger and more complex examples have been supposed to indicate a distinct species;* in the *Mastodon giganteus* the varieties have been seized upon as characters, not only of distinct species, but of distinct genera.† The utmost signification that, in my opinion, can be legitimately assigned to them as distinctive characters, is in relation to difference of sex.

Having thus briefly pointed out the principal characters of each of the seven molars of the *Mastodon angustidens*, I may add that those of the lower jaw are narrower than those of the upper; and that the upper molar teeth are characterised by the slight convex curve, described by the grinding surface in its longitudinal direction, and the lower molars by the corresponding concavity of the same surface. The fore part of an unworn molar is the broadest, and this part of the grinding surface shows first and most the effects of mastication; in the upper molars the inner range of tubercles are most worn, in the lower molars the outer range. By these characters a detached grinder of the Mas-

in the 'Divers Mastodontes,' pl. iii. fig. 4, exemplifies this variety; and the form of the symphysis shows the specimen to have belonged to a female Mastodon.

* See Dr. Kaup's characters of *Mastodon longirostris*, 'Description d'Ossements Fossiles de Darmstadt,' cap. iv. 1835.

† See Dr. Grant in 'Proceedings of the Geological Society,' June 15th, 1842.

todon may with certainty have its place assigned to it in the dental series, and in the jaws supporting them.

I have not thought it necessary to multiply figures of the molars of the *Mastodon angustidens*, which have been at different times discovered in this country; those selected for illustration show the three well-marked grades of size and complication of grinding surface, and will, I hope, suffice, with the descriptions, to enable the collector of fossils to identify subsequent dental remains of this rare British extinct Mammal. The works of Cuvier and of Dr. Kaup above cited, give admirable illustrations of all the teeth of the *Mastodon angustidens.*

The summits of the principal, or normal eminences of the crown, are usually subdivided by shallow clefts into smaller tubercles; a character which is most conspicuous in the incompletely formed small molars at the anterior part of the series. Cuvier has shown this structure in the young *Mastodon's* tooth from Orleans, figured in his 'Divers Mastodontes,' pl. iii. fig. 6. Dr. Kaup has well represented it in his tab. xx. fig. 3, tab. xxi. fig. 1. And Mr. Lyell has given an admirable cut of a fourth lower molar of the *Mastodon angustidens*, from the fluvio-marine crag near Norwich, in the last edition (1841) of his 'Elements of Geology.' I have now before me the germ of a corresponding molar of the same species of *Mastodon*, subsequently discovered by Mr. Robert Fitch in the same stratum and locality.

Captain Alexander has recorded his discovery of a fragment of a young tooth of the *Mastodon*, in the crag at Bramerton, in the third volume of the 'Geological Proceedings.' This tooth belongs to the *Mastodon angustidens*, and the crag is of that fluvio-marine origin which Mr. Lyell has shown to belong to the older pliocene period. In the

Museum of the Geological Society of London, there is a much worn fragment of a large molar of the same species of Mastodon from the fluvio-marine crag at Euston, Suffolk. Fractured and fragmentary molars of the *Mastodon angustidens* have been discovered in the same formation at Horstead, by the Rev. J. Gunn, and at Bramerton by Mr. Samuel Woodward.

Mr. Lyell has recorded other discoveries of Mastodontal remains, in his instructive Memoirs on the tertiary, drift, and boulder formations of Norfolk. "In a crag-pit at Thorpe," he observes,* "Mr. Wigham has obtained a Mastodon's tooth at the bottom of the deposit, near the chalk, associated with pectens and other marine shells." "He also discovered, in 1838, at Postwick, together with the remains of fish and marine shells, part of the left side of the upper jaw of a Mastodon, containing the second true molar, and in the socket the indication of another, namely, the first molar. This fragment was sufficiently perfect to enable Mr. Owen, to whom I submitted it, to refer it to *Mastodon longirostris*, a species also found at Eppelsheim."†

At the period when Mr. Lyell submitted this specimen to my inspection, although I was by no means convinced of the distinction of the *Mastodon longirostris* of Kaup from the *Mastodon angustidens* of Cuvier, I had not entered so fully into the details of the evidence bearing upon this question as to justify me in rejecting the name assigned by the laborious investigator of the Eppelsheim fossils to the species of *Mastodon*, of which certain specimens, figured by Dr. Kaup, bore the closest resemblance to Mr. Wigham's interesting fossil; nor was I then possessed of the rich series of analogical facts in the dentition of the

* Proceedings of the Geological Society, April, 1839, vol. iii. p. 128.

† Mag. of Nat. Hist. 1839, p. 337.

Mastodon giganteus, which now appear to complete the demonstration of the specific identity of the *Mastodon longirostris*, and *Mastodon angustidens.*

Besides portions of jaws and numerous detached molar teeth, fragments of tusks have been discovered in the fluvio-marine crag, exhibiting the characteristic decussating curvilinear impressions of true ivory, and most probably belonging to the *Mastodon angustidens*, which continental fossils prove to have possessed two large tusks in the upper jaw, like the *Mastodon giganteus* of America (see fig. 102). These tusks are less extensively, and less obliquely curved than the tusks of the Mammoth (fig. 85). Certain individuals of the *Mastodon angustidens*, probably the males, have likewise been shewn by continental specimens, to have had small straight tusks in the lower jaw, and I am able to add the testimony of a British fossil in proof of this correspondence between the European narrow-toothed *Mastodon*, and the *Mastodon giganteus* of North America.

Mr. Fitch has this year communicated to me his latest discoveries of Mastodontal and other Mammalian remains in the crag formations of Norfolk, and he has been so obliging as to forward to me the specimens for description. In a note dated "Norwich, September 6th, 1844," he says: "Since you last saw my collection, I have added several very good specimens, which I think you would like to hear about, if not to see. I have obtained several other Mastodon's teeth, one beautifully perfect, and precisely similar to one figured in the Geological Transactions, vol. vii. pl. xxxix. figs. 1, 2, and 3. I have also what I suppose to be the tooth of a Trogontherium, and a tooth I imagine to be of a Bear. I shall be very happy to send them up for your inspection, if you think them of any interest. They are all from our crag-pits in the immediate vicinity of Norwich. I have also a large and curiously flattened portion of a tusk, which is about sixteen inches long; the structure is unquestionably ivory; this our friend Professor Sedgwick said he should like you to see."

The specimen, of which figures are subjoined, is a portion

Fig. 101.

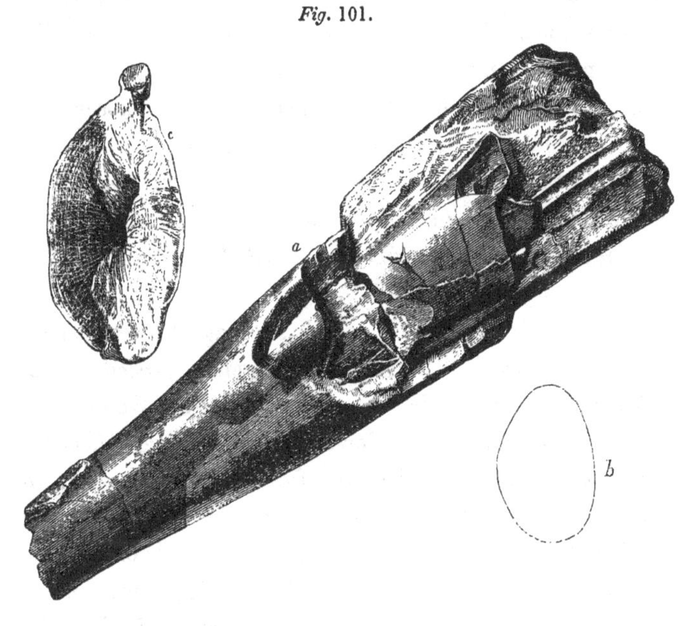

Portion of the lower tusk of the *Mastodon angustidens*, from the fluvio-marine crag, Norwich. *a* ¼ nat. size.

of a straight, subcompressed conical tusk: the base of the fragment is as solid as the apex, and the whole is traversed by a subcentral canal, of nearly the same diameter, which is about three lines, from one end of the fragment to the other. The outer layers of the ivory have been detached, excepting a very small portion near the small end, which retains its thin coating of cement: the decussating curvilinear lines are well displayed at the fractured surface of this end, figured at *c*. The transverse section of this tusk (fig. 101, *b*) gives an irregular oval figure, one side being less convex than the other; and on the lower half of the less convex (outer) side of the tusk, faint traces may be distinguished of longitudinal grooves, about a line in breadth: the slender subcentral canal is nearer the lower than the upper surface of the tusk. In all these characters, the fragment in question agrees with a similar fragment of a tusk, ten inches in length, obtained from the miocene, or older pliocene tertiary deposits at Eppelsheim, and now in the collection of the Earl of Enniskillen; which specimen Dr. Kaup has determined to belong to the lower jaw of his *Mastodon longirostris*, the *Mastodon angustidens* of Cuvier.

The earliest observation of this striking character of inferior tusks, which distinguishes the genus *Mastodon* from *Elephas*, appears to have been made by Dr. Godman, in 1829,* upon a mutilated lower jaw of a young *Mastodon giganteus*, obtained, I believe, from tertiary deposits in Orange County, United States, and at that period in Peale's Museum, New York. The symphysis of this jaw was entire, and contained two short tusks, from four to six inches in length, projecting straight forwards from the extremity of that part of the jaw. As the lower jaws of the mature American Mastodons which were at that time known to science, offered, like those of the species of Elephant, no trace of tusks, Dr. Godman described his specimen as belonging to an extinct animal of a new genus, for which he proposed the name of *Tetracaulodon*. Mr. Cooper of New York, however, "suggested the opinion that the *Tetracaulodon* was nothing but the young of the

* Transactions of the American Philosophical Society, vol. iii, N. S. p. 478.

gigantic *Mastodon*, and that the tusks were merely milk-teeth, which were lost as the animal became adult."* This opinion was opposed by Dr. Hays in an elaborate memoir,† *ad hoc;* and, with regard to a suggestion offered by Mr. Peale, that the tusks on the lower jaw might be only a sexual distinction, Dr. Hays expresses his opinion, "that it is impossible in the existing state of our knowledge, and with our present materials, to confirm or positively refute this suggestion."

Availing myself of the rich accession of evidences of the osseous and dental organization of the *Mastodon giganteus*, collected in the Missouri territory in 1840, and brought to this country in the following year by Mr. Albert Koch, I arrived at the conclusion that the *Tetracaulodon* of Dr. Godman was the immature state of both sexes of the *Mastodon giganteus* of Cuvier, and that in the male, one at least, and usually the right, of the two lower tusks was retained, but that in the female both were lost as she approached maturity.‡ The inferior tusks, with some modifications of the grinding teeth, which I regard as individual varieties, have, nevertheless, been since interpreted as establishing not only the *Tetracaulodon*, but as characterizing six distinct species of that genus.§

Apart from the considerations of the dental characters leading to such opposite conclusions respecting the mastodontal fossils from North America, the theory of the unity of the species, of which the inferior tusks were regarded by me as immature and sexual characters, might have met with a less general reception than has been accorded

* Silliman's Journal, vol. xix. (1830), p. 159, quoted by Dr. Hays in Trans. of Amer. Phil. Soc. vol. iv. (1833.)

† Loc. cit.

‡ Proceedings of the Geological Society, Feb. 1842. The specimens described, and from which the above conclusions were drawn, are now in the British Museum.

§ Proceedings of the Geological Society, June 15, 1842.

to it, had the *Mastodon giganteus* been the sole species of the genus which manifested such remarkable characters.

But an analogous sexual distinction would seem to have characterised the species of Mastodon (*Mastodon angustidens*) most common in Europe, by specimens discovered in the tertiary deposits at Eppelsheim, in Gascony, and in England, as in the example of the inferior tusk above described.

A symphysial extremity of the lower jaw with two sockets, shewing that it had contained tusks slightly inclined downwards, together with portions of nearly straight tusks, from the same formation at Eppelsheim, had been originally assigned by Dr. Kaup to his genus *Dinotherium*; but the subsequent discovery of the remaining part of the same lower jaw as the bi-alveolar symphysis shewed, by the molar teeth, that it was a Mastodon which had possessed the two inferior and almost straight tusks; and upon this specimen, (fig. 96,) which is remarkable for the great prolongation of the symphysis and sockets of the tusks, Dr. Kaup founded his *Mastodon longirostris*; interpreting the character of the lower tusks in the European Mastodon as a specific distinction, just as Dr. Godman had previously interpreted the first discovered American Mastodon's lower jaw with tusks, as evidence of a new genus.

The molar teeth of the Eppelsheim jaw do not, however, differ from those on which Cuvier had previously founded his species called *Mastodon angustidens*, and I have been led by this correspondence, and by the analogy of the *Mastodon giganteus*, to the conclusion* that the lower tusks of the Eppelsheim Mastodon are a sexual character,

* Expressed in my 'Report on British Fossil Mammalia,' in the Report of the British Association, 1843, p. 220.

and that the *Mastodon angustidens* differs from the *Mastodon giganteus* in this, as well as other respects, viz., that both the inferior tusks are retained in the male, instead of one only, as in some of the American specimens. The portion of inferior tusk of the Mastodon from Eppelsheim, in the cabinet of the Earl of Enniskillen, to which reference has already been made, belongs to the left side; whilst that from the crag at Norwich, is from the right side of the jaw. The two sockets in the entire elongated symphysis first discovered at Eppelsheim, are of equal size.

For the reasons above adduced, I assign the fragment of the tusk discovered by Mr. Fitch, to an adult male of the *Mastodon angustidens*, and all the Mastodontal molar teeth which have hitherto been discovered in British strata to the same species.*

From the age assigned to the fluvio-marine crag, and to some of the continental formations, from which remains of the *Mastodon angustidens* have been obtained, it would seem that this species preceded the Mammoth in Europe, and was of older date than the *Mastodon giganteus* of North America.† No remains of the *Elephas primigenius*, at least, have hitherto been discovered in the miocene or older pliocene strata at Eppelsheim which have yielded the most com-

* Dr. Kaup also cites as the character of his *Mastodon longirostris*, that the last molar tooth has five pairs of cones, and a well marked posterior basal ridge (cinq pointes doubles et un talon bien prononcé); but Cuvier refers similar ultimate molar teeth, as the upper one from Trevoux, (pl. i. 'Divers Mastodontes,' fig. 5,) and the lower one from Padua, (ib. pl. iv. fig. 2), to his species 'à dents étroites.' Cuvier likewise figures a last molar tooth *in situ* in the lower jaw with four pairs of cones, and the fifth pair reduced to a talon or basal ridge; but, if this specimen, which was brought by Dombey from Peru, be of the same species with the European 'Mastodontes à dentes étroites,' it will merely illustrate, as I have before shewn, an analogous range of individual variety in the configuration of this complex tooth in the *Mastodon angustidens*, which has been proved to exist in the same tooth of the *Mast. giganteus*.

† See Mr. Lyell's Paper 'On the Geological Position of the *Mastodon giganteus*,' in the Proceedings of the Geological Society, February 1, 1843.

plete specimen hitherto recovered of the bony framework and dentition of the *Mastodon angustidens;* and not a fragment of a bone or tooth of the Mastodon has yet been found in these newer pliocene and post-tertiary deposits of England, which are so rich in remains of the Mammoth.

In other parts of the world the genus Mastodon, under different specific forms from our European *Mastodon angustidens,* has continued to be represented during a later epoch, and to have been contemporaneous with the Mammoth, or other extinct species of Elephant. In certain localities in North America, famous for remains of the *Mastodon giganteus,* as Big-bone Lick, the Mammoths bear to the Mastodons a proportion of one to five.*

A species of *Mastodon,* nearly allied to the *M. angustidens* by the form of the molar teeth, is associated with the Elephantoid Mastodon, and with a true species of *Elephas,* in the tertiary formations of the Sub-Himalayan range. Another species of *Mastodon,* also nearly allied to the *M. angustidens,* if we may judge from the configuration of a molar tooth, has left its remains in the ossiferous caves, and post-tertiary, or newer tertiary deposits of Australia.† From the conformity of the molar teeth, Cuvier regarded a Mastodon, whose remains have been discovered in Peru, as identical in species with the *Mastodon angustidens* of Europe.

We may therefore conclude, that the gigantic proboscidian modification of the Mammalian type was first manifested on our planet under the generic form of the Mastodon, and with teeth which differed less from those of the older tapiroid Pachyderms, than do the grinders of the true Elephants. No genus of quadruped has been more extensively diffused over the globe than the Mastodon. From

* Dekay, 'Fauna of New York,' p. 102. † Annals of Natural History, 1844.

the tropics it has extended both south and north into temperate latitudes; and, in America, remains of the Mastodon have been discovered on the western coast, as high as the 66th degree of north latitude.* But the metropolis of the *Mastodon giganteus* in the United States, like that of the *Mastodon angustidens* of Europe, lies in a more temperate zone, and we have no evidence that any species was specially adapted, like the Mammoth, for braving the rigours of an arctic winter.

The Mastodon unquestionably possessed a long proboscis, the chief office of which, in the Elephant, is to seize and break off the boughs of trees for food. There is nothing in the ascertained organization of the Mastodon, to lead us to doubt that such was also the principal function of the trunk in that genus. Cuvier, however, was of opinion that the Mastodon applied its teeth, as the Hippopotamus and Hog do, to the mastication of tender vegetables, roots, and aquatic plants.† But the large eminences of the grinding teeth, the unusual thickness of the enamel, and the almost entire absence of the softer cement from the grinding surface of the crown, would rather indicate that they had been instruments for crushing harder and coarser substances than those for the mastication of which the more complex but weaker grinders of the Elephants are adapted.

It has been conjectured that the Mastodons were more aquatic, or swamp-haunting quadrupeds than the Elephants; their limbs were, however, proportionally shorter, although constructed on the same type, each foot being terminated by five short and stout toes, which were evidently, by the form of the last phalanx, confined in one common thick hoof. The leg-bones are stronger in proportion than those of the

* Dekay, 'Fauna of New York.'
† 'Ossemens Fossiles,' tom. i. 4to, 1821, p. 225.

Elephant ; the cranium is flatter, and, from the smaller development of the frontal air-cells, it presents a less intelligent character. The almost complete skeleton of the *Mastodon giganteus*, so well known to the public as the "Missouri Leviathan," when exhibited, with a most grotesquely distorted and exaggerated collocation of the bones, in 1842 and 1843, in the Egyptian Hall, Piccadilly, but now mounted, in strict accordance with its natural proportions, in the British Museum, has enabled me to present, in the subjoined cut, as perfect a restoration of the Mastodon, as that of the Mammoth given at the head of the preceding section.

Fig. 102.

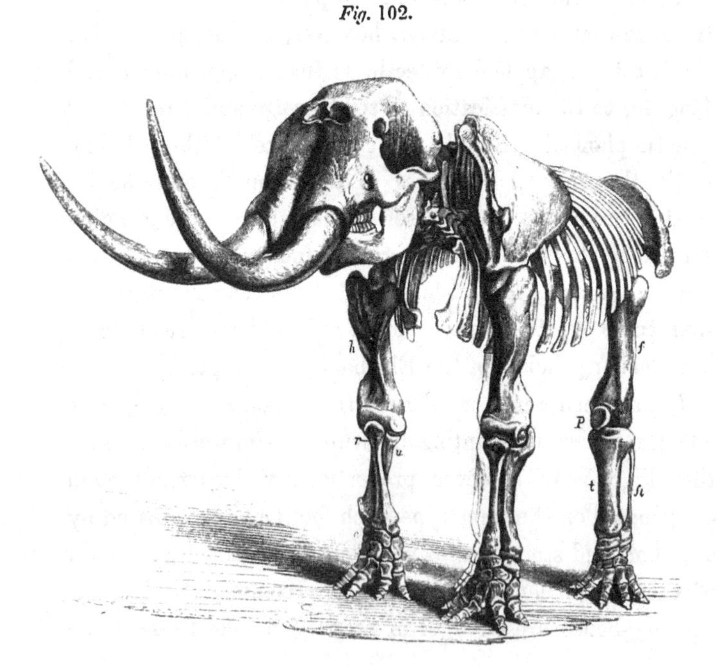

Skeleton of the *Mastodon giganteus*, from Missouri, in the British Museum.

PACHYDERMATA. *TAPIROIDA.*

Fig. 103.

Portion of lower jaw of Coryphodon from eocene clay, Essex coast.
Inner side, $\frac{2}{3}$ nat. size.

CORYPHODON EOCÆNUS. Eocene Coryphodon.

Large Lophiodon, OWEN, Report of British Association, 1843.

IT is not surprising that the rare and extraordinary forms of Mammalia, which supply the transitional links connecting the proboscidian with the tapiroid families of *Pachydermata*, should have escaped observation; if, indeed, they exist in this country, where those tertiary formations, in which alone on the continent their remains have hitherto been found, are sparingly or not at all developed. No remains of Dinotherium, or gigantic

Tapir of Cuvier, for example, have as yet been found in the older pliocene crag of England; although the association of this gigantic Pachyderm with the *Mastodon angustidens*, in the contemporary formations of Eppelsheim and France, has been attested by numerous and well preserved fossils, including the entire cranium.

The molar teeth of the Dinothere had their grinding surface crossed by high and sharp transverse ridges, like those of the *Mastodon giganteus;* but, in most of the teeth, the ridges were restricted to the same number, two, which characterizes the molars of the Tapir. The tusks of the lower jaw, which are early lost in one sex of the Mastodons, were retained in both sexes of the Dinothere with a greater and indeed peculiar degree of downward curvature, yet still manifesting the analogy to the Mastodons by their superior size in the male Dinothere.

These points of resemblance would signify comparatively little in the inquiry into the natural progression of the affinities of the Pachyderms, had the Dinothere been a gigantic Herbivorous Cetacean, as some have conjectured; but, in addition to the arguments in favour of its true Pachydermal character derived by Dr. Kaup * from the texture of the cranial bones, their richly developed air-cells, the deep implantation of the petro-tympanic bone of the organ of hearing, and other particulars of minor import, I may adduce the texture of the dental substances of the molar teeth, and the vertical displacement and succession of the small deciduous anterior molars by true premolars, or "*dents de remplacement*," in support of the view here taken of the position of the genus *Dinotherium* in the Pachydermal series, as a link between *Mastodon* and *Lophiodon*.

* Akten der Urwelt, 8vo, 1841, p. 52.

The large extinct tapiroid Pachyderms have left their remains on the continent in both miocene and eocene formations: in England they are represented by scanty but extremely interesting fossils, which have been obtained from the eocene deposits of the London and plastic clays.

If the specimen, fig. 103, which is a fragment of the right branch of the lower jaw, containing the last and part of the penultimate molar teeth, be compared with the figures which Cuvier has given of the corresponding parts of the *Lophiodon Isselanus*, (Grand Lophiodon d'Issel, 'Ossemens Fossiles,' 4to, 1822, Tapiroids, pl. iii, fig. 3,) or of the *Lophiodon medius*, (loc. cit. fig. 1,) their family likeness will be readily appreciated; but the jaw-bone below the last tooth in the English fossil is deeper in proportion to the size of that tooth, than in the *Lophiodon Isselanus*, and still more so than in the *Loph. medius* which differs by its more slender jaw from the *Loph. Isselanus.* But the more important discrepancies which determine the sub-generic distinction of the large extinct tapiroid lophiodont Pachyderm of our eocene clay, are

Fig. 104.

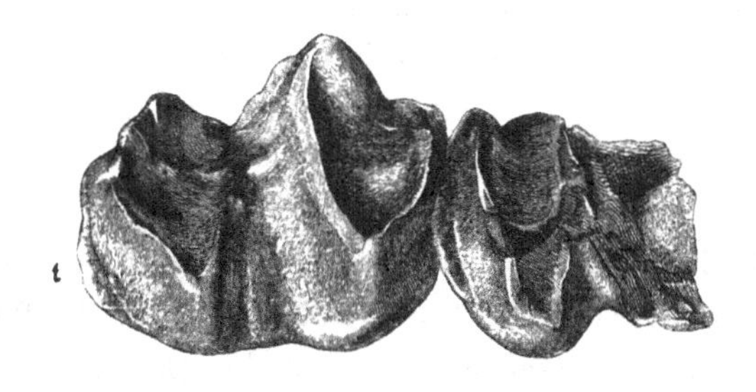

Last molar and part of penultimate molar of Coryphodon eocænus. Upper and outer view; nat. size. Essex coast.

manifested by the last molar tooth, which is fortunately entire.

The crown of this tooth has a smaller antero-posterior diameter in proportion to its transverse diameter, which chiefly depends on the much smaller size of the third or posterior ridge, as compared with the corresponding tooth in the Cuvierian Lophiodons.

From the outer extremity of each of the two principal transverse eminences of the last molar, (fig. 104), a ridge is continued obliquely forwards, inwards and downwards: the anterior one extends to the inner and anterior angle of the base of the crown: the posterior one terminates at the middle of the interspace between the two ridges. The anterior principal transverse eminence, although it has a trenchant summit, as in the known Lophiodons, yet the edge is more concave, the outer and inner extremities rising each into a conical point. The posterior transverse eminence is much lower than the anterior one, and is tricuspid; the trenchant margin connecting the outer and inner points does not extend across the crown parallel with the anterior ridge, as in the Lophiodons, but forms an angle posteriorly, the apex being developed into a third point, which is the highest, and from this point the posterior ridge, or talon, extends downwards and outwards upon the back part of the crown at *t*.

Thus the crown of the last molar of the present species has the two transverse eminences of a Lophiodon's molar so modified that it supports two pairs of points and one single point, like the last lower molar tooth of the fossil jaw from Lot-et-Garonne, described by Cuvier in the 'Ossemens Fossiles,' 1822, tom. iii. p. 404; and like that from the Puy en Velay, described in the posthumous edition of the same work, 8vo., vol. v. p. 480, both of which are referred by Cuvier to the genus *Anthracotherium*. The last molar in the present fossil differs, however, from the teeth above cited, in the height of the connecting ridge of the anterior pair of points, and in the development of the fifth point, not from a third posterior lobe, but from the apex of the angular ridge connecting the posterior pair of points. The typical *Anthracotherium*, of which part of the lower jaw from the lignite beds of Liguria is figured by Cuvier, in the 'Ossemens Fossiles,' 1822, tom. iii. pl. lxxx. fig. 2, differs from the fossil under consideration, in the deep cleft dividing the anterior pair of tubercles; and in the great development of the bifid posterior, or third lobe of the last molar tooth. In the posterior part of the penultimate tooth of the present fossil, it is easy to perceive that the tubercle corresponding with the inner one of the posterior pair in the last molar is obsolete, and represented by a minute eminence near the base of the crown; whilst the tubercle answering to the fifth in the last molar is more elevated, and is nearer the inner side, and the ridge from the outer tubercle terminates there. It is also obvious from the breadth of the fractured part of the anterior fang of the penultimate molar, that its antero-posterior diameter must have more nearly equalled that of the last molar than in the Lophiodons.

The second and third molars of the lower jaw of the 'grand Lophiodon de Buchsweiler,' resemble the last molar of the present fossil, in having the anterior transverse ridge more elevated than the posterior one; but in the fourth molar they are of equal height, (Cuvier, 'Ossemens Fossiles,' loc. cit. p. 202, pl. vii. fig. 1.) The British Tapiroid fully equalled in size that of Buchsweiler, and the fossil in question belonged to a full-grown but not aged individual; for the posterior surface of the anterior ridge of the last molar tooth has been slightly abraded by mastication, and the extent of the fractured jaw behind it proves that there existed no other alveolus posteriorly; but that the perfect tooth *in situ* is the true ultimate molar. From the above described characters of this tooth, we may infer that the whole dental series of the extinct eocene Pachyderm offered modifications of the Lophiodont type of dentition which led towards that of the *Anthracotheria*, more especially of the smaller species from Garonne and Velary.

From the closer resemblance which the fossil presents to the true Lophiodons, it must be regarded as a member of the same family of tapiroid Pachyderms; indicating therein a distinct subgenus, characterised by the want of parallelism of the two principal transverse ridges, and by the rudimental state of the posterior talon in the last molar tooth of the lower jaw. The name *Coryphodon*, which I have proposed for this subgenus, is derived from κορυφὴ a point, and ὀδοὺς a tooth, and is significative of the development of the angles of the ridges into points. The broad, ridged, and pointed grinding surface of the tooth indicates its adaptation to comminute the coarser kinds of vegetable substances; and it is very probable that the habits and food of the Tapir, which is the nearest existing analogue of the Coryphodon, are not very dissimilar from those which characterised of old the present extinct species and the true Lophiodons.

The American Tapir is described as "passing a solitary existence, buried in the depths of the forests and never associating with its fellows; but flying from society and avoiding as much as possible the neighbourhood of man. It rarely stirs abroad from its retreat during the day, which it passes in a state of quiet lethargy, and seeks its food only by night. With the exception of the Hog, it seems to be the most truly omnivorous of the tribe of animals to which it belongs; for scarcely anything comes amiss to its ravenous appetite. Its most common food is vegetable, and consists of wild fruits, buds, and shoots."*

The abundance and variety of the fossil remains of fruits, most of them of a tropical character,† which have been obtained from the same deposits of eocene clay as that which has yielded the subject of the present section, bespeak the extent and nature of those dark and dense primeval forests in which the Coryphodon obtained its subsistence. In size, the ancient British Tapiroid quadruped must have surpassed the largest Tapir of South America, or Sumatra, by one-third. The unique fossil specimen which has led to its determination, was dredged up from the bottom of the sea, between St. Osyth and Harwich on the Essex coast, and now forms part of the interesting and instructive collection of my esteemed friend, John Brown, Esq., of Stanway Green, near Colchester. The specimen is petrified, and heavily impregnated with metallic salts; it presents the usual rich deep brown colour of the fossil bones of the London clay: the pyritic matter which sparkles in the cancelli of the bone, and which lines the pulp-cavity of the broken molar tooth, leaves no room for doubt as to the fossil having been originally imbedded in that eocene tertiary formation of the Harwich coast.

* 'Gardens and Menagerie of the Zoological Society delineated,' 8vo. vol. i. p. 202.

† See Mr. Bowerbank's interesting work on the Fossil Fruits of the London clay, 8vo. Van Voorst.

PACHYDERMATA. *TAPIROIDA.*

Fig. 105.

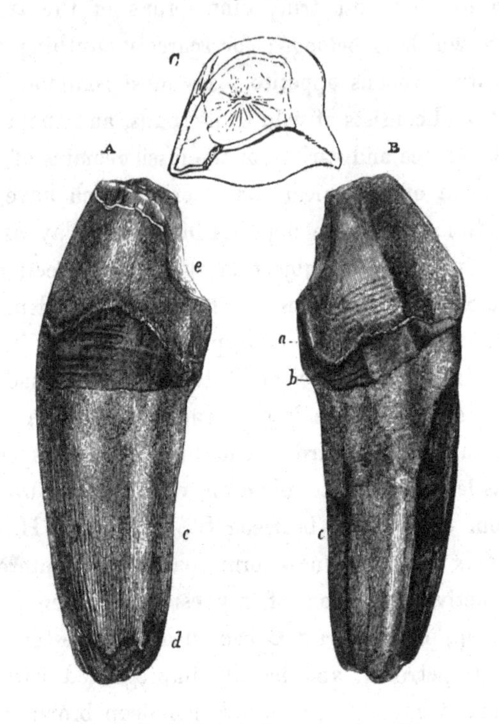

Canine tooth of large Tapiroid Pachyderm, nat. size. From eocene clay.

CORYPHODON, OR LOPHIODON.

THE tooth above figured was brought up from a depth of one hundred and sixty feet out of the plastic clay, during the operations of sinking a well in the neighbourhood of Camberwell; for the opportunity of examining it, I am indebted to Mr. Alport, author of the 'Antiquities and Natural History of the town of Maidstone in Kent.'

The tooth appears to be the right canine of the lower

jaw: the summit of the crown has been abraded, and the posterior part (*e*) excavated by the action of the superior canine upon it, during the lifetime of the animal. The general proportions of this tooth, its degree of curvature, and the relative length of the crown and the fang, accord pretty closely with those of the canines of different species of *Lophiodon* figured by Cuvier in the 'Ossemens Fossiles,' 1822, tom. ii. pt. 1. pl. x., figs. 3 and 12. pl. ix., fig. 11. The crown must have projected but a small distance beyond that of the adjoining teeth, and have been quite concealed by the lips, as in the Tapir, not forming a projecting tusk, and being shorter and thicker than the canine of a carnivorous quadruped. Like the canine of the *Lophiodon tapiroides* in pl. ix. of the volume cited, the growth of the present tooth was completed and the fang terminated by an obtuse solid extremity: but it differs in the fang being less expanded; it is at no part so thick as the base of the enamelled crown: in this respect it resembles more the canine of the *Lophiodon medius*, tom. cit., pl. x. fig. 12, but the crown of the present tooth is proportionally more expanded at the base. The proportions of the crown more nearly resemble those in the *Lophiodon Isselanus*, pl. x. fig. 3; but the fang is ventricose in that species, as in the *Lophiodon tapiroides*. Cuvier does not give a figure of the transverse section of the crown of the canine in any of his specimens: that of the present tooth, (fig. 105, c) is very characteristic, and resembles the transverse section of the crown of the teeth of the great extinct reptile called *Pliosaurus;* the outer surface being nearly flat, and the rest of the crown so convex as to describe a semicircle: a ridge of enamel along each border of the flattened side separates it from the convex side of the crown. The Mammalian nature of the tooth is established by the entire and consolidated fang. The flattened surface, (fig. 105, B,) is gently undulating, convex in the middle and concave at each side near the ridges in the transverse direction: the crown is defended by two layers of enamel: the outer and thicker layer has a minutely wrinkled surface and terminates near the base of the crown by a finely plicated border (B, *a*); extending lower upon the posterior and outer than upon the anterior and inner sides of the crown. The thin and smooth layer of the enamel extends to and defines the base of the crown (B, *b*); the outer layer being coextensive with the inner one only at the two boundary ridges, and the inner layer being extended further upon the tooth at its anterior and inner sides. The length of this tooth must have been three inches when entire; the circumference of the base of the crown is two inches, nine lines. From its close resemblance in the essential characters of its form to the canines of the great extinct Tapiroid Pachyderms, and the apparent specific distinctions from any of the known species of Lophiodon, I strongly suspect it to have belonged to a Coryphodon: its proportions agree with those of the molar teeth of the *Coryphodon eocænus*, and the enamel has the same delicate wrinkled surface; and, although in the question of the specific identity of two fossils from different localities, identity of geological formation would, of itself, be of small moment, it adds to the probability arising from the arguments derived from organic agreement.

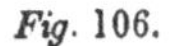

Fig. 106.

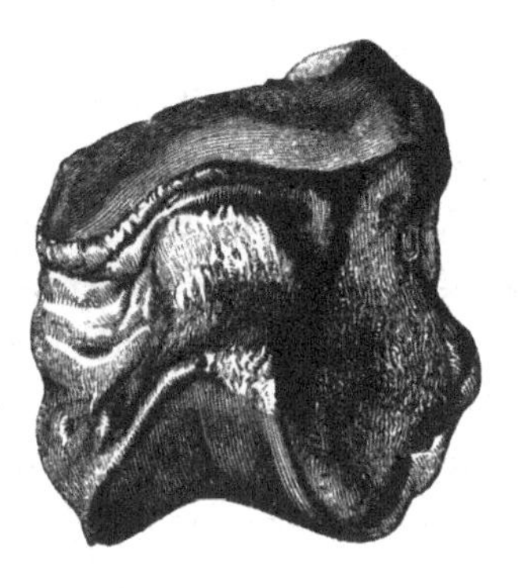

Middle phalanx of right fore-foot, nat. size. Eocene Marl, Isle of Wight.

LOPHIODON, OR PALÆOTHERIUM.

THE fossil bone above figured is a median phalanx of the right fore-foot, and was submitted to me as the bone of an Iguanodon. There is, in fact, a considerable general resemblance between the middle phalanges of this great herbivorous reptile and those of the larger hoofed Mammals; but with respect to the fossil in question, the configuration of the lateral surfaces for the attachment of the ligaments; the production of the inferior border of the distal articulation into a process (*p*) for the insertion of the flexor tendon; and the greater curvature or portion of a circle described by the distal articular extremity, (*e*, *e*,) which indicates a greater extent and freedom of flexion and extension of the toe than the cold-blooded reptiles possess; all combine to prove the fossil to have belonged to the more active, warmer-blooded and higher organized Pachyderm. It agrees most closely with the characters of the corresponding phalanges in the large Tapiroid quadrupeds; and, as it is a little longer in proportion to its breadth than the middle phalanx of the fore-foot of the *Palæotherium magnum*, figured by Cuvier, 'Ossemens Fossiles,' 4to, 1822, pl. xlix, fig. 6, it may probably have belonged to a *Lophiodon* or *Coryphodon*. I owe to Sir Philip Egerton a knowledge of this rare specimen; and to the Marchioness of Hastings, of whose choice collection of Fossil Remains it forms part, the permission to describe and figure it for the present work.

Subjoined is an inside view of the unique specimen of the jaw of the great Tapiroid Pachyderm, from the eocene clay near Harwich.

Fig. 107.

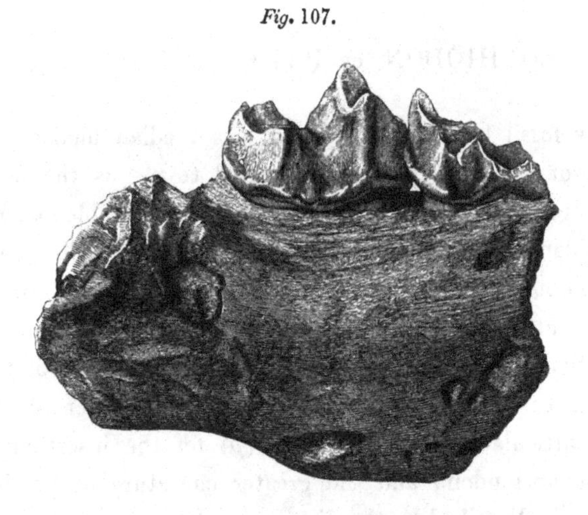

Coryphodon eocænus, $\frac{2}{3}$ nat. size.

PACHYDERMATA. *TAPIROIDA.*

Fig. 108.

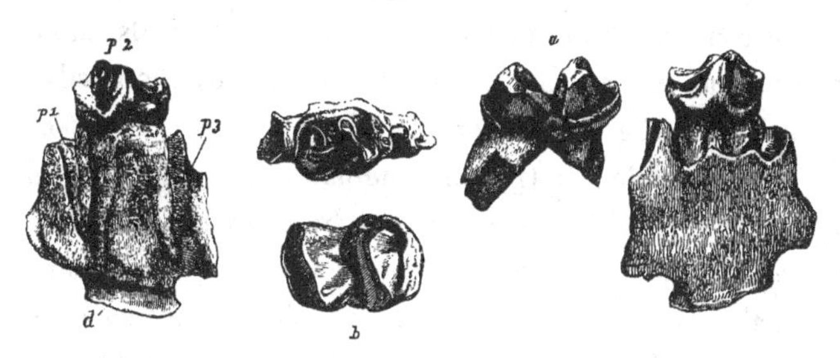

Second and fourth molars, nat. size. Eocene Clay, Bracklesham.

LOPHIODON MINIMUS. Dwarf Lophiodon.

Très-petite espèce de Lophiodon, CUVIER, Ossemens Fossiles, 4to. 1822. Vol. ii. pt. 1. p. 194.

THE first announcement of the former existence of a small Tapiroid Pachyderm, in the eocene deposits of England, on the sole evidence of two molar teeth, will not be received without scrupulous examination of its validity by Comparative Anatomists, who know how many and various Mammalia resemble the Tapirs in the configuration of the crown of the molar teeth, especially those of the lower jaw, to which the fossils in question belong.

The proboscidian Dinothere, the marsupial Nototherе and Diprotodon, amongst the extinct Mammalia; the Kangaroo and the Cetaceous Manatee amongst the living forms of the class, participate with the Tapirs, both recent and extinct, in having the grinding surface of the molar teeth developed into two principal transverse ridges, as shewn in the fossil molar tooth, fig. 107, *a* and *b*. But, setting aside the Dinothere and the large extinct marsupial

Pachyderms by reason of their vastly superior size, the latter also having the two ridges of the grinding surface relatively thinner and higher; and limiting the present comparison to those molar teeth of the animals more nearly equal in size to the species indicated by the fossils; we find that the molars of the Kangaroos, both recent and extinct, differ in the longitudinal ridge which unites together the two transverse ridges, by crossing the middle of the valley; while those of the Manatee have the two principal transverse ridges lower and thicker, their angles are not bent forward, the posterior transverse basal ridge, or "talon," is relatively larger and higher, the anterior talon is wanting, and the fangs descend in parallel lines, or slightly converge near their extremities.

The fossil tooth in question, on the other hand, combines the more obvious and common character of the double-ridged grinding surface with those minor modifications, which distinguish the molars of the Tapir from those of the Kangaroo and Manatee. Both angles of the ridges are slightly bent forwards, making the fore-part of each ridge concave; a secondary ridge is continued from the outer angle of each of the primary ones, from the posterior one to the intermediate valley, from the anterior one to the anterior basal ridge, or talon. From this talon a ridge extends along the outside of the base of the anterior primary ridge, and swells into a small tubercle at the outer angle of the middle valley: this little character is repeated, also, in the molar teeth of the Tapir; but not in those of the Kangaroo or Manatee. Compared with the corresponding tooth, viz. the third or fourth molar, right side, lower jaw, in the American or Indian Tapirs, the following differences are noticeable: the anterior principal transverse eminence is relatively smaller in the fossil, and its exterior basal ridge is stronger: the whole tooth is, likewise, smaller by one third than its analogue in the permanent series of teeth of the existing Tapirs.

The difference in the superior size of the anterior talon and external basal ridge approximates the fossil tooth to the extinct subgenus of Tapiroids, which Cuvier has called *Lophiodon*, as will be evident by comparing the fossil (fig. 108, *a* and *b*) with the figure of the penultimate molar, right side, lower jaw, of the *Lophiodon minimus*,—a species, moreover, precisely corresponding in size with our English fossil,—in the 'Ossemens Fossiles,' 4to. 1822, vol. ii. pt. 1, 'Animaux Fossiles voisins des Tapirs,' pl. x.

More decisive evidence of the special relation of the present fossils to the Lophiodont section of the Tapiroid family, is yielded by the smaller tooth (fig. 108, *p 2*), next to be described. This tooth was found close to the preceding, in the formation of eocene clay, which immediately overlies the chalk at Bracklesham. Compared with the recent Tapirs, it presents the same general modification of the crown, as does the premolar tooth with which the series of the six grinding teeth commences in both Indian and American Tapirs. But, in the fossil, the anterior talon is by no means so large or so much produced: the second eminence is relatively broader: the third transverse eminence, instead of a concavity, presents a prominence with a ridge on each side of its base, and a third intermediate one connecting it with the second eminence of the crown: in all those characters our fossil agrees more closely with the *Lophiodon*, as may be seen by comparing fig. 108, *p 2*, with the tooth *i*, fig. 1. pl. i. of the volume of the 'Ossemens Fossiles,' above cited, representing the jaw of a larger species of *Lophiodon*. But the question of the subgenus of *Tapiroid*, to which the Bracklesham fossils are referable, is more unequivocally decided by the evidence of a smaller premolar tooth anterior to the entire one figured, (fig. 108, *p* 1,) shewn by the remains of the posterior fang, and the socket of the anterior fang, which are fortunately preserved in the fragment of the lower jaw adhering to the entire tooth. In the true Tapirs, both recent and fossil,* only the first of the molar series has the compressed modification of the crown exemplified by the tooth (fig. 108, *p* 2); the second presents the normal quadrate crown with the two transverse ridges. In the Lophiodon, the latter structure is manifested only by the last three teeth, or the true molars; the compressed form being retained by all the three anterior or premolars, the first of which is very small and simple, but implanted by two fangs: the preserved fang and socket of the tooth that preceded the entire premolar in the fossil under consideration, indicate the small size characteristic of the first premolar of a Lophiodon, and the form of a crown of the second premolar in place, leaves only one other question for consideration before deciding upon its reference to *Lophiodon* or *Tapir.*

Both the first and second of the deciduous series of molars of the lower jaw of the young Tapir present the compressed, subtriangular form of crown exemplified by the fossil tooth (fig. 108, *p* 2); and, though there are well-marked differences in the details of configuration, it might be argued that this small fossil tooth was the second deciduous molar of an extinct species of Tapir. This objection, however, is met by the facts, that the second deciduous molar of the Tapir, when its crown has been as much used in mastication as the tooth in fig. 108, has shorter and more divergent fangs; and that in the same extent of jaw as is preserved with the fossil, there would

* *Tapir priscus*, Kaup, *Tapir Avernensis*, Croizet.

be a portion of the reserve-socket for the germ of the succeeding molar tooth, of which there is no trace in the fossil.

The tooth in question must therefore be regarded as a permanent premolar, and as the second in the series; and the fossil is accordingly proved to belong to a species of *Lophiodon*. The premolar (*p* 2) bears the same proportion to the true molar (*a* and *b*), as the premolar of the larger species of *Lophiodon* exhibits in the entire series of the lower jaw figured by Cuvier in the volume above cited, pl. i. Both teeth in fig. 108 belong to the same side of the lower jaw, most probably to the same lower jaw, and they offer no characters by which they can be distinguished from the *Lophiodon minimus*—the "très-petite espèce d'Argenton," described by Cuvier in the volume cited at p. 194.

In the posthumous edition of the 'Ossemens Fossiles,' 8vo. 1834, vol. iii. p. 362, a note is appended to the article on the Lophiodons, in which M. de Basterot, "jeune Naturaliste Anglais," is cited as having maintained an opinion in a paper read to the "Société d'Histoire Naturelle de Paris," that the freshwater marls in central France, from which the remains of the Lophiodons had been derived, belonged to the formation of the plastic clay and lignite, which immediately succeed the chalk.

The determination of the *Lophiodon minimus* in the plastic clay, overlying the chalk at Bracklesham on the Sussex coast, affords satisfactory confirmation of the high antiquity of the epoch of the tapiroid Pachyderms in the tertiary division of geological time.

PACHYDERMATA. *PALÆOTHERIUM.*

Fig. 109.

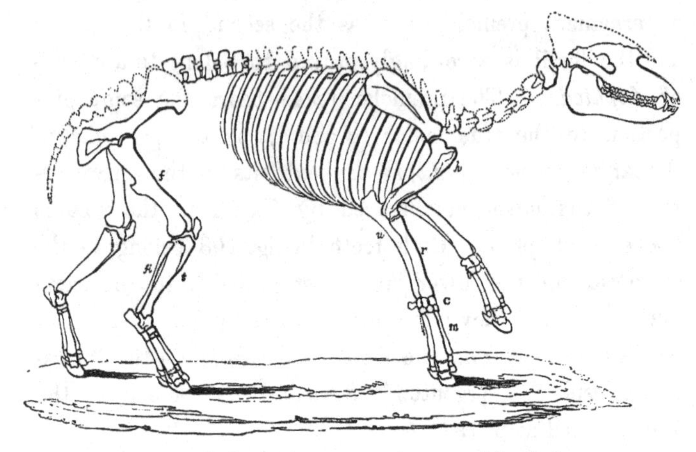

Skeleton of Palæotherium magnum, as restored by Cuvier.
$\frac{1}{20}$th nat. size.

PALÆOTHERIUM MAGNUM. The Great Paleothere.

Palæotherium magnum, CUVIER, Annales du Muséum, iii. pp. 365, 442. vi. p. 265. ix. pp. 15, 29 ; Ossemens Fossiles, tom. iii.

THE genus *Palæotherium* was founded by Cuvier, and several species restored by the masterly determinations of the detached and fragmentary fossils successively discovered in the gypsum quarries of Montmartre, and submitted to the great Anatomist, who may be said to have based the science of Palæontology on the immortal Memoirs descriptive of the present and allied extinct Pachyderms of the Paris Basin, which form the third volume of the 4to. editions of the 'Ossemens Fossiles.'

The dental formula of the genus *Palæotherium*, is:—*i.* $\frac{3-3}{3-3}$, *c.* $\frac{1-1}{1-1}$, *p.* $\frac{4-4}{4-4}$, *m.* $\frac{3-3}{3-3}$, = 44;—that is, there are three incisors in each intermaxillary bone and three on each side of the corresponding part of the lower jaw; one canine tooth, four premolars, and three true molars, on each side of both jaws. The Palæothere has three toes on both the fore and hind feet, and the nasal bones elevated, as in the Tapir, which animal it must have nearly resembled in its general form.

The present species surpassed the largest Tapirs in its size, which equalled that of the Horse; and all the Palæotheres differed from the Tapir in having one toe less upon the fore-foot, and also in the structure of both premolar and molar teeth, which more resemble those of the Rhinoceros.

Of the characters of the true molars of the upper jaw, a clear idea may be gained by the subjoined figure of the first or second of that series, which was discovered in the freshwater eocene marl at Seafield in the Isle of Wight.

Fig. 110.

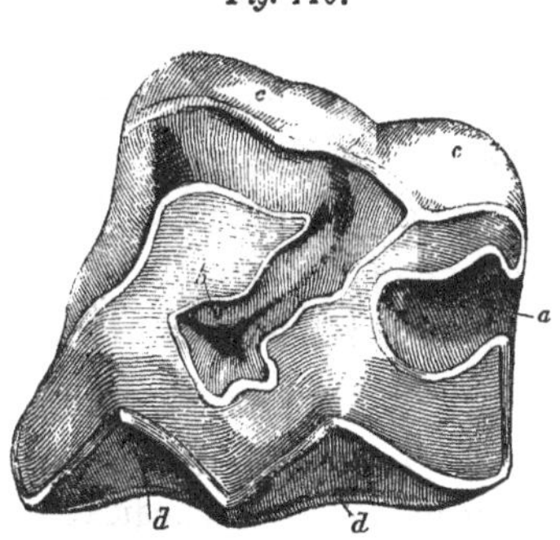

Upper molar tooth, *Palæotherium magnum*.

The crown is almost a cube with a square grinding surface, divided into two lobes by an oblique fissure, *b*, continued from near the middle of the inner surface of the crown obliquely outwards and forwards, two-thirds across the tooth, where it expands into a wide and deep depression. The convex inner sides of the lobes *c*, *c*, are bordered near their base by a ridge. The outer surface of each lobe is gently hollowed out from side to side at *d*, *d*, the hollows being bounded by three longitudinal ridges. The posterior lobe is subdivided by a short and wide fissure *a*, which is expanded and deepened at its extremity, like the fissure *b*. These fissures are formed by folds of the capsule or bag of the formative matrix of the tooth; and as the capsule supports the organ which forms the enamel, the edges of the folds of enamel so formed, are exposed by the wearing away of the grinding surface of the tooth, and being harder than the dentine or central substance of the tooth, they stand up above it like the exterior border of enamel surrounding the tooth. In specimens of fossil molar teeth of aged animals in which the crown has been much worn, the more shallow beginnings of the enamel folds are worn out, and only the deeper terminal depressions remain, forming, as in fig. 113, detached islands of enamel instead of the peninsulas which characterise the grinding surface of the molars of younger animals. It is requisite to bear in mind these changes of the pattern of the grinding surface of the complex molars of the Herbivora in determining the nature of a fossil tooth, lest differences due to age should be mistaken for the distinguishing characters of species or genera.

I have, as yet, seen no other unequivocal relic of the largest species of true Palæothere. For the opportunity of examining the fossil figured, I am indebted to the Rev. T. Darwin Fox, M.A.

PACHYDERMATA. *PALÆOTHERIUM.*

Fig. 111.

Upper molar,
Palæotherium medium.
Binstead, Isle of Wight.

Fig. 112.

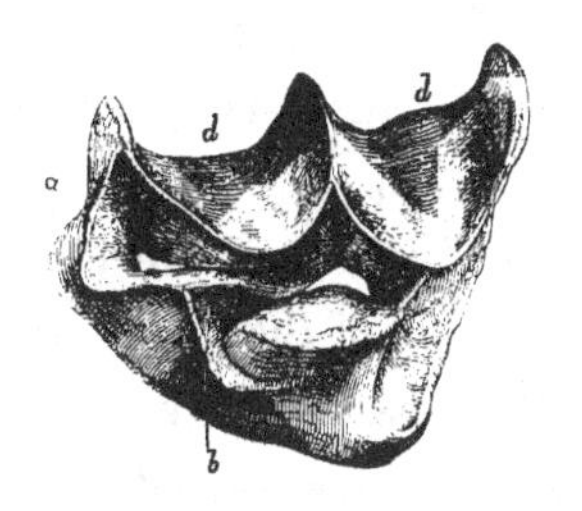

Germ of last upper molar, nat. size, *Palæotherium medium.*
Binstead, Isle of Wight.

PALÆOTHERIUM MEDIUM. Cuv.

The Middle-sized Paleothere.

Most of the Palæotherian fossils that have been collected from the quarries of the hard freshwater marls of the Isle of Wight, belong to the species called by Cuvier *Palæotherium medium.* This animal was about one sixth smaller than the American Tapir, but stood higher on its legs, and had longer and more delicate feet. The grinding teeth are proportionally larger than in the Tapir. In the specimen of the penultimate molar, (fig. 111,) the prominent angles formed by the coronal border of the enamel which covers the outer depressions of the tooth, are more marked than in the molar of the *Palæotherium magnum*, but this is owing to the crown being more deeply worn, which has also reduced the posterior fold (*a*) to its insular form: a small portion, however, of this part of the grinder is broken away in the specimen. The inner principal fold (*b*) retains its peninsular character.

The last molar tooth has a more oblong crown than the preceding ones, the grinding surface of the tooth being more extended in the axis of the jaw; but the posterior margin of the tooth is narrower, and the crown approaches more to the triangular form. In the collection of S. P. Pratt Esq., F.G.S., there is a fine specimen of the germ or newly-formed crown of the last molar tooth, from the right side of the upper jaw, fig. 112.

The two concave enamelled surfaces, *d*, *d*, separated by the three salient ridges, which form the outer wall of the crown, are strongly inclined inwards as they extend downwards; and each concave surface is produced into a point: the enamelled summits of which are entire. Descending folds of the formative matrix of the tooth have left corresponding sinuous depressions on the surface of the crown: one of these (*b*) extends in a sigmoid form from the internal to near the middle of the external wall; a second, (*a*) which begins by a deep fossa at the posterior border of the tooth crosses the preceding, and extends to within a short distance of the anterior border: a ridge is continued from each external angle of the crown first downwards, and then inwards to the opposite internal angle; at the posterior angle it is continued into a prominent conical lobe; the ridge continued to the anterior angle surrounds the base of a larger and higher one, which Cuvier has termed the intermediate lobe. At the commencement of mastication, the dentine is first exposed upon the outer zigzag ridge, and being bound by two parallel lines of enamel, a double crescent is produced, like that on the outer half of the tooth of a Ruminant quadruped; but as mastication proceeds in the Palæothere, the second or internal crescent of enamel is soon obliterated. The field of dentine is widened and bounded by the peripheral ridge of enamel, from which the two oblique peninsular folds are continued into the body of the tooth, corresponding with the primitive depressions on its surface, displayed by the germ of the molar here described (fig. 112).

Fig. 113.

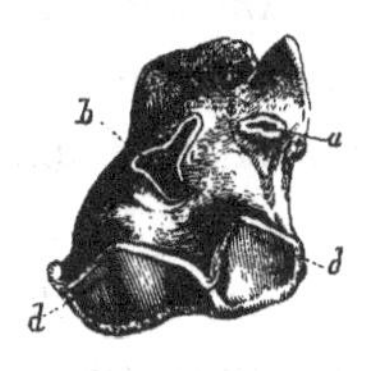

Much worn upper molar, nat. size. *Palæotherium crassum*; Binstead, Isle of Wight.

Fig. 114.

First premolar, nat. size. *Palæotherium medium*; Binstead, Isle of Wight.

The opposite condition of the grinding surface is shewn in a molar tooth of a Palæothere (fig. 113), from the lower freshwater formation at Binstead; the valleys, *a* and *b*, are both reduced to islands of enamel. The specimen is in the Museum of the Geological Society.

The first premolar (fig. 114) is the smallest and most simple of the series: its crown is narrower transversely; the two outer depressions are shallower: there is a longitudinal depression along the inner side of the grinding surface, bounded behind by a prominent ridge: the unworn crown forms an elongate cone; but the surface is soon reduced to an uniform flat tract of dentine, in which state this tooth is commonly found. The specimen figured, from the Binstead quarry, is from a young animal. In the second premolar the internal fold is nearer the anterior border of the crown than in the third and fourth premolars, which differ from the true molars only by a slight inferiority of size.

PALÆOTHERIUM CRASSUM.

THE molar teeth of the lower jaw are, as is usual in the herbivorous Mammalia, narrower transversely, and of a more simple structure than those to which they are opposed above. In the present genus they are seven in number on each side of the lower jaw: the first is the smallest, and has a simple compressed conical crown: the rest have their outer part formed of two half cylinders, except the seventh, which has a third smaller semi-cylindrical lobe. Of such a tooth, which was obtained from the Seafield quarry, Isle of Wight, two views of the recently formed crown are subjoined (fig. 115): *a* is the outer side, shewing the form above described; *b* is the inner side, shewing the longitudinal depressions which penetrate the outer lobes or convexities. This tooth, which is rather less than the corresponding one of the *Palæotherium crassum*, figured by Cuvier, ('Ossemens Fossiles,' 4to, 1822, tom. iii. pl. 1, *q*,) must have belonged to a young animal. It had probably not cut the gum; certainly not come into use, for the margins of the crescentic summits of the three lobes are unworn.

Fig. 115.

a

b

Last lower molar, nat. size. Palæothere. Eocene marl, Isle of Wight.

In the grinding surface of the tooth (fig. 116), which is the fifth of the molar series of the lower jaw, and the first of the three true molars, the two crescents are united by a continuous tract of dentine, the intermediate wall of enamel having been worn down.

Fig. 116.

Fifth lower molar, nat. size. *Palæotherium crassum*; Binstead, Isle of Wight.

PALÆOTHERIUM MINUS.

Of this elegant species, the freshwater eocene deposits of the Isle of Wight have furnished several specimens more entire and better preserved than those of the larger Palæotheres. The collection submitted to my examination by the Rev. Darwin Fox, in 1838,* included a portion of the base of the skull, the right ramus of the lower jaw with all the molars, except the first small spurious one (fig. 117), the proximal end of the right radius, and the shaft and distal end of the right tibia.

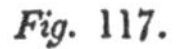

Fig. 117.

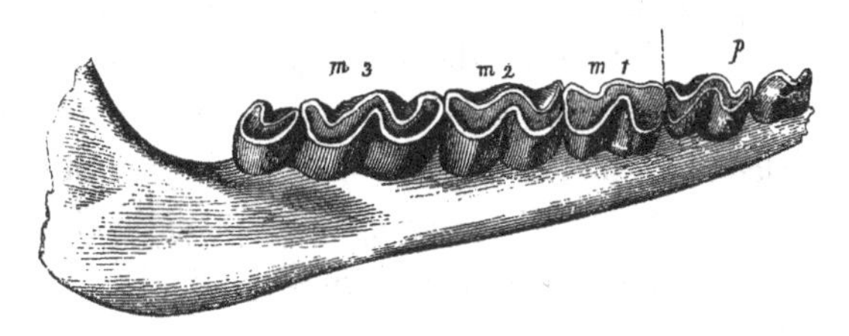

Portion of lower jaw, nat. size. Seafield, Isle of Wight.

Mr. Wickham Flower, F.G.S., possesses the shaft and distal articular end of the humerus of a species of Palæotherium, of the size of *P. crassum*, which was obtained from the eocene clay at Hordwell Cliff, Hampshire: the specimen is black and heavily impregnated with mineral matter. A lower molar tooth, of apparently the same species of Palæothere, was discovered at the same place.

A single incisor, apparently of the lower jaw of the Palæotherium medium, illustrates the identity in form and structure of the cutting teeth, as of the canines and molars of the species of the freshwater eocene deposits of the Isle

* Geological Transactions, vol. vi., second series, p. 41.

of Wight, with that of the gypsum of Montmartre: the trenchant summit of the wedge-shaped crown has been worn down by use in the specimen figured (fig. 118), shewing the great antero-posterior breadth, which increases to the base or commencement of the long subcompressed fang.

Fig. 118.

Incisor; nat. size. Palæothere; Binstead, Isle of Wight.

The more complete remains of the Palæotheria recovered from the gypsum beds of the Paris Basin, revealed to Cuvier that this ancient genus of the Pachyderms had the same number of incisive teeth as the Tapir, viz. six in the upper and six in the under jaw; but they are more equal in size, the outermost of the upper jaw being not so large, and that of the lower jaw not so small in proportion as in the Tapir. The canine teeth of the Palæothere had relatively longer crowns than in the Coryphodon, but were concealed by the lips as in that animal, the Lophiodons, and the modern Tapir.

*Fig.*119.

Palæotherium magnum.

PACHYDERMATA. *RHINOCEROS.*

Fig. 120.

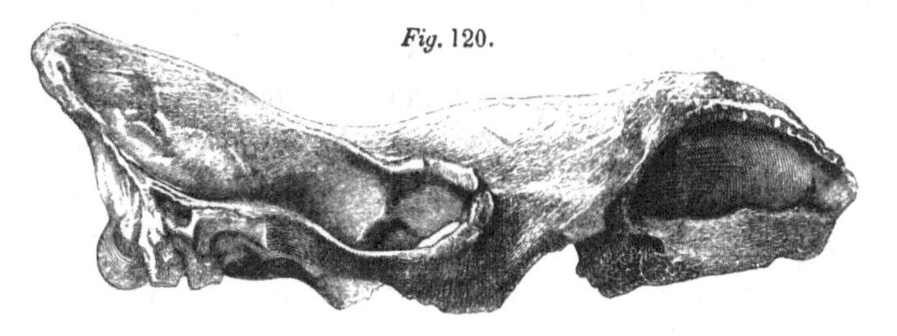

Skull of Rhinoceros tichorhinus. ⅛ nat. size.

Fig. 121.

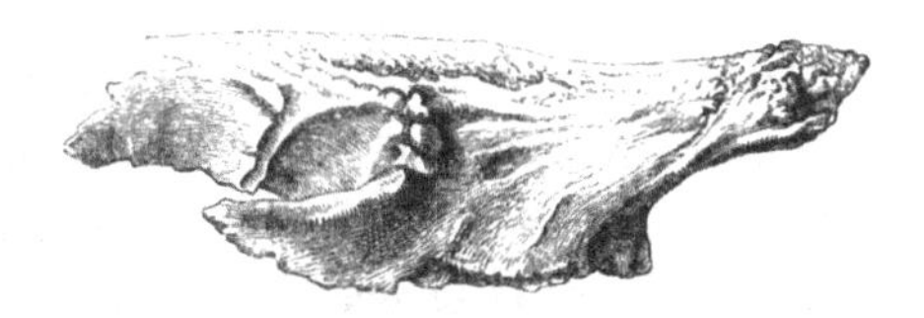

Portion of skull of Rhinoceros, from Newer Pliocene at Chartham, Kent.

RHINOCEROS TICHORHINUS. Tichorhine two-horned Rhinoceros.

Rhinoceros tichorhinus. ou Rh. à narines cloisonnés. Cuvier, Annales du Muséum, tom. iii., p, 46. Ossemens Fossiles, 4to, 1822, vol. ii. pt. i.

The first notice and figure of fossil remains referable to the genus *Rhinoceros*, occurs in a quaint and extremely rare old tract entitled, "Chartham News, or, A Brief Relation of some strange Bones there lately digged up, in some grounds of Mr. John Somner of Canterbury: written by his brother, Mr. William Somner, late auditor of Christ Church, Canterbury, and register of the Archbishop's court there, before his death.—London: Printed for T. Garthwait, 1669." (4to, pp. 10, with a plate.)

"News from Chartham in Kent.—Although it may, and perhaps must be granted, that miracles (strictly understood) are long since ceased; yet in the latitude of the notion, comprehending all things uncouth and strange, (*miranda*, as well as *miracula;* wonders as well as miracles,) they are not so; but do, more or less, somewhere or other dayly exert and shew themselves, *Dies diem docet.*" After a fling at the "New lights that are now-a-days much cried up," and leaving these "spiritual mountebanks and their counterfeit ware,"—a race still far from being extinct,—the worthy 'Register' proceeds "to the matter-of-fact then."

"Mr. John Somner, in the month of September, 1668, sinking a well at a new house of his in Chartham, a village about three miles from Canterbury, towards Ashford, on a shelving ground or bankside, within twelve rods of the river, running from thence to Canterbury and to Sandwich Haven; and, digging for that purpose about seventeen feet deep, through gravelly and chalky ground and two feet into the springs; there met with, took, and turned up a parcel of strange and monstrous bones, some whole, some broken, together with four teeth, perfect and sound, but in a manner petrified and turned into stone, weighing (each tooth) something above half a pound, and almost as big, some of them, as a man's fist."

Alluding to the notices of the remains of giants which were current in the philosophical and other works of the time, the author judiciously remarks:—"And so we must have judged of these teeth and of the body to which they belonged; had not other bones been found with them, which could not be man's bones." "Some that have seen them," he proceeds to say, "by the teeth and some other circumstances, are of opinion, that they are the bones of an Hippopotamus, or Equus fluvialis, that is, a River-horse; for a Sea-horse, as commonly understood and exhibited, is a fictitious thing. Yet Pliny makes Hippopotamum ('mari, terræ, amni communem,') to belong to sea, land, and rivers. But what are the differences and properties of each kind, I leave others to inquire. The earth, or mould about them, and in which they all lay, *being like a sea-earth or fulling earth* has not a stone in it, unless you dig three feet deeper, and then it rises a perfect gravel."

This last passage gives a more exact knowledge of the matrix of the fossils than is usually found in analogous notices: we readily recognise in it the post-pliocene brick-earth and drift which have since yielded, especially in the counties of Kent, Surrey, and Essex, so rich a harvest of the remains of great extinct Pachyderms.

"So have you the story, an account, if you please, of what was found, where, when, and upon what occasion. For more public satisfaction, and to facilitate the discovery; at least to help such as are minded to employ their skill in guessing and judging of the creature, whose remains these are, what it was for kind; we have by and with the help of an able limner, adventured on a scheme or figure of several of the teeth and bones, with their respective dimensions of breadth, length, and thickness."

"No man, we conceive, not willing to be censured of rashness, will be very forward to divine, much less to define or determine what the creature was; and, doubtless, dubious enough it is, whether of the twain, the sea, or the land, may more rightly lay claim unto it."

Mr. Somner having, nevertheless, "taken a large time of consideration of all particulars and circumstances fit to be duly and deliberately weighed and observed in the case," adventures to conjecture it to be "some sea-bred creature ;" and then proceeds to discuss at length the question, "How it possibly came there? Piscis in arido?" with its four following branches:—

"1. Whether the situation and condition, face and figure of the place, may possibly admit of the sea's once insinuating itself thither?

"2. Whether (that possibility being granted or evinced) the sea did ever actually insinuate itself so far as to this place, and when?

"3. How, in probability, and when, this valley or level being once sea-land, should come to be so quite deserted and forsaken of the sea, as it is at this day, the sea not approaching by so many, a dozen, miles or more?

"4. By what means the sea, once having its play there, this creature comes to lodge and be found so deep in the ground, and under such a shelving bank?"

Our limits compel us to terminate here the quotations, and to refer the geologist, interested in such early attempts to solve the problems relating to the changes in the earth's surface, to the pamphlet itself, of which a copy exists in the King's Library in the British Museum, or to the reprint of it in the Philosophical Transactions for 1701, No. 272, p. 882.

With the inquiry into the causes of the sea's progress and retreat in Kent, as evidenced by the supposed "sea-bred monster," we have here, in fact, the less concern, since we shall be able to shew that it belonged to a terrestrial genus of quadruped.

The figures of two of its teeth, "part of what the author intended, if he had lived," are so exact, and the progress of Comparative Anatomy since 1668 has been so immense, that they may now be determined, without much laudable ingenuity or blameable rashness, to have belonged to a Rhinoceros, and to have come from the middle of the molar series of the upper jaw. But we are fortunately enabled to go further, and inquire into the exact species of Rhinoceros to which they belonged: for the identical fossils discovered at Chartham are now preserved in the British Museum. They are noticed by Nehemiah Grew in his 'Catalogue of the Rarities of Gresham College,' p. 254; and were doubtless transferred to their present depository along with the other objects contained in the ancient Museum of the Royal Society.

The annexed cut (fig. 122) is an original figure of the best preserved of the molar teeth from Chartham: it is

Fig. 122.

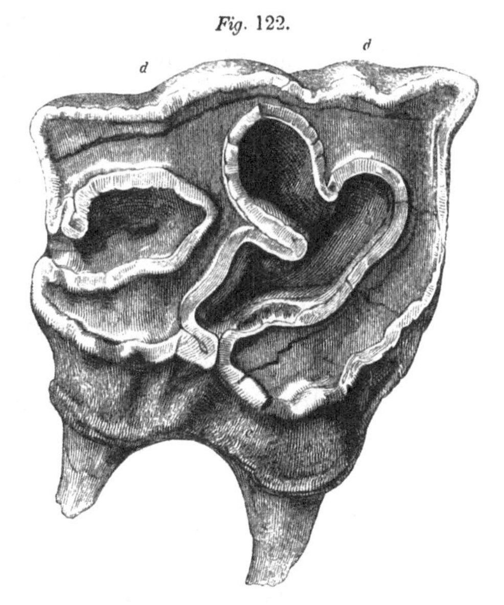

Upper molar tooth of *Rhinoceros tichorhinus*, Newer Pliocene, Chartham, Kent.

the fifth or sixth molar of the right side. It well exemplifies the close analogy of the molars of the Rhinoceros to those of the Palæotherium (see fig. 110). We perceive the same cubical form of the crown; the grinding surface of which is similarly broken by a deep valley, (*a*,) extending from the posterior margin nearly half-way across, and by a deeper and longer valley, *b*, commencing from the middle of the inner side of the crown, and expanding and partly dividing into two deep depressions near its opposite extremity. The principal difference by which the upper molars of the Rhinoceros may be distinguished, independently of their greater size, from those of the Palæotherium, is the much inferior depth of the two longitudinal depressions (*d d*) on the outer side of the tooth, and the feeble development of their boundary ridges. In the Palæotherium, a slight rising may be discerned at the bottom of each of the two deep outer pressions (see fig. 112) : this rising is much increased in the Rhinoceros, and gains the level of the borders of the depressions, giving an undulating character to the outer surface of the tooth. The changes produced by age and progressive wearing away of the grinding surface will be illustrated by subsequent specimens.

One of the "strange and monstrous bones" exhumed with the teeth at Chartham (fig. 121), is described by Grew* as "part of the far cheek, with both the ends and the sockets of the teeth broken off." He compares it with the corresponding part of the Hippopotamus; and, finding "that the orbit of the eye is neither so round nor so big, yet the teeth far bigger;" that the forehead stands higher than the eye, whilst in the Hippopotamus "it lies so low, that it looks like a valley between two hills," he concludes it more likely that it belonged to a Rhinoceros, "for the being whereof in this country we have as much ground to suppose it as of the Hippo-

* Loc. cit., p. 255.

potamus." Of the soundness of Grew's determination, the reader will be able to judge by comparing the figure of the fossil (fig. 121) with that of the entire cranium of the *Rhinoceros tichorhinus*, which is placed above it, at the head of the present section.

Two distinct rough surfaces (*h h*) may be traced on the upper part of the fragment, shewing that the species of Rhinoceros to which it belonged was two-horned; and the anterior surface rises towards its middle part, as if to form the longitudinal ridge, which there characterises the fossil species, and distinguishes it from the African two-horned Rhinoceros, which has a depression at the corresponding part of the skull. But more decisive evidence of the relationship of the Chartham fossil to the extinct *Rhinoceros tichorhinus* is afforded by the remains of the strong and thick bony wall which descended from the bones supporting the horns to form the partition between the two cavities of the nostrils, and give additional strength to that part of the skull.

Cuvier concludes, from this peculiar structure of the most common extinct species of two-horned Rhinoceros of the northern and temperate regions of Asia and Europe, that it bore longer and more formidable nasal weapons than do any of the known existing species with two horns. In the Chartham fossil, a great part of the bony septum is broken away: it remains in the entire skull figured (fig. 120). The skull of the extinct Rhinoceros was relatively longer in proportion, and terminated forwards by a peculiar modification of the nasal bones, which, by the medium of the thickened anterior part of the osseous partition-wall were anchylosed, or joined by a continuous bony mass, with the fore-part of the intermaxillary bones, or those that terminate the upper jaw.

The bony partition-wall, with its peculiar anterior termination,* is well displayed in some of the entire skulls of the tichorhine Rhinoceros, which have been discovered in this country. One of these, figured by Cuvier, 'Ossemens Fossiles,' 4to., 1822, tom. ii., pt. 1., pl. ix., fig. 3, was found in a slate-pit at Stonesfield in Oxfordshire, about four miles from Woodstock. Dr. Buckland possesses fine specimens of the skulls and other bones of the same extinct Rhinoceros, which were discovered, associated with remains of the Mammoth, Hyæna, &c., in the drift on the banks of the Avon, at Lawford, near Rugby.

The most complete skeletons have been found, as might be expected, in caverns or cavernous fissures, where the carcass of the fallen animal has been best protected from external changes and movements of the soil.

Dr. Buckland has recorded one of the most remarkable examples of this kind, which was brought to light in the operation of sinking a shaft through solid mountain limestone (fig. 130, F), in a mining operation for lead-ore near Wirksworth, Derbyshire.† A natural cavern (*ib.* C) was thus laid open, which had become filled to the roof with a confused mass of argillaceous earth and fragments of stone, and had communicated with the surface by a fissure (*ib.* B) fifty-eight feet deep and six feet broad, similarly filled to the top, where the outlet (*ib.* A) had been concealed by the vegetation. Near the bottom of this fissure, but in the midst of the drift (*ib.* D), and raised by many feet of the same material from the floor of the cavern, was found nearly the whole skeleton of a Rhinoceros (*ib.* E), with the bones almost in their natural

* The name imposed by Cuvier on the present extinct species of Rhinoceros has reference to this structure: it is from τεῖχος, a wall, ῥὶν, a nose: *tichorhinus.*

† 'Reliquiæ Diluvianæ,' p. 61.

juxta-position: one part of the skull which was recovered shewed the rough surface for the front horn; the back part of the skull and one half of the under jaw were detached. All the bones were in a state of high preservation. There were no supernumerary bones to indicate the presence of a second Rhinoceros, but a few remains of Ruminants, apparently of extinct species.

The skull of the Rhinoceros, which, with the rest of the bones so fortunately preserved, is now deposited in the Geological Museum at Oxford, shews the bony partition of the nasal cavity characteristic of the *Rhinoceros tichorhinus*, and the lower jaw further illustrates the peculiarities of that extinct species.

As the evidence of a second British extinct species of Rhinoceros will, in the sequel, be established by the characters of the lower jaw, I subjoin two figures of the specimen of that bone from the cave at Wirksworth.

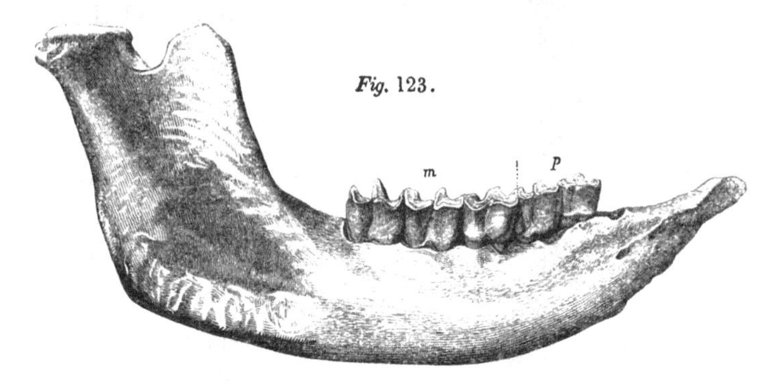

Lower jaw of *Rhinoceros tichorhinus*, Cave, Wirksworth. ⅛ nat. size.

In the side-view of this jaw given above, the extent of the anterior end of the jaw, called the symphysis, in advance of the molar teeth, is shewn: this part is peculiar, in the *Rhinoceros tichorhinus*, both for its length and its small vertical diameter. Pallas believed that he had found remains of the sockets of incisive teeth in the symphysis, and such traces are shewn by one of the specimens from Rugby, in the Geological Museum at Oxford; a structure, as Cuvier justly remarks, which approximates the *Rhinoceros tichorhinus* to the one-horned Rhinoceros of Asia.

Fig. 124.

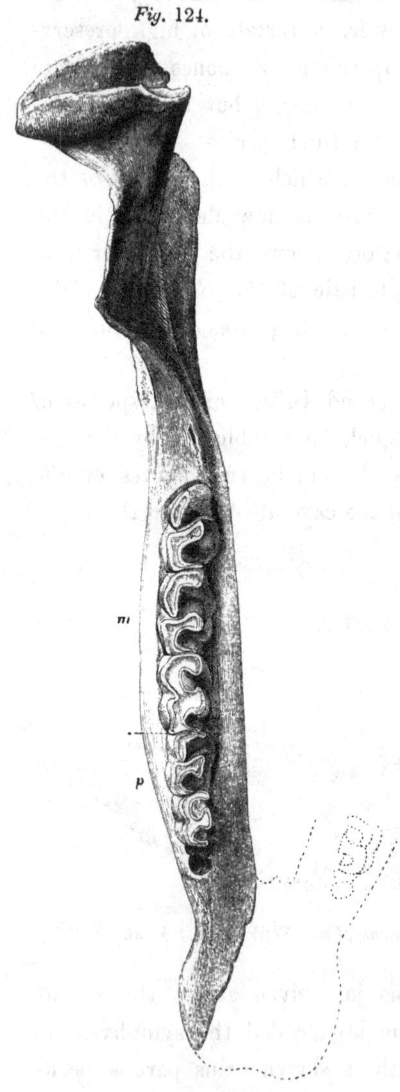

Lower jaw of *Rhinoceros tichorhinus*, Cave, Wirksworth. ¼ nat. size.

Fig. 124 shews the breadth of the symphysis, and the grinding surface of the lower molar teeth; but, before adverting to these, I shall notice the chief modifications of form under which the upper molar teeth of the *Rhinoceros tichorhinus* may present themselves.

It has been already observed, * that, in the cave at

* Ante, p. 259.

Kirkdale, the remains of the large herbivorous quadrupeds were chiefly those of young animals, and such as would most easily fall a prey to the Hyænas, and be dragged by them into their den.

Fig. 125.

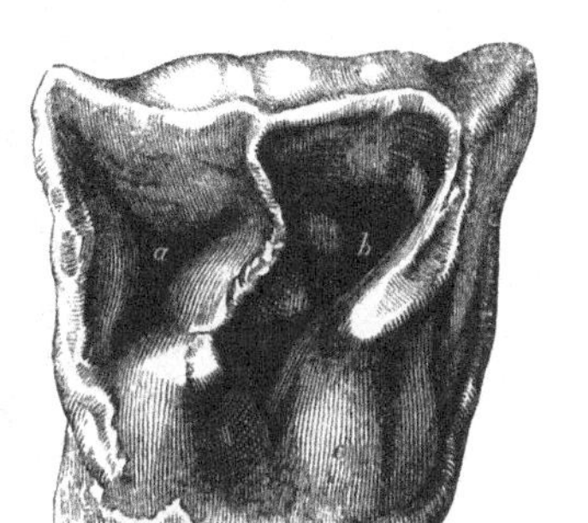

Deciduous upper molar; nat. size; *Rhinoceros tiehcrhinus.* Kirkdale cavern.

Fig. 125 represents the grinding surface of an upper molar tooth of the *Rhinoceros tichorhinus*, which was discovered in the Kirkdale cavern, and is now in the British Museum. This tooth is the third of the series; only the crown had been formed and had not made its appearance above the gum. From its size, it was likewise evidently the germ of a deciduous or milk tooth. The comparison of figure 124, with figure 112, of a similar germ of an upper molar tooth of the *Palæotherium medium*, will illustrate the similarity of plan, and generic modification, of the structure of the teeth of the Rhinoceros, as compared with those of the more ancient Pachyderm. The outer wall of the crown is more even and less deeply indented; the two valleys, *a* and *b*, are wider in the Rhinoceros.

Mastication first exposes the dentine at the summits of the ridges, and produces the two peninsular folds of enamel shewn in fig. 122. The continued wear of the tooth next insulates the posterior division of the transverse peninsula and simplifies it, as at *b* in the molar tooth from the cave of Kent's Hole (fig. 126). As the shorter valley (*a*) is deepest at its extremity, further attrition exposes the dentine at its shallower commencement, and a second island of enamel is produced, as in the molar tooth figured by Cuvier, 'Ossemens Fossils,' 4to., 1822, tom. ii. pt. 1. pl. xiii. fig. 6. In very old Rhinoceroses the first formed island

Fig. 126.

Fourth right upper molar ; nat. size ; *Rhinoceros tichorhinus* ; Cave, Kent's Hole, Torquay.

of enamel, which surrounds the shallowest depression, is worn away, and the grinding surface simplified to the pattern figured by Cuvier in the plate above cited, fig. 5.

The teeth of the lower jaw of the Rhinoceros present the same degree of resemblance to these of the Palæotherium, as exists in the upper jaw. The crown of each molar consists of two vertical crescentic lobes, but these are less regularly curved, are placed more obliquely with regard to each other, and are divided by a deeper cleft. Hence the dentinal substance of the two lobes, when exposed at their summits by attrition, is not so soon blended into one continuous tract as in the Palæothere (fig. 116), but long remains insulated by a complete boundary ridge of enamel in each lobe, as shown in the lower molar tooth of the *Rhinoceros tichorhinus* (fig. 127). This tooth was discovered in the drift gravel, over-lying the London clay, during the operations of digging the Regent's Canal, and is now in the British Museum. It shows also the deeper internal excavation, and the unequal height of the two crescentic lobes, which distinguish the lower molars of the Rhinoceros from those of the largest Palæothere.

Fig. 127.

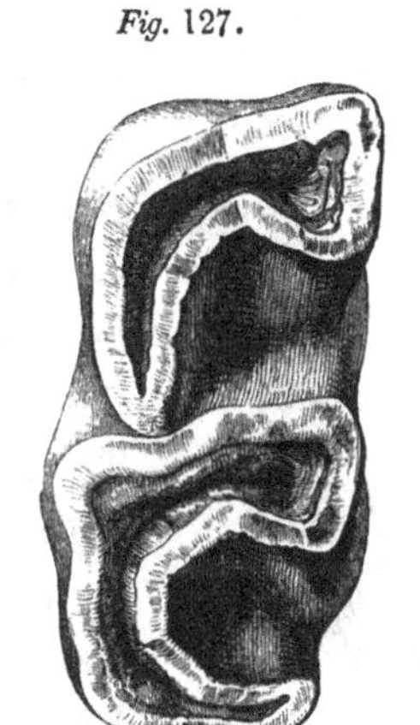

Fifth molar, right side, lower jaw, nat. size; *Rhinoceros tichorhinus.* Drift gravel.

In the lower jaw of the *Rhinoceros tichorhinus*, represented in figures 123 and 124, five molar teeth are shown *in situ*, and the socket of a small premolar in front. The lower jaw, discovered at Montpellier, figured by M. Christol in his Memoir on the species of fossil Rhinoceros, in the 'Annales des Sciences' for 1835, pl. ii. figs. 1 and 2, and referred by that author to the *Rhinoceros tichorhinus*, is described (p. 46) as having all its molars, "munie de toutes ses molaires," of which teeth the figures exhibit six, corresponding in number with those of the specimen from Wirksworth. I have, however, obtained good evidence, from British specimens, of the accuracy of M. Adrien Camper's statement, cited by Cuvier, 'Ossemens Fossiles,' 1822, tom. ii. pt. 1. p. 61, that the tichorhine Rhinoceros had seven molar teeth on each side of the lower jaw, like the existing species; and that the smaller number in the jaws from Montpellier and Wirksworth, is due to the age of the individuals to which they belonged.

The anterior part of the left branch of the lower jaw of a younger Rhinoceros (fig. 128), from the drift at Lawford, near Rugby, now in Dr. Buckland's Museum, contains four teeth, which demonstrate, by their relative position to the broken symphysis, a distinctive character of the *Rhinoceros tichorhinus*, and, at the same time, the existence of a smaller and more simple premolar anterior to that tooth, of which the empty socket is shown in fig. 124. The third tooth, in the present specimen, precisely accords in size and conformation with the second in fig. 124; and the fourth premolar with the third tooth, in fig. 124: the sole differences which the teeth in the younger specimen present, arise from their having been much more recently acquired; the summits of the two crescents, composing the crown of the third tooth, had only just begun to be used in mastication, whilst those of the fourth are entire, and the base of the crown is not quite disengaged from the socket. We have in this instructive specimen the whole series of premolars, or those permanent teeth which succeed and displace the four deciduous molars of the still younger Rhinoceros. The individual to which the fossil in question belonged, must have perished just as it had accomplished this change of its dentition. In fig. 124, it may be observed that the third tooth in place, which is the first true molar, has been more worn than the tooth in advance, from which it is separated by the dotted line; the summits of the two

Fig. 128.

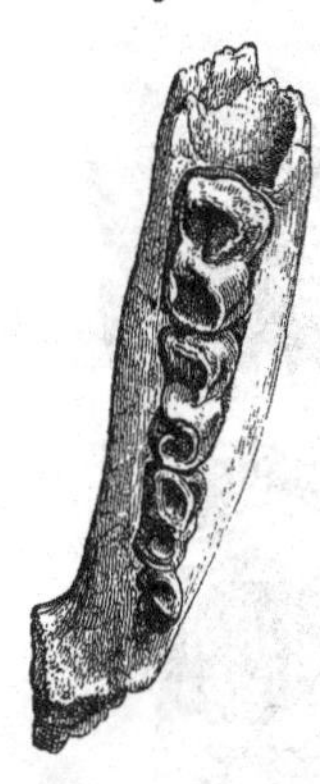

Portion of lower jaw. ¼ nat. size. *Rhinoceros tichorhinus*. Drift, Lawford, Rugby.

crescents are still distinct in the anterior tooth, whilst in that which follows, they are blended by a continuous tract of dentine. This difference arises from the circumstance that the first true molar comes into place immediately behind the deciduous series of four teeth, before these are shed and succeeded by the four premolars shown in fig. 128; it thus assists in performing the essential work of mastication whilst the change of dentition is going on, and is, consequently, worn down to some extent before the fourth premolar has risen into place.

The first premolar in the *Rhinoceros tichorhinus* has a compressed conical crown, the anterior half simple and subtrenchant, the posterior half broader, and impressed by a vertical pit: it is supported by two connate fangs, implanted in a simple alveolus; the antero-posterior extent of the crown is between seven and eight lines, the height of the enamelled part is half an inch: the socket is on the same transverse line as the posterior border of the symphysis. The form and size of the three succeeding molars may be judged of from the figures in cuts 123, 124, and 128. In the last specimen, the longitudinal extent of the series of four premolars is four inches nine lines. The first premolar appears to be shed, in the *Rhinoceros tichorhinus*, before the last true molar rises completely into place.

Similar evidence of the true number of the molar series in the lower jaw of the *Rhinoceros tichorhinus*, is given by another British specimen, to which historical interest is attached, both from its early introduction to the notice of Geologists, and on account of the opinion respecting it which Cuvier has left on record. The specimen in question is that which Douglas has figured in his 'Dissertation on the Antiquity of the Earth,' 4to, 1785, as the "Fossil *animal incognitum* bone from Thame," and which he notices in the Appendix, p. 45, as "the specimen in the Museum of Sir Ashton Lever, No. 20, which was found under ground by digging at Thame, in Oxfordshire."

The original, now in the Geological Museum at Oxford, was kindly pointed out to me by Professor Buckland, who has attached to it the following note:—"In 1829 I purchased this specimen at a sale in London, from the Museum of Mr. Donovan, who probably purchased it at the sale of the Leverian Museum." The extract from the 'Ossemens Fossiles,' 1822, vol. ii. p. 54, is added, as follows. "*Douglas* (loc. cit. App. p. 45,) représente un fragment de mâchoire inférieure contenant trois dents, trouvé en creusant un puits, à *Thame*, dans le comté *d'Oxford*, et conservé alors dans le Musée de *Lever*. Il paroit de l'espèce de Lombardie à narines non cloisonnées."

The distinctive characters of the lower jaw of the species of extinct *Rhinoceros*, called by Cuvier 'non-cloisonné or *leptorhinus*,' are very clearly illustrated by the figures of the Lombardy specimens, which he has given in pl. ix. figs. 8 and 9 of the volume cited, and by the English fossils described and figured in the succeeding section. The lower jaw from Thame manifests as clearly, by the position of the first premolar behind the symphysis, its specific identity with the *Rhinoceros tichorhinus*, and it so closely agrees with the specimen from Lawford (fig. 128), as to render a figure of it unnecessary in this work.

In that which Douglas has given of the natural size, viewed from the inside (the mirror not employed), the second premolar, which was then in place, is behind the symphysis, and the small, partially divided socket for the first premolar has the same relative position to the posterior border of the symphysis as in the lower jaw (fig. 128). Douglas's specimen belonged to an immature

Rhinoceros of nearly the same age as that from Lawford; the summit of the second crescent of the fourth premolar shows that it had just come into use at the period when the animal perished. The anterior of the three ridges, on the inner side of the crown of the third and fourth premolars, supports a small oblong tubercle,* a variety not present in the Lawford specimen. In the *Rhinoceros leptorhinus* of the fresh-water deposits in Lombardy, a species also co-existing of old with the tichorhine Rhinoceros in Britain, the premolar teeth extend forwards much closer to the anterior end of the jaw, and the second premolar is placed in advance of the posterior border of the symphysis (see figs. 132 and 134).

The portion of lower jaw, with two molar teeth, which forms the subject of the first plate in Douglas's 'Dissertation,' and the foundation of much ingenious reasoning, on the supposition that it was part of a Hippopotamus, belongs to a Rhinoceros, and probably to the extinct tichorhine species. It was discovered in "a stratum of drift or river sand, blended with a kind of clay, of a yellowish grey tinge," at the depth of twelve feet, in dig-

* Cuvier, in detailing the discovery at Avary of certain fossils, which he refers to the *Rhinoceros incisivus*, says, "Enfin une dent inférieure, plus usée, est peut-être la cinquième ou la sixième; j'y vois, au deuxième croissant du côté interne, un crochet que je ne retrouve pas dans les autres espèces." 'Ossemens Fossiles,' 1822, tom. iii. p. 391. M. Christol, believing that he had discovered this character in the molars of the lower jaw of the *Rhinoceros tichorhinus*, regards it as distinctive of that species. 'Annales des Sciences,' 1835, tom. iv. p. 62. In the lower molar tooth, which he figures to illustrate this character, it is shown as a minute notch near the upper and posterior part of the middle ridge on the inner side of the crown, which ridge is formed by the posterior and inner termination of the first or anterior crescent; the notch cuts that ridge in a direction downwards and forwards, detaching from it a small conical process. I cannot find a trace of this character in any of the lower molars of the *Rhinoceros tichorhinus* which I have examined; and I have especially compared with the figure given by M. Christol, loc. cit., pl. iii. fig. 1, a molar, the fourth, of the same size and with the same degree of usage. Such small tubercle, notch, or crochet, wherever developed, is most probably an accidental variety.

ging the foundation of a store-house at Chatham, Kent. The figure shows the outer side of the two crescentic or semi-cylindrical lobes, which form the crowns of the lower molars of the Rhinoceros. Douglas presented the specimen to Sir Ashton Lever; and, after the dispersion of the Leverian collection, it was purchased by H. Warburton, Esq., M.P., late President of the Geological Society, and was presented by him to the Museum of the Society.

With regard to other parts of the dentition of the lower jaw of the *Rhinoceros tichorhinus*, allusion has been already made to the traces of sockets of incisive teeth, observed in the expanded symphysis of Siberian and British specimens (p. 334). M. Christol has described and figured the lower jaw of a tichorhine Rhinoceros,* discovered in the post-pliocene marine deposits, ("les sables marins supérieures de Montpellier,") which, like the specimen described by Pallas,† presented four alveoli at the symphysial extremity; the two outer or lateral cavities were two inches deep, and one inch in diameter at the outlet: the left socket contained the base of a fractured incisor; the two middle sockets were reduced to minute circular pits, not exceeding three lines in depth, and four in diameter. The last true molar is not quite in place, and its anterior crescent is very little worn, indicating that the individual with the above-described condition of the lower incisors was scarcely full grown, certainly not an

* Annales des Sciences Naturelles, 1835, tom. iv. pl. 2, fig. 1 and 2. The second premolar (the first in the specimen figured by M. Christol) seems to me to be proportionally too large, and too much advanced, for the species to which this lower jaw is referred.

† The words of Pallas are, "In apice maxillæ inferioris, seu ipso margine, ut ita dicam, incisorio, dentes quidem nulli adsunt; verumtamen apparent vestigia obliterata quatuor, alveolorum minusculorum æquidistantium, è quibus exteriores duo, obsoletissimi, sed intermedii, satis insignibus fossis denotati sunt." Novi Commentarii Petropol., t. xiii. p. 600.

aged animal. The upper incisors appear to be earlier lost; and the traces of those below are generally obliterated in specimens of *Rhinoceros tichorhinus* with the molar series complete.

The characters of other enduring parts of this species, as defined by Cuvier, have been satisfactorily confirmed, not only by the discovery of the almost entire skeleton of the same individual tichorhine Rhinoceros, in the Cave at Wirksworth, but by other not less extraordinary and instructive instances.

In 1816 a considerable portion of the skeleton of a Rhinoceros was discovered by Mr. Whidbey, engineer of the Plymouth Breakwater, in one of the cavernous fissures of the limestone quarries at Oreston, near Plymouth: the following parts, most of which were determined by Mr. Clift, were recovered and preserved:—

Two molar teeth of the upper jaw.
Four do. do. lower jaw.
Portion of the first vertebra, atlas.
Portions of four dorsal vertebræ.
Portions of two caudal vertebræ.
Portions of four ribs.
The symphysial end of an os pubis.
Portions of the right and left scapulæ.
Both articular extremities of the left humerus.
Do. do. right ulna.
Do. do. left radius.
The right os unciforme.
The middle metacarpal bone of the right fore-foot.
A phalanx of the same toe.
Both articular extremities of the right femur.
Part of both extremities of the left femur.
The left patella.

A fragment of the left tibia.

Two portions of metatarsal bones of the right hind-foot.

The size and form of the teeth, and the thick and strong proportions of the remains of the bones of the extremities, indicate them to have belonged to an animal of the same species as that still more entire specimen discovered in the Derbyshire cavern.

The state of the epiphyses of the long bones proves that the animal had not quite reached maturity; but in the same cavernous fissure, at Oreston, there was found part of the right humerus of an older individual of the *Rhinoceros tichorhinus*.

The broken bones have suffered from clean fractures; none of them are gnawed or waterworn: the cavern containing them was fifteen feet wide, twelve feet high, forty-five feet long; it was filled with solid clay, in which the bones were imbedded: they were situated about three feet above the bottom of the cavern.*

In similar and adjoining caverns (fig. 50, A and B) detached bones and teeth of the same extinct species of Rhinoceros were found; they were associated in one of the fissures with remains of a large species of Deer, and of the *Ursus spelæus*; in another with fossil bones of *Equus*, *Bos*, *Cervus*, *Ursus*, *Canis*, *Hyæna*, and *Felis spelæa*. None of the bones exhibit marks of having been gnawed or broken by the teeth of the great cave-haunting Carnivora; but both these

* Philosophical Transactions, 1817, p. 176: the specimens are now preserved in the Museum of the Royal College of Surgeons, London. One of the bones was analyzed by Mr. Brande, who found it to consist of

Phosphate of lime	60
Carbonate of Lime	28
Animal matter	2
Water and loss	10
	100

and the herbivorous species appear to have perished by accidentally falling into the cavernous fissures before they were filled up by the mud, clay, and drift.

The remains of the Rhinoceros discovered in the cave at Kirkdale, tell a very different story: they manifest, as Dr. Buckland has demonstrated, abundant evidence of the action of the powerful jaws and teeth of the Hyænas, whose copros and vestigia prove that ancient cavern to have been a place of refuge to those Carnivora.* The fossil bones of the Rhinoceroses found in this cavern, as well as in that near Torquay, called Kent's Hole, belonged to animals which inhabited England during the period immediately preceding the deposition of the unstratified drift, and they coexisted with the Mammoth, Hippopotamus, huge Aurochs, Ox and Deer, which likewise became the occasional prey of the Hyænas, whose dwelling-place was thus converted into a kind of charnel-house of the large Herbivora.

The circumstances under which remains of the Rhinoceros have been discovered in the limestone caves of the Mendips, and in those on Durdham Down, lead to similar explanations of their mode of introduction.

The humerus of a Rhinoceros was discovered, associated with remains of the *Hyæna spelæa*, in one of the caves in the carboniferous limestone at Cefn in Denbighshire, at a height of about one hundred feet above the present drainage of the country.† The Rev. Mr. Wilson, of Leyton, has kindly submitted to my examination a collection

* Ante, pp. 141—147.

† These caves were described by the Rev. Edward Stanley, now Bishop of Norwich, in the proceedings of the Geological Society, vol. i. p. 402. Mr. Murchison remarks (Silurian System, p. 552,) that the evidence produced is scarcely adequate to sustain the inference that the cave was inhabited, though it affords satisfactory proof that such wild animals then existed in an adjacent region.

of bones, discovered by the Rev. R. Greaves in a fissure of a limestone rock in Caldy Island, off Tenby, most of which proved to belong to the *Rhinoceros tichorhinus*. A femur of the same species was discovered by Dr. Lloyd in a fissure of the Aymestry limestone. Mr. Murchison, who cites Dr. Lloyd's discovery, proceeds to say, (loc. cit. p. 554):—

"That quadrupeds of extinct species inhabited this (silurian) region, is proved by the contents of certain gravel heaps on its eastern limits. In a pit, south of Eastnor Castle, where the fragments consist exclusively of silurian rocks and syenite of the adjacent hills, the remains of the Elephant and other animals have been found, and at Fleet's Bank, near Sandlin, the bones of a Rhinoceros and Ox. The latter were found by Mr. J. Allies, who has also collected the bones of the Horse, Rhinoceros, Elephant, &c., at Powick, and those of a Rhinoceros at Bromwich Hill, near Worcester."

Remains of the Rhinoceros were discovered by Mr. Strickland, associated with those of the Elephant and Hippopotamus, in the fluviatile deposits of the valley of the Avon, near Cropthorn, Worcestershire. These deposits appear to form part of the same series which he has traced from Defford, in that county, to Lawford, in Warwickshire, where they have yielded bones of the Rhinoceros in great abundance and perfection. Remains of this Pachyderm were likewise associated with those of the Elephant and Hippopotamus in the analogous fresh-water deposits of the valley of the Thames. The tooth, figured in Mr. Trimmer's Memoir on those at Brentford (Philosophical Transactions, 1813, pl. ix. fig. 2), is an upper molar of a Rhinoceros, not of the Hippopotamus, as there stated.

The fresh-water formations, exposed on the cliffs of our

eastern coast, have yielded very fine remains of more than one extinct species of Rhinoceros.

The *Cambridge Advertiser*, for the 26th of February, 1845, contains the following announcement :—

"FOSSIL REMAINS; CROMER.—The late high tides have partly uncovered the lignite beds along the base of the cliffs, and among the fossil remains of that stratum have been found a fine specimen of the lower jaw of a Rhinoceros, with the seven molar teeth in good preservation; together with molars of the Elephant, Hippopotamus, and Beaver."

The jaw of the Rhinoceros has been obligingly transmitted to me for examination by its present possessor, Robert Fitch, Esq., F.G.S. It is the left ramus of a young, but nearly full-grown individual of the *Rhinoceros tichorhinus.* The socket of the first small premolar is not obliterated; the second and third premolars, the last deciduous molar, and the first and second true molars, are in place: the crown of the last true molar is just about to emerge from its alveolus; the last premolar is concealed in the substance of the jaw, beneath the third much worn tooth in place. This interesting specimen, which exemplifies one of the later stages of the dental changes of the extinct Rhinoceros, will be again adverted to in comparison with a corresponding fossil of the *Rhinoceros leptorhinus.*

With regard to the most instructive remains of the Rhinoceros from Lawford near Rugby, Cuvier (loc. cit. p. 80) expressly refers the cubitus to the 'espèce cloisonnée;' and again, with regard to the 'os innominatum,' he says, that it seems to belong to the species with the osseous septum, viz. the *Rhinoceros tichorhinus:* in reference to the tibia and the cervical vertebræ, Cuvier confines his observations to their differences as compared with the recent *Rhinoceros indicus* (p. 84), or to their want of sufficiently distinguishing characters (p. 76).

Dr. Buckland possesses some very fine and perfect specimens of the humerus of the *Rhinoceros tichorhinus*, from Lawford, of one of which Cuvier has given figures in pl. xv. figs. 5 and 6, of the volume above cited. The humerus is remarkable in the Rhinoceros, and especially in the great extinct tichorhine species, for its strength and the enormous thickness of the upper end; in one of the Lawford specimens the circumference at that end is two feet, the entire length of the bone being one foot, seven inches. The great tuberosity is developed into a strong curved plate, which bends over the broad and deep bicipital groove: the deltoid crest, continued downwards from the tuberosity also manifests prodigious strength. Cuvier remarks that the trochlear articular surface for the radius is more oblique, and its lower crest longer, in the fossil, than in the recent Rhinoceros of India.

Fig. 129.

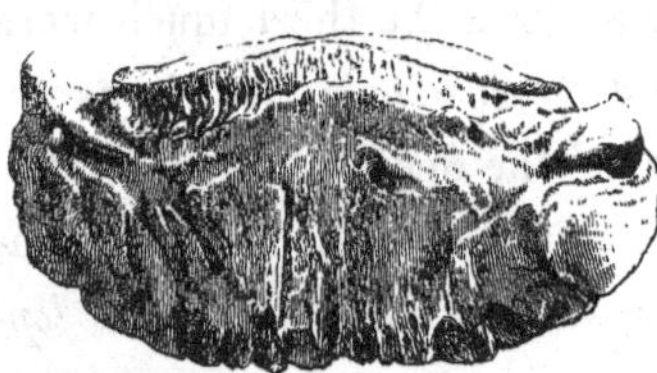

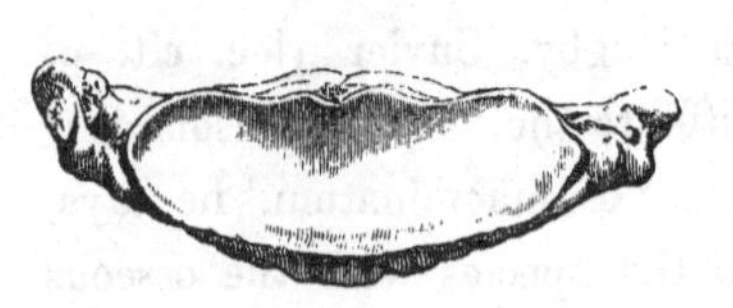

I subjoin two views of an ungual phalanx of a Rhinoceros (fig. 129), which was obtained from the brick marl, at Gray's Thurrock, Essex; an opportunity of examining this fossil, and of giving these illustrations, having been kindly afforded me by Mrs. Mills, of Lexden Park, near Colchester. The upper figure shows the rough anterior surface of the bone, sculptured by the canals for the blood-vessels, supplying the secreting organ of the thick hoof which once adhered to it: the under figure shows the smooth articular surface which played upon that of the second phalangeal bone.

Of the bones of the hind extremity Dr. Buckland's collection at Oxford contains a rich series, from which, indeed, Cuvier derived much of his knowledge of the anatomical distinctions of this part of the skeleton of the *Rhinoceros tichorhinus.* He figures a fine specimen of the os innominatum, or haunch-bone, (tom. cit. pl. xiv. figs. 1 and 2,) which, compared with that bone in the existing one-horned and two-horned Rhinoceros, exhibits a narrower and longer "foramen ovale:" the lateral borders of the iliac bones are more oblique and more concave towards the neck; the anterior border is less convex, especially towards the external angle; and this angle is narrower, more pointed, and not forked; the external angle of the tuberosity of the ischium is also more pointed. The femur or thigh-bone of the Rhinoceros may be distinguished from that of the Hippopotamus, Aurochs, and other large herbivorous quadrupeds of similar size, by a flattened process extending outwards from near the middle of the outer part of the shaft: this process is termed the "third trochanter." The shaft is broad and flat, especially at the upper end. I have compared the proximal part of the thigh-bone of the young Rhinoceros from Oreston, in which the hemispherical articular head and the great trochanter were in the state of detached epiphyses, with the femur of a young *Rhinoceros indicus* in the same state, and found the depression for the ligamentum teres shallower in the fossil: the post trochanterian depression is also shallower, and the third trochanter smaller. The shaft is thicker in proportion to the lower condyloid expansion than in the African, Indian, or Sumatran Rhinoceros; and the fore part of the shaft, above the joint for the patella, or knee-pan, is more excavated than in the other fossil species found in Britain, viz., the *Rhinoceros leptorhinus.*

Although the remains of the great tichorhine Rhinoceros have not been found in such abundance in the caves, the unstratified drift, and the post-pliocene fresh-water deposits of Britain, as those of its more gigantic contemporary the Mammoth, the two-horned Pachyderm seems to have been as extensively distributed over the land which now constitutes our island. The works of continental palæontologists demonstrate that this Rhinoceros was similarly associated with the Mammoth in the more recent deposits of France, Germany, and Italy.*

But the most abundant as well as the best preserved specimens of the tichorhine Rhinoceros have been discovered in the northern latitudes of Asia, which appear to have been the regions most frequented by it; and where the same evidence has been obtained of its special adaptation to colder climates than those inhabited by existing Rhinoceroses, as that which has been previously detailed in reference to the Mammoth.

The very remarkable discovery of the extinct Rhinoceros preserved in ice was made nearly twenty years before the analogous one of the frozen Mammoth, noticed in a foregoing section;† and is narrated by Pallas in the 4th volume of his 'Voyages dans l'Asie Septentrionale,' (4to., 1793, pp. 130—132), as follows:—

"I ought here to mention an interesting discovery,

* Cuvier showed that the famous fossil Morse of Monti, discovered at Mont Blancano, near Bologna, was the lower jaw of the *Rhinoceros tichorhinus*, (tom. cit. p. 73.)

† Ante, p. 263.

which I owe to M. le Chevalier de Bril. Certain Jakoutzki hunting this winter" (1771-2) "near Viloui, found the body of a great unknown beast. The Sieur Ivan-Argounof, Inspector of Zimovia, caused to be transmitted to the Prefecture of the province of Jakoutzk the head, a fore-foot, and a hind-foot of the animal, the whole of which were in an excellent state of preservation.* He says in his Memoir, dated the 17th of last January," (1772) "'that they found, in the month of December, the animal dead, and already much decomposed,† at about forty versts above Zimovié de Vilouiskoe, on the sand of the bank, at the distance of one toise from the water and four toises from another higher and more precipitous escarpment: it was about half buried in the frozen sand. They took its dimensions on the spot: it was three and three quarters Russian ells' (aunes de Russie, about eleven and a half English feet) 'in length, and they estimated its height at three and a half ells. The body of the animal, still retaining its corpulency,' (encore dans toute sa grosseur,) 'was clothed with its skin, which resembled leather; but it was so far decomposed that they were unable to bring away more than the head and the feet. These I saw at Irkoutsk; they seemed to me, at the first view, to belong to a Rhinoceros, which had been in full vigour. The head, especially, was very recognisable, because it was covered by its skin. The skin had preserved all its exterior organization, and one could see upon it many short hairs,' (*on y appercevoit plusieurs poils*

* Pallas, in a more elaborate account of the same discovery which he communicated to the Imperial Academy of Sciences at Petersburg, states, "Reliquum vero cadaver, corruptum valde, licet corio naturali adhuc obvolutum, in loco relictum, periit:" Novi Commentarii Petropol., 1773, tom. xvii. p. 587.

† In his Memoir in the Petersburg Transactions, Pallas observes, "fœtorem spirabant non recens corruptarum carnium, sed latrinis prorsus antiquis comparandum, quasi ammoniacalem." Loc. cit. p. 589.

courts). 'The eyelids and eyelashes even had not entirely fallen into decay. I saw a substance in the cavity of the skull; and here and there, beneath the skin, were the remains of the putrified flesh. I remarked on the feet the very obvious remains of the tendons and cartilages, where the skin was wanting. The head had lost its horn,* and the feet their hoofs. The situation of the horn, the fold of integument which surrounded it, and the separation' (of the toes?) † 'which existed in the fore-feet and hind-feet are certain proofs of the animal being a Rhinoceros.' I have given an account of this singular discovery in the Memoirs of the Academy of Petersburg, and refer my readers to that work to save repetition. They will there see the reasons in proof that a Rhinoceros has been able to penetrate near the Lena in high northern latitudes, and the circumstances that have led to the discovery in Siberia of the remains of so many strange animals."

In this Memoir, Pallas specifies the short hairs, strongly implanted in pores of the skin covering the vertex, and growing in tufts (fasciculatim nascentes) from the sides of the mandibular region, of rigid texture and cinereous grey colour, with here and there a black hair longer and stiffer than the rest. The hairs adhered to many parts of the skin of the legs, from one to three lines long, of a dirty cinereous colour. So much hair as grew from the parts of the frozen Rhinoceros observed by Pallas, he never

* "La tête étoit dégarnie de sa corne," are the words of the French translator and editor Peyronie; but Pallas, in his Memoir, expressly mentions the two horns: "*Cornua* cum capite adlata non fuerunt, prius forte abrupta et a flumine vel transeuntibus gentilibus, qui venationi operam navant, ablata. Apparent autem *cornu nasalis* pariter atque *frontalis* evidentissima vestigia." Novi Comment. Petropol., tom. xvii. p. 588.

† In the Memoir, "De Reliquiis animalium exoticorum," Pallas, speaking of the feet, says, "In quibus non solum divisura ungularum, Rhinocerotis characteristica, sed corium pariter," &c.

observed on any living species; and he asks whether it does not indicate the Rhinoceros of the Lena to have been an aboriginal of the temperate latitudes of Asia?

It must not be inferred from the observations which Pallas was able to make on the hair of the legs of the frozen Rhinoceros, that its body was less warmly clad than that of the Mammoth. No naturalist, unacquainted with the woolly covering of the arctic Musk Ox, could have inferred it from an inspection of the legs only, which are clothed with short, dull, brownish-white hair, unmixed with wool.

Of the subsequent discoveries of carcasses of Rhinoceroses in the frozen soil of Siberia, I can only learn that they prove the hide to have been destitute of those singular folds which characterize that part in the existing one-horned Rhinoceros; and that one of the horns, probably the first or nasal horn, has been obtained, which measures nearly three feet in length, and thus confirms the deductions of Cuvier from the osseous septum supporting the nasal bones, as to the size of this formidable weapon: it is preserved in the Museum of Natural History at Moscow.

Although the molar teeth of the *Rhinoceros tichorhinus* present a specific modification of structure, it is not such as to support the inference that it could have better dispensed with succulent vegetable food than its existing congeners; and we must suppose, therefore, that the well-clothed individuals who might extend their wanderings northwards during a brief but hot Siberian summer, would be compelled to migrate southward to obtain their subsistence during winter. Plants might then have existed with longer periods of foliation than those which now grow. This, at least, is a less extreme hypothesis than the sudden change from a tropical to an arctic climate, which has been proposed to account for the preservation in ice of entire Elephants and Rhinoceroses; and Mr. Darwin has well remarked that "as there is evidence of physical changes, and as the animals have become extinct, so may we suppose that the species of plants have likewise been changed." But, admitting the more probable necessity of migration, we may derive some insight into the habits of the Siberian Rhinoceros by inquiring into those of existing large Herbivora of Arctic climes, which were represented by species coeval with those extinct Rhinoceroses. Pallas describes and figures in the same Memoir "De reliquiis animalium exoticorum" in which he describes the frozen Rhinoceros, the fossil remains of a Musk Ox (*Ovibos*, De Bl.), which seems to be not more satisfactorily distinguishable from the existing species* than is the *Urus priscus* from the great Lithuanian Aurochs: the Musk Ox is remarkable at the present day for its geographical position in high northern latitudes, and its adaptation to such by its peculiarly fine woolly clothing, and its periodical migrations have been noticed by experienced naturalists. The appearance of the Musk Ox in the month of May on Melville Island in latitude 75°, was one of the phenomena ascertained in Captain Parry's first voyage, and "is interesting," Dr. Richardson observes,† "not merely as part of their natural history, but as giving us reason to infer that a chain of islands lies between Melville Island and Cape Lyon, or that Wollaston's and Banks's Lands form one large island, over which the migrations of the animals must have been performed."

* Cuvier, 'Ossemens Fossiles,' 4to. 1823, tom. iv. p. 156.

† 'Fauna Boreali-Americana, Mammalia,' p. 276.

Fig. 130.

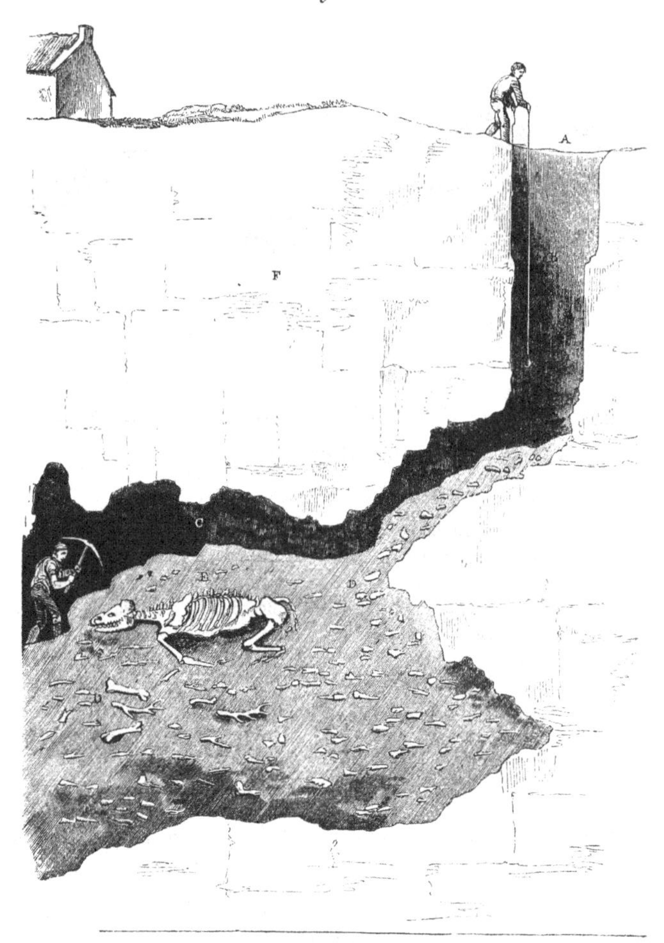

Section of the Dream Cave at Wirksworth, (Buckland, 'Reliquiæ Diluvianæ,') showing the position of the fossil Rhinoceros, E.

PACHYDERMATA. *RHINOCEROS.*

Fig. 131.

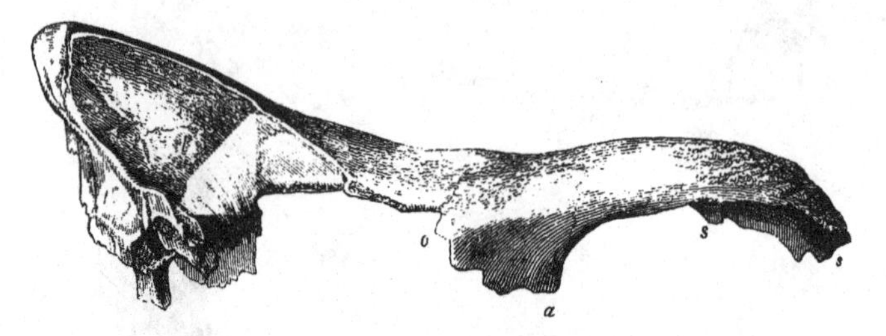

Upper part of skull, *Rh. leptorhinus*, ⅛ nat. size. Clacton, Essex.

RHINOCEROS LEPTORHINUS. Leptorhine Two-horned Rhinoceros.

Rhinoceros leptorhinus, ou Rh. à narines non-cloisonnées et sans incisives, — CUVIER, Ossemens Fossiles, 4to., 1822, tom. ii., pt. 1, p. 71, pl. ix. figs. 8 and 9; pl. xiii., figs. 4 and 5.

„ *Kirchbergense,* — JÄGER, Die Fossilen Sæugethiere, Wurtemberg, fol. 1839, p. 179, tab. xvi., figs. 31, 32, 33.

„ *Merckii,* — KAUP, Akten der Urwelt, 8vo., 1841, p. 6, tab. i., figs. 1, 3, 4, and 5; tab. ii.

WHILST the catalogue of extinct European Rhinoceroses has been augmented, since the time of Cuvier, by a few well-determined and many nominal species, one, which the great Palæontologist had himself inscribed there by the name of *Rhinoceros leptorhinus*, has been almost blotted out and lost sight of, through the defective character of part of the evidence on which he founded the species.

The name '*leptorhinus*' and its French synonym '*à narines non-cloisonnées*,' more commonly applied by

Cuvier to the species in question, were suggested by the characters of the fossil skull of a Rhinoceros discovered by M. Cortesi in a fresh-water upper tertiary deposit at Plaisance, as they appeared in a drawing transmitted to Cuvier, who had not had an opportunity of studying the original, which is preserved in the 'Musée des Mines' at Milan. Confiding in the drawing, which is engraved in the 'Ossemens Fossiles,' 4to., 1822, tom. ii., pt i., Rhinoceros, pl. ix., fig. 7, Cuvier was led to conclude that the Rhinoceros of Plaisance differed from that of Siberia and northern Europe in having "the cerebral part of the skull less prolonged and less inclined backwards; in the position of the orbit above the fifth molar tooth; in the anterior termination of the nasal bones by a free point, and in the absence of any attachment of them to the intermaxillaries by a vertical osseous septum; in the minor degree of prolongation of the intermaxillary bones, which were of a totally different form, presenting, in short, as little as the nasal partition, any of those characters for which the skull of the *Rhinoceros tichorhinus* was so remarkable." (Tom. cit. p. 71.) From these apparently broad distinctions, Cuvier did not hesitate to admit the specific difference of M. Cortesi's Rhinoceros; and he even ventured to state that it incontestably approached nearer to the *Rhinoceros bicornis* of the Cape than to any other known species. (Tom. cit. p. 71.)

This summary of the cranial characters of the *Rhinoceros leptorhinus* is repeated without modification in the posthumous 8vo. edition of the 'Ossemens Fossiles,' 1834, tom. iii., p. 136.

In the following year, however, M. de Christol communicated to the 'Annales des Sciences,' 2de série, tom. iv., p. 44, a more accurate figure (pl. ii, fig. 4) of the cranium of the Rhinoceros discovered at Plaisance, and the results of a careful comparison of three large drawings of that fossil, made at his request by MM. de la Marmora and Gené at Milan; from which he was led to conclude that the drawing published by Cuvier was very defective in one of the most essential points, and had led the great Anatomist into the error of creating a species which had never existed.*

M. Christol found, in fact, that the bony septum of the nose had been omitted in the sketch engraved in the 'Ossemens Fossiles,' whilst a considerable portion of it actually existed in the fossil; and that the anterior extremity of the nasal bones, represented as projecting freely forwards in the Cuvierian figure, were evidently broken off in the actual fossil, according to the large drawings transmitted to him by Prof. Gené. (Loc. cit. p. 70.)

The discrepancies between the figures published by Cuvier and M. Christol are obvious enough; and one can scarcely avoid conceding to the later observer, that he has established the fact of the existence, in M. Cortesi's fossil, of the chief character, viz., the bony partition of the nose, the absence of which was mainly depended on by Cuvier as the distinctive feature of his *Rhinoceros à narines non-cloisonnées*. Since, however, this species rests not only upon M. Brongniart's drawing of the skull at Milan, but upon characters deduced, by Cuvier's own observation, from lower jaws obtained from fresh-water deposits in Italy, M. Christol, who had not any more than Cuvier

* "Cuvier n'a pas eu occasion de la voir, il n'a pu en décrire la tête que d'après un dessein qui, tout en retraçant assez exactement les contours généraux de cette tête, est très incomplet dans le point le plus essentiel, et me paraît avoir induit Cuvier en erreur en le portant à créer une espèce qui n'a point existé." Christol, loc. cit. p. 47.

personally inspected or compared M. Cortesi's fossil, expects too much when he demands the entire suppression of the *Rhinoceros leptorhinus* from the catalogue of extinct species.

I shall be able, indeed, to show that the partial bony septum, and its confluence with the extremities of the nasal bones, inferred by M. Christol to exist in the skull of the Rhinoceros at Milan, do not, of themselves, give proofs of its identity with the species called *Rh. tichorhinus*; and although, in the absence of direct inspection of the fossil in question, I cannot presume to question the accuracy of M. Christol's determination of it, I may observe that the points above cited, upon which he chiefly grounds his opinion, are not incompatible with the characters which I have ascertained to belong to the skull of the *Rhinoceros leptorhinus.*

Before adverting to these, I shall first adduce evidence of the existence, in British fresh-water newer-pliocene deposits, of a Rhinoceros, having the same characters of the lower jaw and teeth which Cuvier has ascribed to his *Rhinoceros leptorhinus.*

The specimens described and figured in the 'Ossemens Fossiles,' tom. cit. pl. ix., figs. 8 and 9, were discovered in Tuscany, and are the most common kind of Rhinoceros jaws in that part of Italy, where, however, the lower jaw of the *Rhinoceros tichorhinus* has likewise been found. From this the jaw of the *Rh. leptorhinus* differs "by the continuation of the series of molar teeth close to the anterior end of the jaw, which is short and not prolonged into a prominence, or expanded part;" and these characters Cuvier correctly cites as evidence of the close resemblance of the leptorhine Rhinoceros to the two-horned species of the Cape. (Tom. cit. p. 72.) The fossil specimen (fig. 132), which, in like manner, differs as much from the lower jaw of the *Rh. tichorhinus* (fig. 123) as it resembles that of the *Rh. bicornis*, was discovered by John Brown, Esq., F.G.S., in the fresh-water pliocene deposits

Fig. 132.

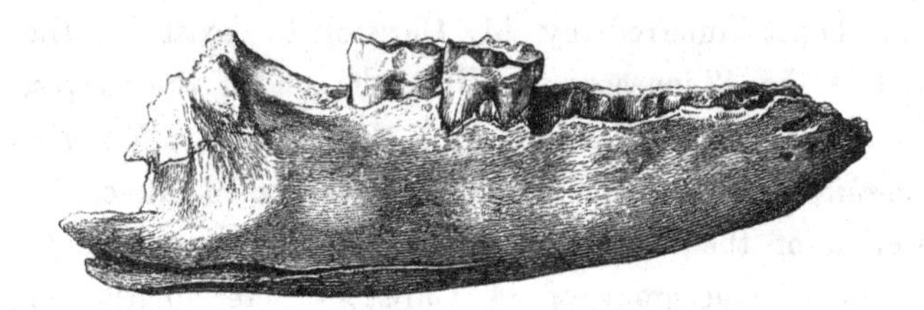

Lower jaw, *Rh. leptorhinus*, ⅛ nat size. Clacton, Essex.

at Clacton on the Essex coast. It consists of the right branch of the lower jaw, wanting the angle and coronoid ascending process and the end of the symphysis, and it contains the last and penultimate molars, and the sockets of four molars anterior to these. The entire length of the specimen is one foot six inches and a half; the depth of the jaw behind the last molar tooth is four inches nine lines; its depth behind the third molar tooth is three inches four lines. The extent of the molar series, from the front of the second socket to the back of the last socket, is ten inches. I assume the anterior alveolus, (fig. 133, *p* 2,) which lodged a two-fanged premolar, exceeding one inch in antero-posterior extent, to have been the second of the series; the deep depression, exposed on the broken part of the symphysis anterior to this socket, is the dental canal; it is shown at *v*, fig. 133, in which a view of the alveolar border of the jaw is given on the same scale as that of the figure of the lower jaw of the *Rhinoceros leptorhinus* in the 'Ossemens Fossiles' (tom. cit. pl. ix., fig. 9), which appears to be the same scale as that on which Dr. Kaup's specimen of the lower jaw of the *Rhinoceros Merckii* is figured in the 'Akten der Urwelt,' tab. ii.

The socket of the second molar, (*p* 2,) or the sixth, counting from behind forwards, is entirely in advance of the transverse line drawn across the back part of the symphysis, and the molar series is consequently extended much closer to the end of the jaw than in the *Rhinoceros tichorhinus*. This part of the symphysis also is rounded inwards towards its anterior termination in the present specimen, producing a very different contour from that produced by the swelling out of the same part to form the flattened spatulate extremity, characteristic of the lower jaw of the *Rh. tichorhinus* (fig. 124). The lower border of the jaw is less curved in the *Rh. leptorhinus*, and the depth less suddenly diminished at the symphysis. The fore-part of the base of the coronoid process is more prominent externally in the *Rh. leptorhinus* than in the *Rh. tichorhinus*. The molar teeth are larger, and the series occupies a greater extent in the jaw of the leptorhine species.

Fig. 133.

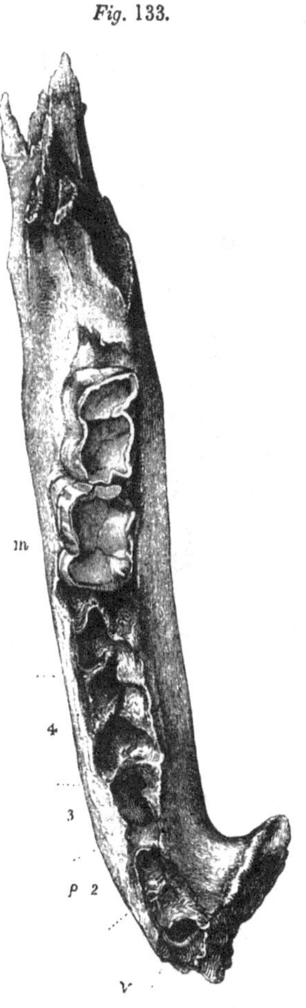

Lower jaw, *Rhinoceros leptorhinus*. $\frac{1}{4}$th nat. size. Clacton, Essex.

	Rh. leptorhinus.		Rh. tichorhinus.	
	In.	Lin.	In.	Lin.
Antero-posterior extent of last two molars .	4	3	3	9
" " penultimate molar .	2	0	1	9
Transverse diameter of base of crown of penultimate molar	1	6	1	2

In the present specimen of the jaw of the leptorhine Rhinoceros (fig. 133), the worn state of the last two molars shows that it had belonged to an old individual: but the difference of size is equally manifested by the specimen of a fragment of the left branch of the lower jaw of the *Rhinoceros leptorhinus* (fig. 134), also obtained by Mr. Brown from the fresh-water deposits at Clacton, and containing the last three molars, in the same state of attrition as those in the jaw of the *Rhinoceros tichorhinus* (fig. 124). There is a difference also in the proportional size of the posterior lobe of the last molar tooth, which is greater in the *Rh. leptorhinus.* The lower terminations of the internal depressions of the molars are less angular and less narrow in the *Rh. leptorhinus*; and the three inner columns or prominences of the molars are less flattened.

Fig. 134.

Rhinoceros leptorhinus. ¼ nat. size. Walton.

The specimen of the fore part of the lower jaw of a somewhat younger leptorhine Rhinoceros, obtained by Mr. Brown from the fresh-water deposits at Clacton, Essex, and containing the second, third, and fourth premolars *in situ* (fig. 135), yields a specific character in the larger proportional size of the second premolar; which will be recognized by comparing the annexed figure with fig. 128, and is demonstrated by the following admeasurements:

	Rh. leptorhinus.		*Rh. tichorhinus.*	
	In.	Lin.	In.	Lin.
Antero-posterior extent of second, third, and fourth premolar	4	3	4	0
" " second premolar	1	3	1	0
" " fourth premolar	1	7	1	7

Fig. 135.

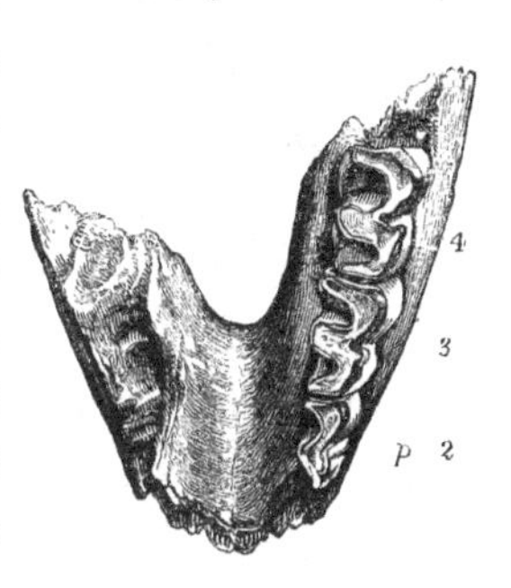

Rhinoceros leptorhinus, rather less than ¼ nat. size. Clacton.

There is a still more marked distinction of form, which, as it is rarely manifested in the lower molars of the Rhinoceros genus, I have here illustrated by two cuts of the natural size; fig. 136, showing the inner side of the second premolar of the *Rhinoceros leptorhinus*, and fig. 137, *p* 2, that of the *Rh. tichorhinus*, the latter being from Lawford, near Rugby.

Fig. 136.

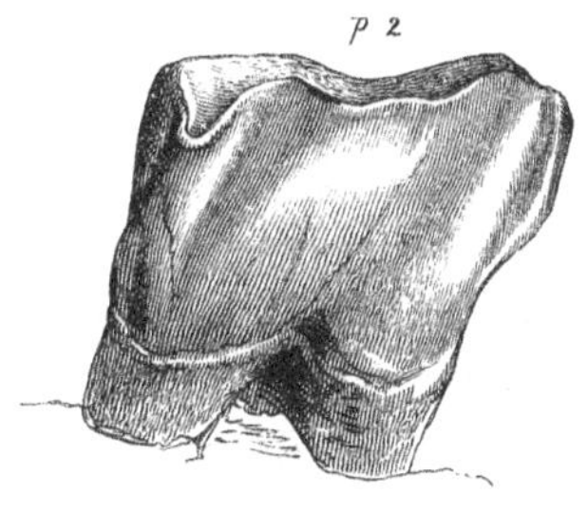

2nd premolar, nat. size. *Rhinoceros leptorhinus.* Clacton.

Fig. 137.

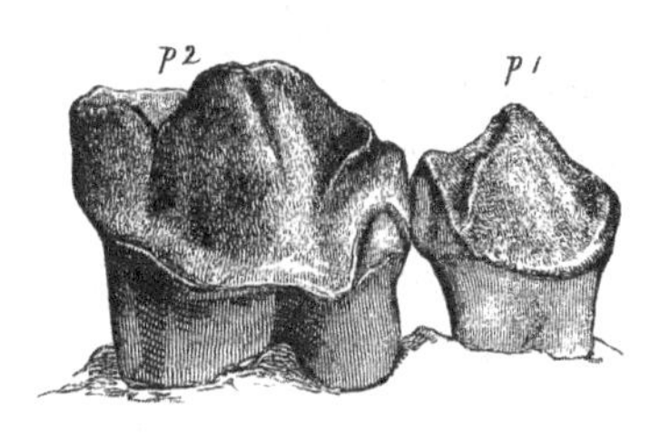

1st and 2nd premolars, nat. size. *Rhinoceros tichorhinus.* Lawford.

In Dr. Buckland's collection at Oxford there is a specimen of a considerable part of the right ramus and symphysis of the lower jaw of a young, but nearly full grown, *Rhinoceros leptorhinus.* The last molar tooth has half risen above its alveolus, and the summit of the anterior crescent had just begun to be used in mastication: the penultimate grinder is in place, the sockets of the antepenultimate molar, and of the three adjoining premolars, vacant; that of the first premolar is obliterated: the whole of the socket of the second, and part of that of the third premolar are in advance of the back part of the symphysis. Besides this well-marked distinctive character, the present fossil displays the more convex curvature of the lower border of the jaw, its greater thickness in proportion to its depth below the premolar series. These differences are well brought out in contrast with the portion of jaw from the fresh-water beds of the Cromer Cliff, which belonged to a younger individual, and of which comparative admeasurements are subjoined:—

	Rh. leptorhinus.		*Rh. tichorhinus.*	
	In.	Lin.	In.	Lin.
Depth of jaw below the middle of third premolar	2	0	3	0
Greatest thickness of the same part of the jaw	1	7	1	8
Depth of the jaw below middle of the penultimate molar	3	0	3	5
Antero-posterior breadth of penultimate molar	2	0	1	9
" " of last molar . .	2	3	1	8

The last two admeasurements show the characteristic superior size of the molar teeth in the *Rh. leptorhinus.*

Dr. Kaup has described and figured a portion of a lower jaw of a Rhinoceros discovered in the Rhine formations ("im Rheine gefunden"), the left ramus of which, according to the figure,* contains the fourth, fifth, and sixth molars, the roots of the third and second, and the anterior root of the seventh molar; the second molar being in advance of the posterior commencement of the symphysis, as in the lower jaw of the *Rhinoceros leptorhinus* of Italy, figured by Cuvier (loc. cit. Rh. pl. ix., fig. 9), and as in the specimen from Clacton, figs. 132 and 133.

Dr. Kaup, believing that in his Rhenish specimen the

* 'Akten der Urwelt,' tab. ii., fig. 1, p. 6.

teeth occupy a greater space, and that the edentulous end of the symphysis is broader than in the jaw of the *Rh. leptorhinus*, figured by Cuvier, refers it to a distinct species, which he calls *Rhinoceros Merckii.* The symphysis is not, however, entire in either of the specimens compared, according to the figures, from which I can by no means satisfy myself of their specific distinction. The length of the alveolar series, from the sixth to the second molar, inclusive of the specimen from Clacton (fig. 133), is 0·205 in French millemetres, or eight and a quarter inches English; in the Italian specimen, and also in that from the Rhine, if Dr. Kaup's figure be, like Cuvier's, one-fourth the natural size, the same dimension gives 0·225 millemetres, or nine inches: but different specimens of the lower jaw of the *Rhinoceros tichorhinus* have presented as much variety of size. I conclude, from the foregoing comparisons, that the lower jaw of the Rhinoceros from the Rhenish deposits, as well as that from Essex, are specifically identical with the lower jaws from Tuscany, which Cuvier has referred to his *Rhinoceros leptorhinus.*

But what are the characters of the rest of the cranium, and in what degree do the proportions of the nasal bones accord with the name imposed upon the species which the lower jaw incontestably proves to be distinct from all other species known at the period of its first description? M. Christol has shown that the answers given to these questions on the authority of the cranium discovered by M. Cortesi are unsatisfactory. No portion of the upper jaw or cranium was associated with the Rhenish specimen of the lower jaw of the *Rhinoceros leptorhinus* described by Dr. Kaup. But the discoverer of the corresponding portion of the same species in our own freshwater deposits was so fortunate as to obtain, by his own personal exertions, at the same time and place, the whole of the upper portion of the cranium, with a considerable proportion of the occiput, and a fragment of the upper jaw with the last molar tooth *in situ*; other upper molars being found detached, but in close proximity with the cranium. The side-view of this portion of cranium (fig. 131), reduced to the same proportion as that of the *Rh. tichorhinus* (fig. 120), shows the minor degree of elevation of the interorbital platform supporting the second or frontal horn, the minor degree of concavity between this surface and the cranium proper, the greater length of the nasal aperture, and the less prominent or convex contour of the anterior and rougher surface for the nasal horn: the limited extent of the bony partition-wall (*s s*), dividing the nasal cavity, and supporting the nasal bone, is also shown in this view, the lower part of the wall being broken away, but not the posterior margin, which terminates by a smooth rounded border. The bony partition-wall extends, in fact, from the anterior end of the nasal bones, only half-way towards the posterior boundary of the nasal apertures (*a a*), the view across the posterior half of which is uninterrupted. In the *Rhinoceros tichorhinus* the bony septum extends from the fore-part of the nose to the vomer behind, and serves to support not only the nasal, but the frontal horn. That the well-marked but interesting transitional character of the partial bony septum is not a fallacious appearance due to accidental loss or fracture, is demonstrated by the under or inner surface of the nasal platform, of which a reduced view is given in fig. 138. This surface, behind the bony septum (*s s*), is quite smooth and free from any marks of sutural attachment of an unanchylosed prolongation of a bony vomer; the surface is slightly convex transversely, concave longitudinally, with the free lateral margins bent down. The short septum is firmly anchylosed,* and gradually increases in thickness to the anterior deflected extremities of the nasal platform, where the appearance of the fractured surface of the confluent bones indicates that, when entire, they had been united by continuous ossification to the intermaxillaries, as in the *Rhinoceros tichorhinus*.† Very clear evidence of the distinction of the two species is obtained by comparing the upper surfaces of their skulls; and the reader may pursue the same comparison by means of the subjoined figure (139), and the

Fig. 138.

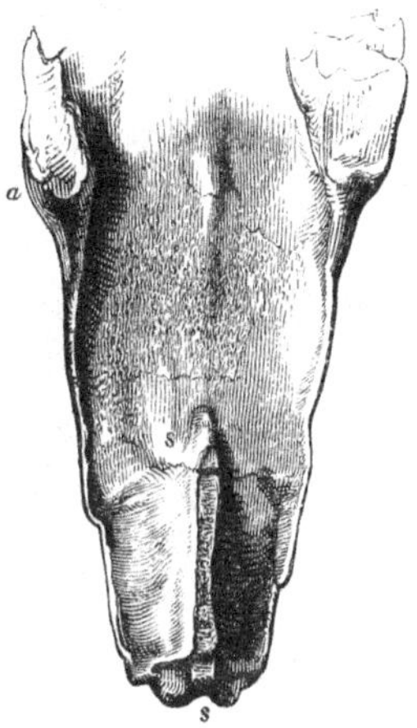

Under surface of nasal bones of *Rhinoceros leptorhinus*. $\frac{1}{6}$ nat. size. Clacton.

* This fact shows that the limited extent of the bony septum in the present cranium is not a consequence of immature age; not only the size of the skull, but the obliteration of the cranial sutures, proves it to have belonged at least to a fully mature individual. In the tichorhine Rhinoceros the bony septum is not anchylosed to the nasal platform until the animal has quite attained its maturity. In the young but full-grown specimen discovered in the frozen sand at Viloui, the bony septum was still free at its upper border. Pallas says, "Os scutiforme, quod cornu nasalis firmamentum præstat cum subjecto fulcro osseo, crassissimo vomeri comparando nondum evaluit; sed harmoniâ tuberculosâ totius plani, ut epiphyses ossium juniorum solent, inarticulatur." Novi Comment. Petropol. xvii. (1773), p. 590.

† When I first saw this specimen at Stanway during a tour of inspection of collections of British Fossils, preparatory to drawing up the Report on that subject for the British Association, I was induced, from the prevalent belief in the osseous septum anchylosed to the nasal bones as the peculiar characteristic of the *Rhinoceros tichorhinus*, to refer the Clacton cranium with those characters to that species; this error in the 'Reports of the British Association,' 8vo., 1843, p. 222, I am now able to correct.

Fig. 139.

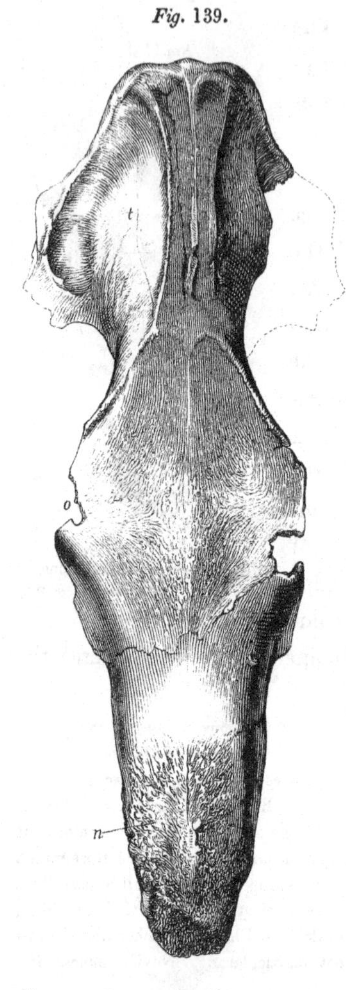

Upper surface of the skull of *Rhinoceros leptorhinus.* ⅛ nat. size. Clacton.

corresponding view of the skull of the *Rh. tichorhinus*, given by Cuvier in the 'Ossemens Fossiles,' 4to., 1823, tom. iii., pl. lxxix., fig. 5.

So compared, the Clacton specimen will be seen to be narrower in proportion to its length, especially at the cerebral and nasal regions: the confluent nasal bones (*n*) are not only more slender, but are more attenuated anteriorly, and thus vindicate the appropriateness of the name *leptorhinus* originally applied to the present species by its first discoverer.* The interorbital surface (*f*) for the frontal horn is not only less elevated, but is much less rugose, and is separated by a smooth space of some extent from that (*n*) for the

* The French name, *Rhinoceros à narines non cloisonnées*, more commonly applied by Cuvier to this species, is now proved to be inapplicable; the more accurate term would be *à narines demi-cloisonnées*; but, as the nasal bones notwithstanding their partial osseous supporting wall, are actually more slender than those of the *Rh. tichorhinus*, there is no objection to the Latin nomen triviale *leptorhinus*, and every reason for retaining it.

nasal horn. We may therefore infer, from the latter character, that the second horn was smaller in the leptorhine than in the tichorhine Rhinoceros, and connect in physiological relationship with this indication the non-extension of the bony supporting wall beneath the second platform.* Another distinction is the narrower interspace between the curved ridges (*t t*) which indicate the extent of origin of the temporal muscles upon the sides of the cranium: and this is not due to any difference of age; for the skull of the tichorhine Rhinoceros, with which I compared the Clacton specimen, belonged to an old individual, and yet exhibited the same superior width between the temporal ridges as is shown in the Cuvierian figure above referred to. The plane of the occiput is less inclined from below upwards and backwards than in the *Rh. tichorhinus*, and this region of the skull of the leptorhine species differs more strikingly in its form (fig. 140): it is narrower in proportion to the length of the skull, and especially at the upper part, which gives it a triangular figure with the apex cut off. In the *Rh. tichorhinus* it is more square-shaped, and the upper overhanging ridge is thicker and more rugged, indicating more powerful ligamentous and muscu-

Fig. 140.

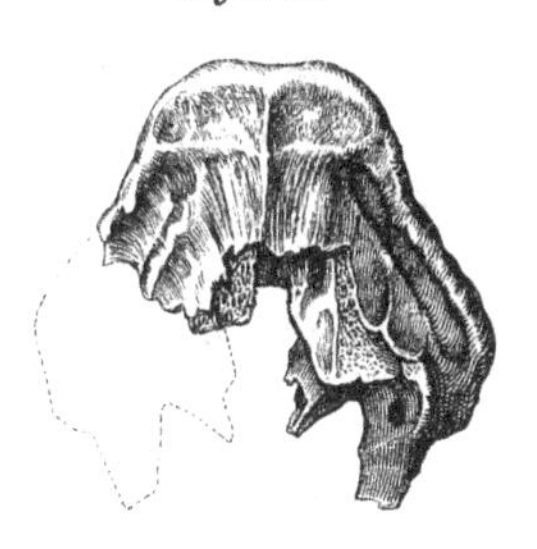
Occiput of *Rhinoceros leptorhinus*. $\frac{1}{8}$ nat. size. Clacton.

* The existing species of two-horned Rhinoceroses of Africa present the same difference in the proportions of their horns, as was manifested by the two extinct European species above compared. The *Rh. Keitloa* of Dr. Smith has both horns of equal length; the *Rh. simus* has the frontal horn much shorter than the nasal one.

lar attachments in relation to the stronger and heavier horns.*

The true characters of the skull of the *Rhinoceros leptorhinus*, and its distinction from that of the *Rh. tichorhinus* being established, there remains only to compare it with the skulls of other known species of Rhinoceros. The descriptions by Cuvier, 'Ossemens Fossiles,' 4to., 1824, tom. v., pt. ii., p. 502; by Dr. Kaup, 'Ossemens Fossiles de Darmstadt,' 4to., 1834, 3me cahier, p. 39, and by M. Christol, 'Annales des Sciences Nat.,' 1835, p. 76, of the extinct two-horned species, referred by Cuvier to his *Rh. incisivus*, but first accurately determined by Dr. Kaup under the name of *Rh. Schleiermacheri*, and shortly after by M. Christol under that of *Rh. megarhinus*, leave no room for doubt as to its specific distinction from the *Rh. leptorhinus.* Cuvier and Dr. Kaup are silent as to the presence or otherwise of a bony nasal septum in the *Rh. Schleiermacheri*, and the excellent figure of the skull of that species in Dr. Kaup's work shows no trace of it. It is equally absent in the original figure of the Montpellier specimen, given by M. Christol in the 'Annales des Sciences,' tom. cit., pl. ii., fig. 5; and the latter author expressly states "that

* These specific distinctions of the *Rh. tichorhinus*, and *Rh. leptorhinus*, will be readily appreciated by the subjoined table of comparative dimensions:—

	Rh. tichorhinus.		*Rh. leptorhinus.*	
	In.	Lin.	In.	Lin.
Length of the skull (in a straight line) .	31	0	28	0
Least breadth between temporal ridges .	3	6	1	5
Breadth of nasal bones opposite the hind border of the nasal aperture . . .	6	0	5	10
Breadth, opposite middle of nasal aperture .	6	6	4	9
Breadth of the anterior extremity of nasal platform	4	0	2	9
Length of nasal aperture	8	0	8	10
Breadth of upper part of occiput . . .	8	0	4	0
Do. of middle of occiput . . .	9	6	6	9

the nasal bones are broad, long, straight, horizontal, not massive, but strong and 'élancés,' without a septum below ('sans cloison en dessous'), abruptly bent down near their free extremity, which terminates in a point directed downwards and a little forwards," *ib.* p. 77. The marked difference in the form of the cranium of the *Rh. leptorhinus*, besides that essential structural one in the presence of the osseous septum, will be appreciated by comparing the contour of the nasal platform in fig. 131 with M. Christol's figure and accurate description of the same part in the *Rh. Schleiermacheri.* Cuvier deemed the skull of this species to resemble that of the Sumatran two-horned Rhinoceros more than any other, but to be proportionally shorter, with the nasal platform broader and less pointed, its convexity more prominent, and the temporal ridges more approximated, so as to form a sagittal crest. (Tom. cit. p. 502.) Now, in each of these particulars, the *Rh. Schleiermacheri* equally departs from the *Rh. leptorhinus;* which, by its proportionally longer cranium, with a narrower and more gradually attenuated nasal platform (fig. 139, *n*), presenting a more gradual and less elevated convex curve (fig. 131), and with the flat space intervening between the less approximated temporal ridges, still more nearly resembles the skull of the *Rh. Sumatranus* than does that of the *Rh. Schleiermacheri.* The *Rh. leptorhinus* differs, nevertheless, from the *Rh. Sumatranus* (see Cuv., op. cit., tom. iii., pl. lxxix., fig. 3) in its proportionally longer and narrower cranium, in the more backward production of the occipital ridge, and still more essentially by the ossified septum and its confluence with the fore-part of the nasal bones. From the skull of the *Rhinoceros incisivus*, to which Cuvier erroneously supposed that of Schleiermacher's species to belong, our present specimen is readily distinguished by both its shape, its partial bony septum, and the surfaces for the attachment of the horns; which surfaces are shown, by Dr. Kaup's beautiful discovery, to be wanting in that accordingly hornless extinct Rhinoceros, which, by way of compensation, was provided with unusually large incisive tusks. (Kaup, loc. cit., p. 109, pl. x.) By the absence of incisors, and by the form of the lower jaw, the *Rh. leptorhinus* resembled the incisorless *Rhinoceros bicornis* of the Cape; but, by the form and proportions of the cranium, it much more nearly resembled the two-horned Rhinoceros of Sumatra, and thus combined in its own organization characters now distinct, and shared between two existing Rhinoceroses, the habitats of which, in the present geographical distribution of Mammalia, are divided by a thousand miles of ocean.

Our chief information of the extent of the range of the extinct species of Rhinoceros is derived from the discoveries of their fossil teeth, which are the most common and the most recognizable remains of these great Pachyderms.

Cuvier expresses his regret that he had had no opportunity of examining the superior molar teeth of the *Rhinoceros leptorhinus*, so that he knew not whether they presented characters analogous to those which distinguish the molars of the existing species. He appeals to the Italian naturalists to supply this hiatus; and to this desirable object the specimens which were obtained by Mr. Brown in the same deposits at Clacton, with the cranium and lower jaws of the leptorhine species, have greatly contributed.

The upper molars from Clacton consist of the last and penultimate ones of the left side, and the ante-penultimate molar of the right side. If this tooth (fig. 141) be compared with the upper molar of the *Rhinoceros tichorhinus* (fig. 122), which has been worn down to about the same degree, it will be seen that, in fig. 141, the valley, *b*, is wider at its commencement, and that the termination, where the letter is placed, is smaller and of a triangular

Fig. 141.

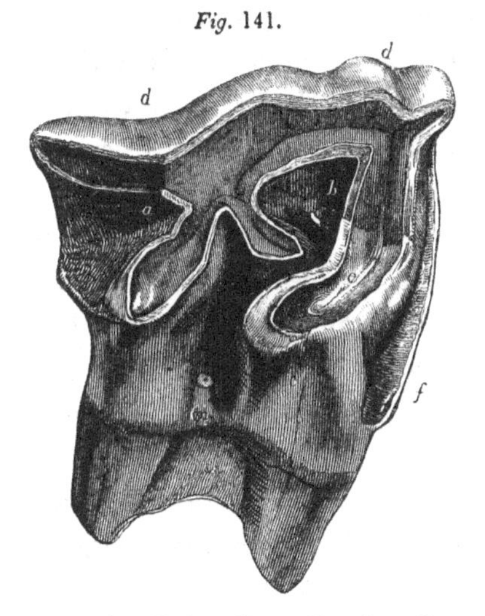

5th upper molar, *Rh. leptorhinus.* Nat. size. Clacton.

form: in the tichorhine molar it is much more expanded and bilobed by its extension towards the middle of the outer surface of the crown. The valley between these two terminal divisions, in the tichorhine molar, is so shallow, that the outer lobe is soon separated as an island of enamel, according to the pattern shown in fig. 126, and the valley then preserves an almost uniform width to the termination marked by the letter *b*. In the upper molar of the leptorhine Rhinoceros, the valley is either divided by the wearing away of the shallow fold of enamel between the end of the narrow process entering the valley and the opposite bank, *e*, whereby the end of the valley, *b*, is wholly insulated, which change is shown in the molar of the *Rhinoceros leptorhinus*, from Crozes,

Department du Gard, figured, but not recognized as such, in the 'Ossemens Fossiles,' tom. ii., pt. 1, Rhinoceros, pl. xiii., fig. 4; or the whole valley is gradually diminished in depth, without the separation of an enamel-island, but continuing to manifest its characteristic wide beginning, as is shown in the upper molar from the same locality in France, figured by Cuvier, tom. cit. pl. xiii., fig. 5. These varieties depend on the varying depth of the narrow part of the valley at the end of the small intruding promontory, and they are exemplified in two of the molars from Clacton: but neither of the patterns of the grinding surface of the upper molars of the *Rh. leptorhinus*, produced by the effects of mastication upon the valley, *b*, are presented by the molars of any of the recent Rhinoceroses, except the two-horned species of Sumatra. In this the valley, *b*, very closely resembles in its form and intruding promontory that in the upper molars of the leptorhine Rhinoceros; but the ridge on the outer side of the tooth, corresponding to that marked *d'* in fig. 141, is much more produced, and the adjoining convexity at the middle of the outer surface is flatter.

But to proceed with the comparison between the upper molars of the extinct tichorhine and leptorhine Rhinoceroses; the lateral valley, *a*, is wider and deeper at its commencement, and shallower at its termination in fig. 141 than in figs. 122 and 126; it is not so soon, therefore, worn down into a second island of enamel, like that shown in the molars of the tichorhine Rhinoceros figured by Cuvier, loc. cit., pl. xiii., figs. 1 and 6: the inner termination of the lobe, *c*, is broader and more bulging in the leptorhine Rhinoceros, the outer longitudinal ridge, *d'*, is more produced, and the anterior basal ridge, *f*, is longer and better developed. The small tubercle, *m*, is commonly, but not constantly present at the entry of the valley, *b*. I have never seen it in an upper molar of the tichorhine Rhinoceros.

Professor Jäger has figured an upper molar tooth from the opposite side of the jaw to that in fig. 141, in the Second Part of his 'Fossilen Sæuge-thiere Würtembergs,' fol., 1839, tab. xvi., fig. 31. It was discovered in a sand-pit ("sand-grube") at Kirchberg, in Wirtemberg, and exhibits about the same amount of attrition, the same characteristic form of the principal valley, the anterior basal ridge, the prominent longitudinal ridge (*d'*), and the expanded convex bases of the inner lobes, separated by the wide beginning of the valley, as in the Clacton leptorhine molar. Professor Jäger notices the latter character,* and the little tubercle (*m*) at the base of the valley, which is likewise present in our Clacton leptorhine molars;† but he does not allude to the more important character, which his figure represents, of the simple termination of the valley (*b*).

The zealous investigator of the Wirtemberg Fossils appears not to have perceived the specific resemblance between the molars from Kirchberg and that from Crozes (Gard), figured by Cuvier, tom. cit., pl. xiii., fig. 4. And, as Cuvier had not obtained evidence to connect these specimens with his *Rh. leptorhinus*, nor, indeed, appears to have appreciated their difference from the molars of the tichorhine Rhinoceros,‡ Professor Jäger had no clue to the

* Professor Jäger, after noticing the general resemblance of the fossil tooth with a corresponding one of the African two-horned Rhinoceros, observes, "allein er unterscheidet sich von ihm ausser der Grösse durch die mehrere Rundung und Trennung der innern Abtheilungen, wodurch er sich noch insbesondere von demselben Zahne von Cannstadt, tab. xvi. fig. 10, unterscheidet, so wie durch den kleinen höcker in der Mitte zwischen beiden. p. 180.

† This is more strongly developed in the molar teeth of the *Rhinoceros incisivus* (*Acerotherium*, Kaup). The Constadt tooth above cited is a molar of the *Rh. tichorhinus*, closely agreeing with that from Chartham, fig. 122.

‡ The molar tooth of the tichorhine Rhinoceros, figured in the 'Ossemens

discovery that the molars of the Rhinoceros from Kirchberg belonged to a distinct species which had already received its appropriate name; and he therefore proposes to denominate it "*Rhinoceros Kirchbergense*"* (sic, p. 179).

Dr. Kaup has given a reduced and reversed view of the same molar tooth in his 'Akten der Urwelt,' 8vo., 1841, taf. i., fig. 4; he equally appreciates the distinction of its structure from the corresponding molars of the *Rhinoceros tichorhinus*, and at the same time recognizes its specific identity with the molars from Crozes. The means of identifying it with the *Rh. leptorhinus* were equally wanting to the Palæontologist of Darmstadt, who, notwithstanding a name had been already attached to the species by Professor Jäger, proposes to call it *Rhinoceros Merckii*. The last molar tooth of the left side, which is retained in a portion of the upper jaw from the fresh-water deposits at Clacton, closely resembles the corresponding less worn molar of the right side from Kirchberg, figured by Professor Jäger in the work cited, pl. xvi., fig. 32, and, like it, differs from the corresponding tooth of the *Rh. ticho-*

Fossiles,' tom. ii. pt. 1. pl. vi. fig. 5, in which the enamel island is formed by the insulation of one lobe of the expanded termination of the valley (*b*), is thus described: "On y voit aussi très-bien la fossette, résultant de l'union du crochet postérieur avec la colline antérieure, et l'echancrure postérieure commence à être cernée."—P. 57. The molar tooth of the leptorhine Rhinoceros, figured in pl. xiii. fig. 4, in which the enamel island is due to the insulation of the entire unexpanded end of the valley (*b*), is thus described, "Le trou antérieure y est dejà distinct par l'union du crochet de la colline postérieure avec la colline antérieure, mais l'echancrure postérieure n'y est point encore cernée."—Ib. p. 58.

* The *nomina trivialia*, formed by latinizing German names of individuals or places, grate harshly upon the ear. One regrets the obligation to adopt such a name as *Schleiermacheri* in place of *megarhinus*, but the law of priority is absolute. With regard to names derived from particular localities, they are obnoxious to the graver objection of indicating very partially and imperfectly the geographical range of the extinct species to which they are applied.

rhinus in the relatively thicker and more bulging base of the inner and anterior lobe, in the more even and less undulating surface, which extends from the anterior external to the posterior internal angle of the crown, and in the absence of the infundibular cavity at the posterior angle of the crown.

The only portion of the vertebral column of a Rhinoceros discovered at Clacton was the os sacrum ; this bone, by the anchylosis of five vertebræ, and the broad, thick, rough plate of bone extending horizontally from the confluent ends of the spines of the first three vertebræ must have belonged, like the cranium, to a fully mature individual. It is of an almost equilateral triangular form, six inches nine lines across the base, and six inches in length; it differs from the os sacrum of the *Rh. Sumatranus* in the oblique truncation of the lower angles of the transverse processes of the fourth vertebra, and the less elongated form of the articular surfaces on the forepart of those of the first vertebra. I have not had the opportunity of comparing this sacrum with that of the *Rhinoceros tichorhinus;* but it most probably belongs to the same species as the other fossils from the fresh-water deposits at Clacton.

Cuvier, having obtained evidence that a fossil humerus of a *Rhinoceros*, discovered by Professor Nesti in the Val d'Arno, differed from the humerus of the tichorhine Rhinoceros by its longer and more slender proportions, by its longer and less prominent deltoid crest, and by some minor characters, suspected it to belong to the *Rh. leptorhinus.* The association with the unquestionable remains of that species in the fresh-water deposits at Clacton, of a considerable portion of a humerus of a Rhinoceros, participating in all the distinctive features of that from the

Val d'Arno, and closely agreeing with the figures given by Cuvier in the 'Ossemens Fossiles,' Rhinoceros, pl. x., figs. 1 and 2, confirms the accuracy of the reference of the Val d'Arno remains to the *Rhinoceros leptorhinus*. The humerus now before me, discovered by Mr. Brown at the same time and place with the leptorhine cranium, presents a most striking contrast with the proportions of the humerus of the tichorhine Rhinoceros before cited, from Lawford.

I subjoin the following comparative dimensions:

	Rh. leptorhinus.		*Rh. tichorhinus.*	
	In.	Lin.	In.	Lin.
Length, from the head to the beginning of the anconal depression	10	0	10	6
Length of the deltoidal crest	7	3	8	0
Circumference of the proximal end	19	0	26	0
Smallest circumference of the shaft	7	9	10	6
Breadth of the proximal end	7	0	9	6

In Mr. Brown's specimen the distal end is broken off.

An ulna, slightly mutilated, from the till at Walton, near Essex, in like manner agrees in its proportions with that from the Val d'Arno, figured by Cuvier in the plate cited, fig. 13.

The long and slender proportions of the femur of the Italian Rhinoceros are noticed in the 'Ossemens Fossiles;' the third trochanter is thrown more forward, and the great trochanter does not descend to join the third.

I have had no means of applying these characters to the identification of the leptorhine species as an English fossil; the only part of the femur found associated with the skull and teeth of the *Rh. leptorhinus* at Clacton being the distal extremity, on the characters of which the text is silent, and the reduced figures inexpressive in the 'Ossemens Fossiles.' This fragment having been kindly transmitted to me by Mr. Brown, together with the other specimens of *Rhinoceros leptorhinus* from Clacton, I have compared it with the corresponding part of the femur of a *Rhinoceros tichorhinus*, obtained from the drift near Moscow.

The first and most obvious distinction of the Clacton femur is the narrower, shallower, and more oblique surface of the shaft, immediately above the articular surface for the patella; the convex ridge continued upwards from the internal and more prominent boundary of that surface is broader, more rounded, and more gradually blended with the shaft of the femur; the whole surface exterior to this ridge slopes more suddenly to the outer side of the bone, and there is a much deeper excavation below the rotular articulation. In the femur of the tichorhine Rhinoceros, the transverse exceeds the antero-posterior diameter of the shaft six inches from the lower end; in that of the leptorhine species, these proportions are reversed at the same part of the shaft. The outer side of the femur behind the outer ridge is more concave in the Clacton specimen, which measures, from the fore to the back part of the external condyle, eight inches; it most probably belongs to the leptorhine species.

In Mr. Brown's collection there are specimens of upper molar teeth of the *Rhinoceros leptorhinus* from the till at Walton in Essex. One of these is the last molar, which had just come into use when the animal perished. Another specimen is a third upper molar, worn down to its base. The same Geologist also possesses the germ of the ante-penultimate molar of a *Rhinoceros leptorhinus* from Grays, in Essex, in which many smaller processes are sent off into the principal valley (*b*), in addition to the larger promontory. A similar modification of a superior molar tooth of the leptorhine Rhinoceros from Tuscany is noticed in the addition to the paragraph on that species in the 8vo. edition of the 'Ossemens Fossiles,' tom. iii., p. 138: I am not disposed, however, to place much stress upon this as a specific character.

Mr. Parkinson appears to have been the first to recognize remains of the Rhinoceros in the formations on the Essex-coast. He says: — "From several fragments of bones, which I met with in the Essex bank, I was led to suppose that the remains of some other very large animal, besides those of the Elephant and Elk, had been there imbedded."—'Organic Remains,' vol. iii. p. 371. The upper part of an os femoris, which differed from that of any animal with whose skeleton Mr. Parkinson was acquainted, induced him to be more particular in his research, and led to his discovery of the tooth of the Rhinoceros, which he has represented in Plate xxi. fig. 3. (op. cit. p. 372.) "This tooth," he proceeds to say, "is an upper molar of the left side, is pretty much worn, and must have belonged to a small animal, since it is not one half the size of the teeth which are found at Chartham." The figure shows all the essential characters of the upper molars of the *Rhinoceros leptorhinus*.

A part of a fossil lower jaw, discovered in the tertiary marine deposits of Monte Blancano, near Bologna, which had obtained notoriety through Professor Monti's description of it, in 1719, as part of the skull of a Morse, was not only proved by Cuvier to be part of a Rhinoceros, but the great Anatomist congratulated himself on being able to determine, by the prominent symphysis, that it had belonged to the *Rhinoceros tichorhinus*. "This discovery," he remarks, "is one of great importance, since it shows that the two species" (the tichorhine and leptorhine) "had inhabited Italy," op. cit. p. 143.

The identification of the fossil teeth respectively referred, in the works of Cuvier, Jäger, and Kaup, to the *Rh. tichorhinus*, *Kirchbergensis*, and *Merckii*, with the *Rh. leptorhinus*, demonstrates a further range of that species, which we now know to have been associated with *Rh. tichorhinus* in France, in Germany, and also, by the instructive specimens obtained by Mr. Brown, in our own island.

Mr. Fitch of Norwich possesses specimens of upper and lower molar teeth of the *Rh. leptorhinus* from the fresh-water (lignite) beds on the Norfolk coast near Cromer, which demonstrate the occurrence of this species in the same deposit with the *Rh. tichorhinus.*

I have not, hitherto, met with any specimens of the *Rhinoceros leptorhinus* from the ossiferous caves of England, nor does the species appear to have extended its range to Siberia, where the tichorhine Rhinoceros most abounded. In this country, as in Wirtemberg, Darmstadt, Central France, and Italy, the remains of the leptorhine Rhinoceros have been left in tranquil deposits of fresh-water lakes or rivers.

Mr. Brown informs me, that at Clacton these deposits line a basin of the London clay, upon which they immediately rest. The deepest part of the basin is twenty feet below the surface, and is covered by a stratum about six inches thick, of red sand, with marine and fresh-water shells; above this, by a deposit five feet thick of peaty matter, with interrupted beds containing marine and fresh-water shells: above this is another thin layer of red sand, with marine and fresh-water shells; then comes another bed of peaty matter four feet thick, overlaid by a thin bed of red sand, with fresh-water shells; and this is covered by a stratum of flinty gravel, four to five feet thick, which supports the superficial vegetable mould. The remains of the Rhinoceroses, with associated Mammoths and Aurochs, were discovered in the deepest part of the basin; but in the space of three hundred yards towards the north, it rises to the surface and is capped by the gravel. Mr. Brown, in concluding his account of the ancient lacustrine basin, which formed the grave of the huge pachyderms and ruminants that once roamed upon its banks, or wallowed in its muddy shallows, says, "As the bones and teeth which I have now much pleasure in sending you, were all collected by myself, I can vouch for their being marked correctly, as to locality."

The habits of the less robust and less formidably armed species no doubt differed from those of the tichorhine Rhinoceros, which is more extensively distributed over England; some Naturalists have recognized different habits in the three or four species of Rhinoceros now living in Africa, and which differ from each other in form and structure much less than did the extinct leptorhine and tichorhine Rhinoceroses of Europe.

Although the number of species, now extinct, which ranged over the Europæo-Asiatic continent equalled or surpassed that of the existing species of *Rhinoceros*, no fossil remains referable to this genus have ever been discovered in America or Australia. This peculiar form of horned Pachyderm appears to have been confined, from its first introduction into our planet, to the same great natural division of the dry land—the Old World of the geographers—to which the existing representatives of that form are still peculiar.

PACHYDERMATA. *SOLIPEDIA.*

Fig. 142.

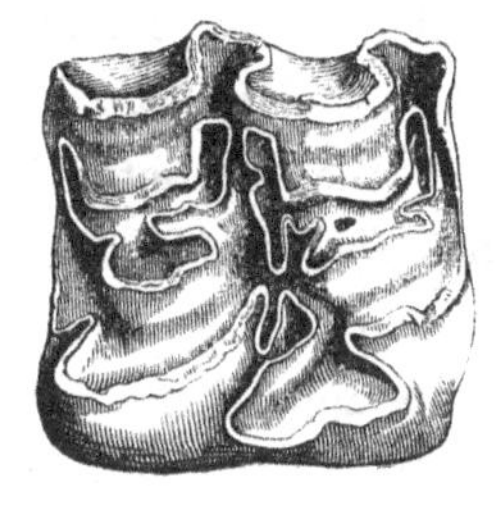

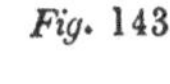

Fig. 143

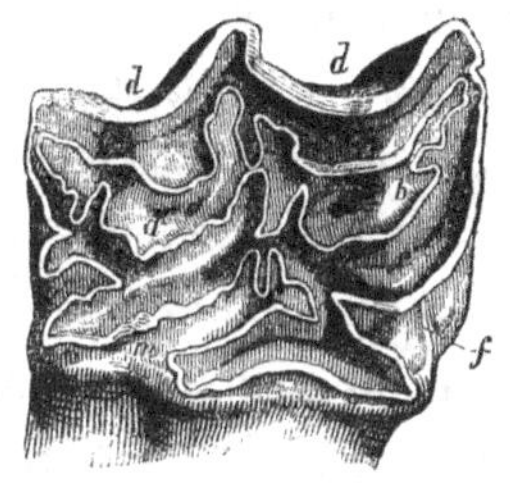

Nat. size.

3rd upper molar, recent Horse. 3rd upper molar, *Equus fossilis*. Kent's Hole.

FOSSIL HORSE. *Equus fossilis.*

Cheval fossile, Cuvier, Ossemens Fossiles, 4to, 1822, tom. ii. pt. i. p. 109.
Equus fossilis, V. Meyer, Palæologica, 8vo. 1832, p. 79.

In England, as on the Continent, remains of the genus *Equus* attest that a species equalling a middle-sized Horse, and one of the size of a large Ass, or Zebra, have been the associates of the Mammoth, Rhinoceros, and other extinct quadrupeds, whose remains are so generally dispersed in the drift formations, bone-caves, and the post-pliocene tertiary deposits. Almost every geological author, who has had occasion to notice the Mammalian fossils of these recent periods, has made mention of such a combination. It has been observed by Dr. Mantell* in the "Elephant-bed" at Brighton; by Mr. Clift† in the cavernous fissures at Oreston; by Dr. Buckland‡ in the ossiferous caves at Kirkdale, in the Mendips, and at

* 'Fossils of the South Downs,' 4to., 1822, p. 283.
† 'Phil. Trans.' 1823, p. 86.
‡ 'Reliquiæ Diluvianæ,' pp. 18, 75.

Paviland; by Mr. Lyell * in the tertiary deposits on the Norfolk coast; by Col. Hamilton Smith † in the bone-caves near Torquay; and by Mr. Morris ‡ in the Mammaliferous deposits in the valley of the Thames, as at Wickham, Ilford, Erith, Grays, and Kingsland. Cuvier records many instances of the like association of a Horse with the undoubted extinct species of Mammals in the corresponding formations on the continent.

No critical anatomical comparison appears hitherto to have been instituted with regard to the relations of these British equine fossils with the existing species. That the fossils vary in size amongst themselves has been more than once noticed; and Dr. Buckland makes a remark § expressive of his suspicion that they belonged to more than one species.

The largest-sized fossil *Equus* from British strata is indicated by molar teeth, obtained by Mr. Lyell from a bed of laminated blue clay, with pyrites, eight feet thick, overlying the Norwich crag at Cromer, where they are associated with remains of the *Mammoth*, *Hippopotamus*, *Rhinoceros*, *Bos*, *Cervus*, and *Trogontherium*. The antero-posterior diameter of one of these teeth, the second in the lower jaw, was one inch four-tenths, equalling that of the largest dray-horse of the present day: other corresponding fossil teeth of *Equus* have measured in the same diameter, some one inch two-tenths, and some one inch. The intermediate size, which equals that of the teeth of a horse of between fourteen and fifteen hands high, is the

* 'Phil. Mag.' vol. xvi. (1840.) pp. 349, 362.

† 'Naturalist's Library,' Horses, p. 63.

‡ 'Mag. of Nat. History,' 1838, p. 539.

§ Loc. cit. p. 75, with respect to the equine remains discovered in the Oreston caverns:—"Horses about twelve, of different ages and sizes, as if from more than one species."

most common one presented by fossils. Several of the equine molar teeth from Kent's Hole, Torquay, indicate a horse as large as that from the blue clay at Cromer; but the size of the fossil species would be incorrectly estimated from the size of the teeth alone. Although the equine fossils are far from rare, yet they have hitherto in England been always found more or less dispersed or insulated, and no opportunity has occurred of ascertaining the proportions of one and the same individual by the comparison of an entire skeleton with that of the existing species of *Equus*.

The best-authenticated associations of bones of the extremities with jaws and teeth, clearly indicate that the fossil Horse had a larger head than the domesticated races; resembling in this respect the Wild Horses of Asia described by Pallas,* and in the same degree approximating the Zebrine and Asinine groups.

It is well known that Cuvier† failed to detect any characters in the bones or teeth of the different existing species of *Equus*, or in the fossil remains of the same genus, by which he could distinguish them, except by their difference of size, which yields but a vague and unsatisfactory approximation. MM. H. v. Meyer and Dr. Kaup have, however, pointed out well-marked distinctive characters in the fossil *Equidæ* of the older pliocene and miocene tertiary deposits of the Continent.

The second and third molars of both jaws in most of the equine fossil specimens of the teeth from our more recent deposits and caverns which I have examined, are narrower transversely in comparison with their antero-posterior diameter than in the existing Horse; and a

* 'Zoographia Rosso-Asiatica,' tom. i. p. 255.

† 'Ossemens Fossiles,' 4to. tom. ii. pt. i. p. 111.

similar character appears to have been recognized by M. H. v. Meyer in some of the fossil equine teeth from the Eppelsheim sand, since he cites the *Equus angustidens* as a synonyme of the species which he subsequently described under the name of *Equus asinus primigenius*.*

The upper molar teeth of the Horse resemble those of the Palæotherium in the two deep longitudinal channels (*d d*, fig. 143,) which traverse their outer side, but the enamel-linings of those channels are not produced into points on the grinding surface. This surface of the equine molar also presents a close analogy with that of the Rhinoceros; to aid in tracing which, the corresponding but modified folds and islands of enamel in the complex molar of the Horse (fig. 143) are marked with letters corresponding with those on the upper molar of the *Rhinoceros leptorhinus* (fig. 141). For the details of the characteristic structure of the teeth of the genus *Equus*, which would be unsuited to the present work, I must refer the reader to the 'Ossemens Fossiles,' and to my 'Odontography'; and here merely add, that the character by which the Horse's molars may best be distinguished from the teeth of other Herbivora, corresponding with them in size, is the great length of the tooth before it divides into fangs. This division, indeed, does not begin to take place until much of the crown has been worn away; and thus, except in old Horses, a considerable proportion of the whole of the molar is implanted in the socket by an undivided base. In an old molar with roots, the pattern of the grinding surface, as it is shown in the figs. 142 and 143, is a little changed by partial obliteration of the enamel folds, but enough generally remains to serve, with the form of the tooth, to distinguish it from the rooted molar of a Ruminant.

* 'Palæologica,' p. 80.

Figure 143 shows the grinding surface of the third molar, right side of the upper jaw, in a fossil from the cave of Kent's Hole, Torquay. It presents the same fossilised condition as the bones and teeth of the extinct Rhinoceros and the great *Carnivora* from the same depository. The upper molars of the Horse are slightly curved; and a fossil species, contemporary with the Megatherium in South America, differs from the existing Horse by the greater degree of that curvature: but there is no such difference in the present fossil, which is of equal length with a large Horse's tooth compared, viz., three inches and a quarter; neither is there any modification of the pattern of the enamel folds on the grinding surface deserving to be regarded as specific. This degree of difference is indicated only by the smaller transverse as compared with the antero-posterior diameter; and the same difference of proportion, as compared with the teeth of the common existing Horse, is shown in the figure of the upper molar from the cave at Kirkdale, in the 'Reliquiæ Diluvianæ,' pl. vii., fig. 7. In general, I have found that

Fig. 144.

3rd lower molar, recent horse, Nat. size.

Fig. 145.

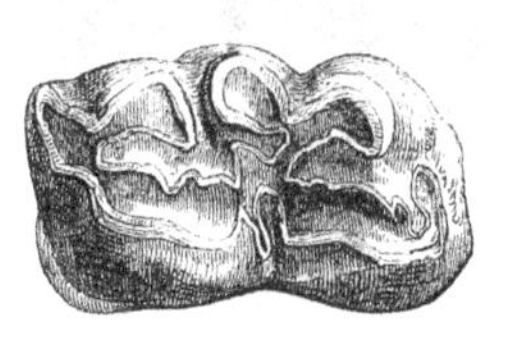

3rd lower molar, *Equus fossilis*, Oreston, Nat. size.

the lower molar teeth of the fossil *Equus* present the same difference in their narrower transverse diameter: this character is shown in the cut of the grinding surface of one of these teeth (fig. 145) from the cavernous fissure, A, at Oreston (fig. 50, p. 132); and it is illustrated by contrast with the same view of the corresponding lower molar of a common Horse of about fourteen hands high (fig. 144). Some of the numerous fossil equine teeth of large size, from the cave at Kent's Hole, do not manifest this character; but the large-sized molar teeth of the Horse, from the newer pliocene blue-clay at Cromer,* are as much narrower transversely, compared with the teeth of the large varieties of the existing Horse, as are the somewhat smaller molars from Kent's Hole, Kirkdale, and Oreston. One of the Cromer fossil teeth, from the lower jaw, with a grinding surface measuring one inch five lines in long (antero posterior) diameter, and eight lines in short (transverse) diameter, presented a swelling of one lobe, near the base of the implanted part of the tooth (fig. 146).† To ascertain the nature and cause of this enlargement, I divided it transversely, and exposed a nearly spherical cavity, large enough to contain a pistol-ball, with a smooth

Fig. 146.

Diseased lower molar, *Equus fossilis*, Cromer.

* Lyell 'On the Boulder Formation and Fresh-water Deposits of Eastern Norfolk,' Philosophical Magazine, 1840, p. 361.

† I am induced to cite one of the curious examples of disease in an extinct animal from the rarity of its occurrence in the tissue which is the subject of it.

inner surface (fig. 147, *c*). The parietes of this cavity, composed of dentine and enamel of the natural structure, were from one to two lines and a half thick, and were entire and imperforate. The water percolating the stratum in which this tooth had lain, had found access to the cavity through the porous texture of its walls, and had deposited on its interior a thin ferruginous crust, but the cavity had evidently been the result of some inflammatory and ulcerative process in the original formative pulp of the tooth, very analogous to the disease called 'spina ventosa' in bone. The incisors of the Horse are distinguishable from those of the Ruminants by their greater curvature, and from those of all other animals by the fold of enamel, which penetrates the body of the crown from its summit, like the inverted finger of a glove. When the tooth begins to be worn, the fold forms an island of enamel, inclosing a cavity partly filled by cement, and partly by the discoloured substances of the food, and is called the 'mark.' In aged Horses the incisors are worn down below the extent of the fold, and the 'mark' disappears. In the incisor tooth (fig. 148) from drift gravel, overlying the chalk at Hessle, near Hull, the mark (*m*) still remains, showing that the tooth had belonged to a Horse not aged.

Fig. 147.

Section of diseased lower molar. *Equus fossilis.* Cromer.

Fig. 148.

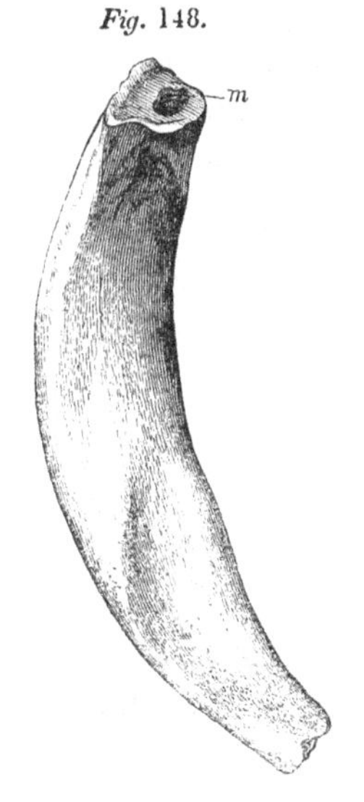

Incisor of fossil Horse, Drift, Hessle. Nat. size.

Cuvier has given most excellent figures of the principal bones of the existing Horse, in the volume cited of the 'Ossemens Fossiles,' plates i. and iii. Amongst the most recognisable ones are the astragalus (fig. 153), and the last phalanx, which is enveloped by the hoof, called by farriers the 'coffin-bone' (fig. 154). The thigh-bone is distinguished from that of any Ruminant of the same size by the flattened process from the outer side of the shaft below the great trochanter; and the Horse thereby manifests its affinity to the Rhinoceros and Palæothere.

Mr. Fitch of Norwich has a lower molar tooth of a Horse three and a half inches in length, and a metacarpal or cannon bone ten inches long, with one of the splint-bones anchylosed to it; both are from Pliocene deposits in Norfolk.

The largest bone of an extremity of a fossil Horse which I have seen, is a second phalanx from the upper pliocene deposits at Walton-on-Naze, Essex, where it was discovered by Mr. Brown of Stanway; it measures two inches eight lines in extreme breadth, and two inches four lines in length. The corresponding bones from Oreston are smaller. Mr. Brown has also found remains of a Horse associated with those of the *Rhinoceros leptorhinus*, *Elephas* and *Urus*, in the fresh-water deposits at Clacton, in Essex. Remains of the *Equus fossilis* have been discovered, similarly associated with larger extinct Pachyderms, in the pliocene formations at Audley End, by the Hon. R. C. Neville. The wide distribution of the fossil Horse over the surface of this island, in the pliocene and later deposits, is indicated by the citations at the commencement of this section.

I have been favoured with the following notes of the discovery of fossil teeth of a species of *Equus* in Ireland, by

John Thompson, Esq., of Belfast. In sinking a well near Downpatrick, in the county of Down, two teeth were found in a stratum of gravel far below the present surface. A tooth was found at Newry under similar circumstances. In the county of Antrim teeth of the Horse have been found four feet below the surface in drift gravel near Belfast, and at the bottom of a turf-bog near Broughshane.

The more common species of fossil Horse from the drift formations and ossiferous caverns, which differs from the existing domestic Horse in its larger proportional head and jaws, resembling in that respect the Wild Horse, but apparently differing in the transversely narrower form of certain molar teeth, may continue to be conveniently indicated by the name of *Equus fossilis*, the Latin synonyme of the 'cheval fossile' of Cuvier.

Fig. 149. *Fig.* 150.

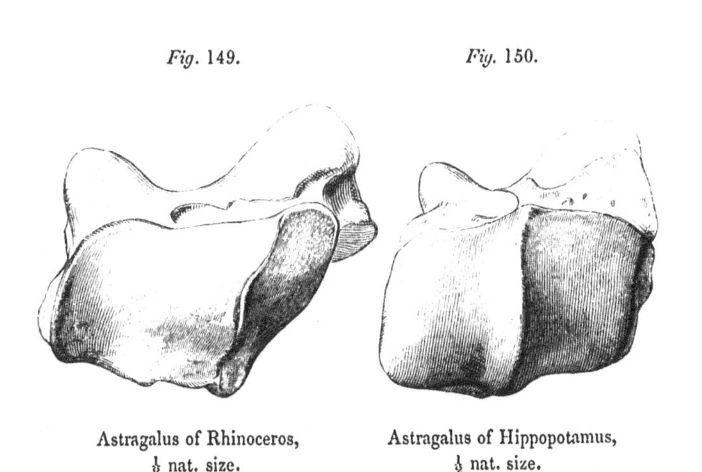

Astragalus of Rhinoceros, ½ nat. size. Astragalus of Hippopotamus, ½ nat. size.

PACHYDERMATA. *SOLIPEDIA.*

Fig. 151.

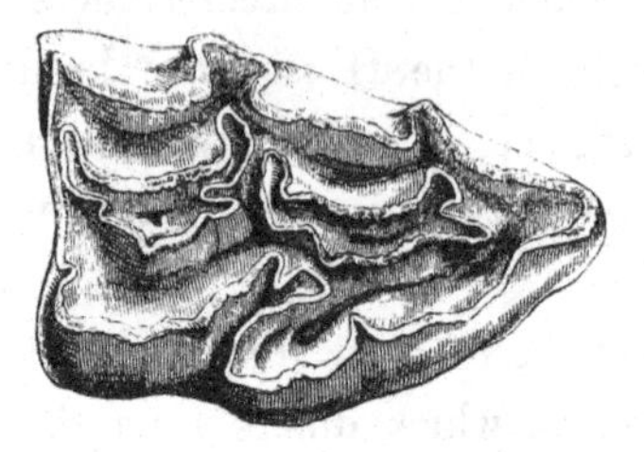

1st upper molar, nat. size, *Equus caballus.*

Fig. 152.

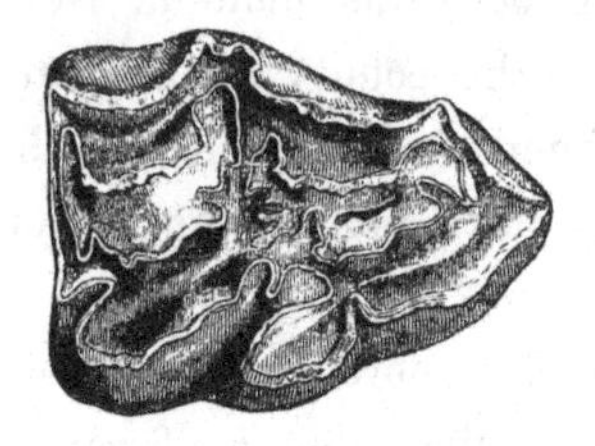

1st upper molar, nat. size, *Equus plicidens*, Oreston.

FOSSIL HORSE, WITH THE ENAMEL-FOLDS OF THE MOLAR TEETH PLICATED. (Equus plicidens.)

Amongst the numerous teeth of a species of *Equus*, as large as a horse fourteen hands and a half high, collected from the Oreston cavernous fissures, I have found specimens clearly indicating two distinct species, so far as specific differences may be founded on well-marked modifications of the teeth.

One of these, like the ordinary *Equus fossilis* of the drift and pleistocene formations, differs from the existing *Equus caballus* by the minor transverse diameter of the molar teeth, and is noticed in the preceding section; the other, in the more complex and elegant plication of the enamel, and in the bilobed posterior termination of the grinding surface of the last upper molar, more closely approximates to the extinct Horse of the miocene period, which M. H. v. Meyer has characterised under the name of the *Equus caballus primigenius.** The Oreston remains differ, how-

* 'Nova Acta Acad. Nat. Curios.' tom. xvi. p. 448.

ever, from this in the form of the fifth or internal prism of dentine (*m*) in the upper molars, and in its continuation with the anterior lobe of the tooth; the fifth prism being oval and insulated in the *Equus primigenius* of v. Meyer.

The Oreston fossil teeth, which in their principal characters manifest so close a relationship with the miocene *Equus primigenius*, differ, like the later drift species (*Eq. fossilis*), from the recent Horse in a greater proportional antero-posterior diameter of the crown of the second upper molar, and also in a less produced anterior angle of the first molar, as shown by the tooth figured in cut 152, as contrasted with the corresponding one of the recent Horse (fig. 151.)

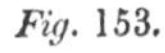

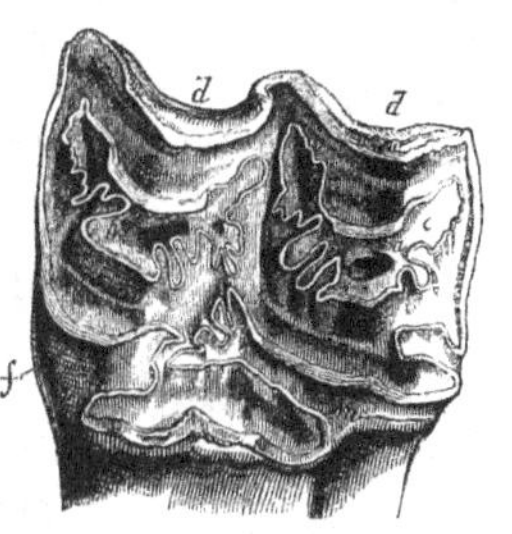

2nd molar, upper jaw, nat. size, *Equus plicidens*, Oreston.

Fig. 153 illustrates the character, above adverted to, of the complex plication of the enamel, as it appears on the grinding surface of a partially worn upper molar tooth, the second of the right side: the length of this tooth is three inches four lines, and the fangs had not begun to be formed. One cannot view the elegant foldings of the enamel in the present fossil teeth, and in those of the more ancient primigenial species (*Hippotheria*) of the continental miocene deposits, without being reminded of the peculiar character of the enamel of the molar teeth of the Elasmotherium, in which it is folded in elegant festoons. This extinct pachyderm, which surpassed the Rhinoceros in size, resembled that genus very closely in the general disposition of the folds of enamel in the grinding teeth, but agreed with the genus *Equus* in the deep implantation of those teeth by an undivided base. The Elasmothere appears, therefore, to have formed one of the links, now lost, which connected the Horse with the Rhinoceros, and it is interesting to observe that some of the extinct species of Horse, in the analogous complexity of the enamel folds, more closely resembled the Elasmothere than do the present species.

Fig. 154.

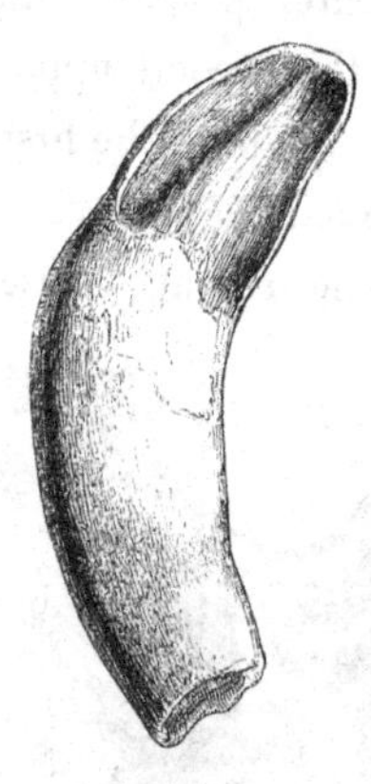

Lower canine tooth, nat. size, *Equus plicidens*, Oreston.

The canines are small in the Horse, and rudimental in the Mare. I figure here the fossil right lower canine of a colt, found in the same cavernous fissure (B) as the plicident molars, and probably, therefore, belonging to the same species: the view of the inner side, given in fig. 154, shows the folding in of the anterior and posterior margins of the crown, characteristic of the canines of the genus *Equus*, and which is very well marked in the present specimen. The incisors associated with the plicident molars offered no distinctive characters.

Some of the bones of the extremities of the fossil Horse from the same fissure (B) of the Oreston Caves, indicate an animal about thirteen hands and a half high. The astragalus, reduced in fig. 155, one third the natural size, is a very characteristic bone of the present genus; the upper articular surface, which is here represented, is oblique, and the two convex ridges are divided by an unusually deep, almost angular, valley; the articular pulley, or trochlea, in the lower end of the tibia has, of course, a corresponding form —the cavities and eminences being reversed; by the depth and obliquity of these, the tibia and astragalus of the Horse may readily be distinguished from those bones in any other quadruped of similar size. The last phalanx, or hoof-bone, is an equally characteristic bone ; a reduced view of the upper and anterior surface of one of these, obtained with

Fig. 155.

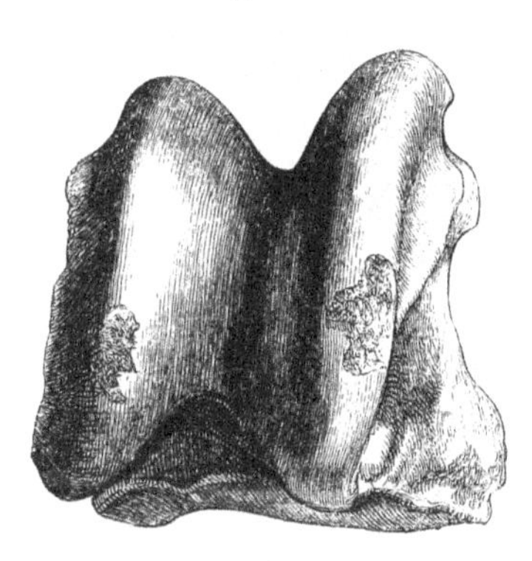

Astragalus of fossil Horse, ⅓ nat. size, Oreston.

Fig. 156.

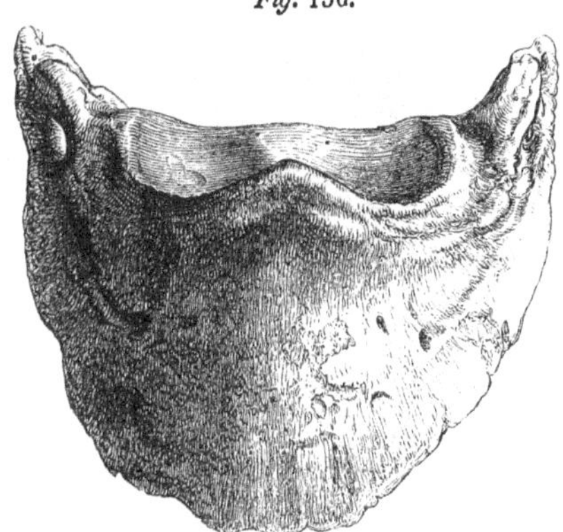

Hoof-phalanx of fossil Horse, ⅓ nat. size, Oreston.

the other bones of the hind-foot from the same fissure at Oreston, is given in fig. 156

The contemporary species associated with the *Equus fossilis* in the Oreston Caverns, but indicated to be distinct by the structure of the molar teeth above described, I have called, in my 'Report on British Fossil Mammalia,'* *Equus plicidens*, on account of the characteristic plications of the enamel. I have not yet seen any teeth from British strata having the well marked characters of those of the *Hippotherium* of Dr. Kaup (*Equus caballus primigenius* of M. H. v. Meyer) ; but the teeth of the extinct slender-legged Horse, or Hippothere, transmitted by Capt. Cautley to the British Museum, are identical with those of the above species from the European miocene.

* 'Trans. Brit. Association,' 1843, p. 231.

Fig. 157.

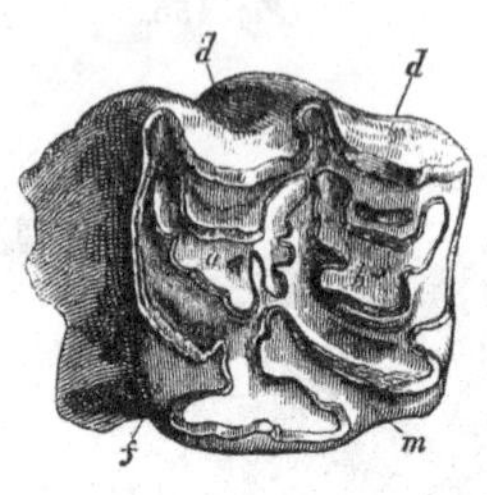

Upper molar, nat. size, *Asinus fossilis*, Oreston.

Fig. 158.

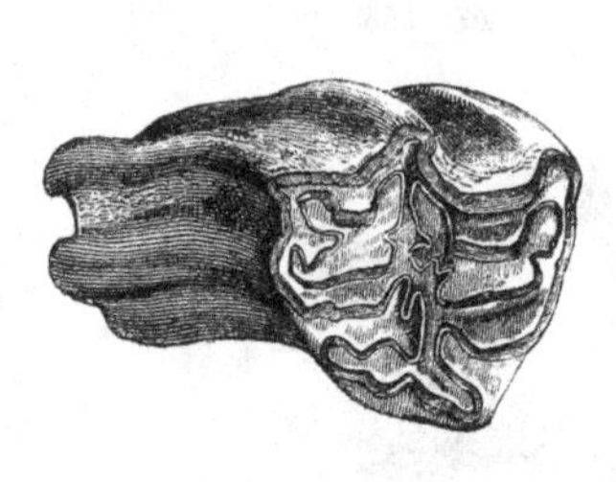

Last upper molar, nat. size, *Asinus fossilis* Oreston.

FOSSIL ASS, OR ZEBRA. Asinus Fossilis.

IN the more recent or diluvial formations a fossil species of *Equus*, smaller than either of the preceding and about the size of the Wild Ass, is indicated by molar teeth. Of these I have examined a middle molar of the left side of the upper jaw, from the drift overlying the London clay at Chatham; a corresponding molar from the opposite side of the upper jaw (fig. 157), from the drift at Kessingland in Suffolk; the last upper molar (fig. 158), from the same deposit and locality; and a fifth molar, left side of lower jaw, from a cavernous fissure at Oreston: all these teeth were in the same fossilized condition as the associated remains of extinct Mammals with which they had clearly been contemporaneous.

In the collection of Miss Gurney of Northrepps Cottage, near Cromer, I saw a fossil second phalanx, or pastern bone, of a small species of *Equus*, about the size of the Zebra, from the pliocene crag at Thorpe. Dr. Mantell states that teeth and bones of an *Equus*, from the supercretaceous drift deposits, which, on account of the abundant

Mammoth's remains, he has called the "Elephant Bed," on the Brighton cliffs, "are referable to a small species, about the size of a Shetland Pony."* If we admit the subgeneric separation of those species of the genus *Equus*, Cuv., that have callosities on the fore-legs only, the tail furnished with a terminal brush of long hair, and a longitudinal dorsal line, the last-indicated fossil species may be named *Asinus fossilis*.

Several bones of a large Ass have been found with remains of the Beaver and the Wild Boar in the marl beneath the peat-formation at Newbury, Berks.

In reviewing the general position and distribution of the fossil remains of the genus *Equus*, we find that this very remarkable and most useful form of Pachyderm made its first appearance with the Rhinoceros during the miocene tertiary periods of geology.

From the peculiar and well-marked specific distinction of the primogenial or slender-legged Horses (*Hippotherium*), which ranged from central Europe to the then rising chain of the Himalayan mountains, it is most probable that they would have been as little available for the service of civilized man as is the Zebra or the Wild Ass (*Equus hemionus*) of the present day; and we can as little infer the docility of the later or pliocene species, *Equus plicidens* and *Equus fossilis*, the only ones hitherto detected in Britain, from any characters deducible from their known fossil remains.

There are many specimens, however, that cannot be satisfactorily distinguished from the corresponding parts of the existing species, *Equus caballus*, which, with the Wild Ass, may be the sole existing survivors of the numerous representatives of the genus *Equus* in the Europæo-

* 'Medals of Creaton,' 1844, vol. ii. p. 40.

Asiatic continent during the pliocene period. The species of *Equus* which existed during that geological period in both North and South America, appears to have been blotted out of the Fauna of those continents before the introduction of Man. The aborigines whom the Spanish Conquestadors found in possession of Peru and Mexico, had no tradition or hieroglyphic indicative of such a quadruped, and the Horses that the invaders had imported from Europe were viewed with astonishment and alarm.

The researches of Mr. Darwin and Dr. Lund have, however, indisputably proved that the genus *Equus* was represented in South America during the pliocene period by a species (*Equus curvidens*) which I have shown to be distinct* both from the European fossils and the existing species. Fossil remains of the Horse have likewise been discovered in North America. The geographical range of the genus *Equus* at the pliocene period was thus more extensive than that of *Rhinoceros*, of which both the extinct and existing species are confined to the continents of the Old World of geography. The Horse, in its ancient distribution over both hemispheres of the globe, resembled the *Mastodon*, and appears to have become extinct in North America at the same time with the *Mastodon giganteus*, and in South America with the *Mastodon Andium* and the Megatherium. Well may Mr. Darwin say, "It is a marvellous event in the history of animals, that a native kind should have disappeared, to be succeeded in after ages by the countless herds introduced with the Spanish colonist!"†

* 'Catalogue of Fossil Mammalia in the Museum of the Royal College of Surgeons,' 4to., 1844, p. 235.

† 'Voyages of the Adventure and Beagle,' vol. iii. p. 150.

PACHYDERMATA. *HIPPOPOTAMUS.*

Fig. 159.

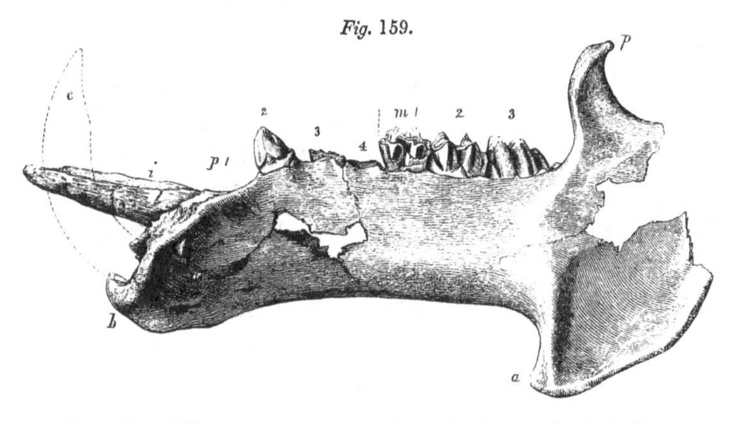

Lower jaw of Hippotamus major, $\frac{1}{7}$ nat. size. Fresh-water clay-beds, Cromer, Norfolk.

LARGE FOSSIL HIPPOPOTAMUS. Hippopotamus Major.

Grand Hippopotame Fossile,	CUVIER, Ossemens Fossiles, 4to., 1821, tom. i. p. 310, pl. i.—vi.
Hippopotamus,	PARKINSON, Organic Remains, 4to., 1811, vol. iii. p. 374, pl. xxi. fig. 1.
"	TRIMMER, Philosophical Transactions, 1813, p. 131, pl. ix. figs. 1 and 3,* pl. x.
"	BUCKLAND, Reliquiæ Diluvianæ, pp. 18, 42, 176, pl. vii. figs. 8, 9, 10.
Hippopotamus major,	OWEN, Report of British Association, 1843, p. 223.

IN glancing retrospectively towards the dawn of the scientific investigation of Fossil Remains, one is struck with the early introduction of the idea that the Hippopotamus had contributed to those found in the temperate latitudes of Europe: this amphibious quadruped seems, in fact,

* Fig. 2. pl. ix. is the upper molar of a Rhinoceros, but I am unable, from the position in which it is figured, to determine the species.

to have been the first to which large fossil bones and teeth were referred, after the notion that they were the relics of giants of the human species began to be exploded.

Thus the learned Saxon scholar, Somner, acquaints us that some who had seen the Chartham fossils were of opinion that they were bones of a River-horse;* and the antiquarian Douglas misinterpreted in like manner the jaw and teeth of a Rhinoceros, much of the ingenious speculations in his 'Dissertation on the Antiquity of the Earth' being based on the assumption that the fluviatile deposits at Chatham, in the instance which he describes, had yielded "hippopotamic remains." "When we consider," he says, "the great distance of the Medway from the Nile, or other rivers near the tropics, where these kinds of animals are now known to inhabit, and when we have no authority from the Pentateuch to conclude that any extraordinary convulsion of nature had impelled animals at that period from their native regions to countries so remote, so we have no natural inference for concluding that the deluge was the cause of this phenomenon." Taking into consideration the geological features of the stratum of the river soil, he concludes "that as the Hippopotamus is known to be the inhabitant of muddy rivers like those of the Nile and the Medway, it should therefore argue that this animal was the inhabitant of those regions, when in a state of climature to have admitted of its existence."†

This conclusion is essentially correct, though based in the present instance on wrong premises; neither the or-

* Ante, p. 326.

† 'A Dissertation on the Antiquity of the Earth,' by the Rev. James Douglas, 4to., 1785, pp. 9, 11.

ganic remains from Chatham, any more than those from Chartham, having appertained to a "river or sea bred creature." The genus of land-quadrupeds, to which these fossils actually belonged, is nevertheless, at the present day, as much confined to the tropics as is the Hippopotamus.

No long time elapsed before true Hippopotamic remains were discovered in the same deposits which had yielded the bones and teeth of Rhinoceroses. It was most probably from fresh-water marl that the entire skull of the Hippopotamus was obtained, which is stated in Lee's 'Natural History of Lancashire' to have been found in that county under a peat-bog, and from which work Dr. Buckland has copied the figure given in plate xxii., fig. 5 of the 'Reliquiæ Diluvianæ.' From the indication of the second premolar in this figure we may, I think, discern the greater separation of that tooth from the third premolar, which forms one of the marks of distinction between the fossil and recent Hippopotamus.

Mr. Parkinson, in the third volume of his 'Organic Remains,' 4to., 1811, p. 375, treating of the Hippopotamus, says, "In my visits to Walton, in Essex, I have been successful in obtaining some remains of this animal." These fossils are now in the Museum of the Royal College of Surgeons, and are referable to the extinct species subsequently determined by Cuvier in the second edition of the 'Ossemens Fossiles,' under the name of *Hippopotamus major*. The first specimen, cited by Mr. Parkinson as "an incisor of the right side of the lower jaw," is the great median incisor, which, when entire, must have been eighteen inches in length. It has lost much of its original animal matter, and is considerably decomposed. This tooth may be distinguished from the straight

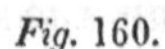

Fig. 160.

Structure of ivory of Hippopotamus tusk in transverse section.

inferior tusk of the Mastodon by its partial investment of enamel, —or when this is lost, as in the decayed specimen from the till at Walton, by the fine concentric lines on the fractured surface of the ivory (fig. 160), the corresponding surface in the tusk of Mastodon presenting the decussating curvilinear striæ as shown in fig. 101 *c*. The second specimen from Walton is thus described by Mr. Parkinson :—" The point of an inferior canine tooth or tusk, measuring full nine inches in circumference, and having seven inches in length of triturating surface (fig. 161). From the great size of this tooth, it is very likely to have belonged to the same animal to which the preceding tooth belonged. Besides the longitudinal striæ and grooves observable in the enamel of its sides and inferior part, it is characterised by strong transverse rugous markings, which are placed at nearly regular distances of about two inches, and are observed to exist in the same manner on the fragment which joins to it."*

Fig. 161.

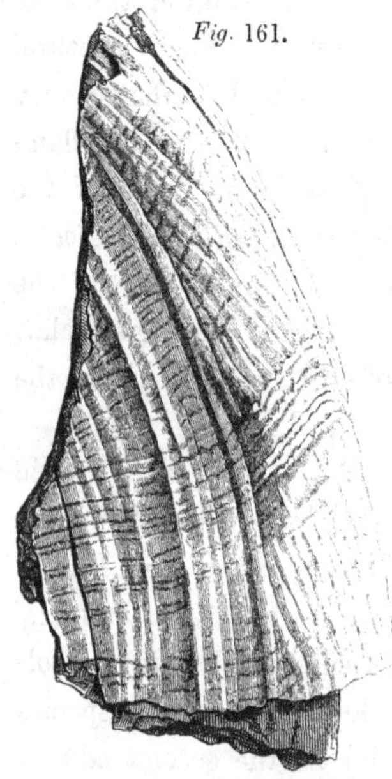

End of lower tusk of Fossil Hippopotamus, ½ nat. size. Walton.

* 'Organic Remains,' vol. iii., p. 375. In pl. xxi., fig. 1., there is a good figure of the penultimate molar of the lower jaw of the fossil Hippopotamus from Walton.

The third specimen described in that work is a fragment of the left lower canine tusk of a young Hippopotamus; it had scarcely come into use, and the pulp-cavity extends to near the apex of the conical and unworn crown. From the absence of the transverse rugous markings in the enamel, and the roundness of the circumference of this first-formed portion of the tusk, Mr. Parkinson was induced to suspect that it might have belonged to the small Hippopotamus;* but similar modifications are observable in the recently protruded tusk of the young African Hippopotamus, and are doubtless due to the immaturity of the individual of the fossil species which yielded this small tusk.

Mr. Parkinson says, "Remains of the Hippopotamus have been found, I am informed, in some parts of Gloucestershire:"† and prior to the publication of the third volume of the 'Organic Remains,' Sir Everard Home had deposited in the Museum of the College of Surgeons a tooth — the third premolar, right side, upper jaw — of the *Hippopotamus major*, Cuv., which had been dug up in a field called Burfield, in the parish of Leigh, five miles west of Worcester. Mr. Strickland's valuable observations‡ on the fluviatile deposits in the valley of the Avon, have confirmed these indications of the remains of the Hippopotamus in that locality, and have thrown much light on the conditions under which the extinct species of that now tropical genus of Pachyderm formerly existed in the ancient waters that deposited those sands.

* *Hippopotamus minor*, a small extinct species determined by Cuvier in the first edition of the 'Ossemens Fossiles,' but of which I have not yet met with any authentic remains from British strata.

† Op. Cit. p. 375. ‡ 'Proceedings of the Geological Society,' vol. ii. p. 111.

Mr. Parkinson lastly cites the remarkable discovery by Mr. Trimmer of the remains of the Hippopotamus in the fresh-water deposits at Brentford, an account of which Mr. Trimmer afterwards communicated to the Royal Society,* with excellent figures of the principal fossils of the Hippopotamus, and of those of the Mammoth, Rhinoceros, and large Deer therewith associated. These specimens were collected in two brick-fields; the first about half a mile north of the Thames at Kew Bridge, and with its surface about twenty-five feet above that river at low water. The strata here are,—first, sandy loam, from six to seven feet, the lowest two feet slightly calcareous; this yields no organic remains. Second, sandy gravel a few inches thick, with fluviatile shells and a few bones of land animals. Third, loam, slightly calcareous, from one to five feet; between this and the next stratum peat frequently intervenes in small patches of only a few yards wide and a few inches thick: here bones and horns of Ox and Deer occur, with fresh-water shells. Fourth, gravel containing water; this stratum varies from two to ten feet in thickness, and is always deepest in the places covered by peat: in it were found the remains of the Mammoth, teeth of the Hippopotamus, and horns and teeth of the Aurochs. This stratum, like the fresh-water deposits at Clacton with similar Mammalian fossils, rests upon the eocene London clay, the fossils of which, with a few exceptions are, as Mr. Trimmer correctly observes, "entirely marine." The first stratum in the second brick-field is a sandy loam, calcareous at its lower part, eight or nine feet thick, in which no organic remains were observed. In the second stratum, consisting of sand, becoming coarser towards the lowest part, and ending in

* 'Philosophical Transactions,' 1813, p. 131.

sandy gravel from three to eight feet, were found, "always within two feet of the third stratum, the teeth and bones of the Hippopotamus, the teeth and bones of the Elephant, the horns, bones, and teeth of several species of Deer and Ox, and the shells of river fish. The remains of Hippopotami are so extremely abundant, that, in turning over an area of one hundred and twenty yards in the present season," (1812) "parts of six tusks have been found of this animal." (Op. cit. p. 135.) Mr. Trimmer adds, that "the gravel-stones in this stratum do not appear to have been rounded in the usual way by attrition, and that the bones must have been deposited after the flesh was off, because in no instance have two bones been found together which were joined in the living animal; and further, that the bones are not in the least worn, as must have been the case had they been exposed to the wash of a sea-beach." (Ib. p. 136.)

When the flesh and ligaments of a dead Hippopotamus, decomposing in an African river, have been dissolved and washed from its bones, these will become detached from one another, and may be separately imbedded in the sedimentary deposits at the bottom without becoming much waterworn in their course previous to entombment. Although, therefore, the bones of the Brentford Hippopotamus were imbedded after the flesh was off, the individual to which they belonged might not have been transported from any great distance, the phenomena being perfectly in accordance with the fact that the animal had lived and died in the stream with the fresh-water mollusks, the shells of which characterize the sedimentary deposit in which its bones were subsequently buried. All the well-observed phenomena attending the discovery of Hippopotamic remains have concurred in establishing the truth of the conjecture of Douglas, that such animals, though now tropical, were formerly inhabitants of these regions.

Additional arguments, as novel as ingenious, in support of the same conclusion have been deduced by Dr. Buckland from his examination of the cave at Kirkdale and of the remains of the quadrupeds, including the Hippopotamus, which he discovered in that remarkable depository of organized fossils. Of the great amphibious Pachyderm he cites six molar teeth and a few fragments of canine and incisor teeth, "the best of which are in the possession of Mr. Thorpe, of York."* Fig. 10, in plate vii., represents a much-worn last deciduous molar of the upper jaw of a young Hippopotamus, and figs. 8 and 9 two permanent molars which had just cut the gum, and had not had their fangs completed when the animal perished: the tooth in pl. xiii. fig. 7, is the last deciduous molar of the lower jaw. These teeth of the Hippopotamus, therefore, like the teeth of the Mammoth† associated with them in the Kirkdale Cave, prove that they were young and inexperienced individuals that had fallen into the clutches of the co-existing predatory Carnivora which made that cave their lurking-place, and perfectly coincide with the conclusions which Dr. Buckland thus enunciates:—"The facts developed in this charnel-house of the antediluvian forests of Yorkshire demonstrate that there was a long succession of years in which the Elephant, Rhinoceros, and Hippopotamus had been the prey of the Hyænas, which, like themselves, inhabited England in the period immediately preceding the formation of the diluvial gravel; and if they inhabited this country, it follows as a corollary that they also inhabited all those other regions of the northern hemispheres in which similar bones have been

* 'Reliquiæ Diluvianæ,' p. 18.

† Ante p. 259, 334.

found under precisely the same circumstances, not mineralized, but simply in the state of grave-bones imbedded in loam, or clay, or gravel, over great part of northern Europe, as well as North America and Siberia."*

Fossil remains of Hippopotamus have been found in some abundance, and in a more perfect state than those in the fluviatile deposits of the valleys of the Thames and Avon, in the formations of clay and sand with lignite beds, also of fresh-water origin, that overlie the Norwich crag upon the eastern coast of Norfolk.†

The fine example of the ramus or half of the lower jaw of the *Hippopotamus major*, represented in figures 159 and 162, was obtained from this pliocene formation near Cromer. It forms part of the rare and instructive series of fossils which Miss Anna Gurney, in the exercise of a beneficence which is combined in her noble character with an enlightened appreciation of whatever tends to promote science, has caused to be rescued from the destructive operations to which the sea-coast in the vicinity of her residence is peculiarly exposed. The fishermen and other poor inhabitants of the coast have been encouraged by her judicious bounty to collect and preserve the specimens that, by the action of high and stormy tides, become detached from the cliffs; and the evidences of the ancient beings of this island thus saved from destruction, have proved of essential service in the present attempt to record the extinct species of British *Mammalia*.

The half-jaw, of which the side-view is given in fig. 159, measures two feet in length, and one foot one inch and a half from the summit of the coronoid process, *p*,

* 'Reliquiæ Diluvianæ,' p. 42.

† See Mr. Lyell's Memoir on the Geology of this coast in the 'Philosophical Magazine' for May, 1840.

to the precurved angle at the base of the ascending ramus *a*. Both these processes are broken off in the fossil lower jaw of the *Hippopotamus major* preserved in the Grand Ducal cabinet at Florence, and figured by Cuvier in pl. iv. of his chapter on the Fossil Hippopotamus. (Op. cit.) Our English specimen fully confirms the difference, which the Florentine jaw left somewhat doubtful, in the degree of forward curvature of the process, *a*, which curvature is more rapid and extensive in the recent than in the fossil Hippopotamus. The lower contour of the horizontal ramus begins to be convex almost immediately anterior to the above curvature in the recent species; in the fossil it continues concave to the alveolus of the canine tusk, *b*. The coronoid process, *p*, is more vertical in the fossil; it inclines forwards before curving back in the recent species. Of the narrower interspace between the two rami and the sharper angle at their anterior union, so well marked in the Florentine jaw, the Norwich specimen does not afford evidence; but it shows the same equality of breadth of the jaw along the outside of the molar series. The swelling-out to form the socket of the canine, commences, as in the Italian specimen, anterior to the premolar tooth, *p* 2, and not, as in the African Hippopotamus, opposite the middle of the molar series.*

Traces of the socket of the first premolar, *p* 1, still remain in the fossil; the second premolar, *p* 2, is relatively larger, and is separated by a wider interspace from the third than in the recent Hippopotamus; and the oblique ridge on the inner surface of the crown is more developed in the fossil. The first true molar, *m* 1, presents a basal ridge on the outside of the hinder lobe, and a tubercle at the base

* Compare fig. 162 with the same view of the lower jaw of the recent Hippopotamus in the 'Ossemens Fossiles,' tom. cit., 'Hippopotame vivant,' pl. ii. fig. 4.

of the inner division of the two lobes, which I have not found in the corresponding tooth of the recent species. The last lower molar, *m* 3, which is characterized by a third accessory lobe, has a longer antero-posterior diameter in comparison with its transverse than in the recent species; but the agreement in the size and shape of the molar teeth is very close.

Fig. 162.

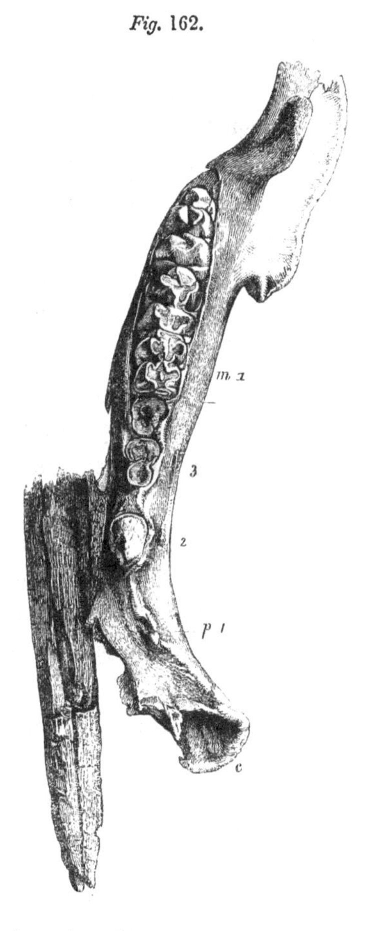

Lower jaw of *Hippopotamus major*, $\frac{1}{6}$ nat. size; fresh-water clay-beds. Cromer, Norfolk.*

The antero-posterior diameter of the base of the crown of this tooth is one inch six lines (·0039 millimetres).

The antero-posterior diameter of the last molar of the *Hippopotamus major* from Walton is three inches three lines; the transverse diameter of the base of the first lobe one inch and a half.

The great straight incisive tusk of the lower jaw is commonly found in a state of decomposition, with the ivory separating into a series of superimposed cones. In

* In this figure, the incisive tusk, *i*, is drawn too much inclined outwards; it should be parallel with the molar series.

a water-worn tusk of this kind from the beach at Cromer, the bases of certain of these cones at intervals of about an inch form slightly projecting ridges, encircling the tusk rather obliquely, and causing an undulation of the surface.

The canines are wanting in the lower jaw from Cromer; but a portion of an inferior canine of a larger specimen of the *Hippopotamus major*, in the Museum of Miss Gurney, measures three inches and a half in diameter across the flattened side: fig. 161 gives a reduced view of the inner side of the extremity of a lower tusk from the fresh-water deposits at Walton, of nearly equal dimensions. Mr. Brown of Stanway possesses a portion of a smaller tusk of the fossil Hippopotamus, from the same formation and locality.

In the Norwich Museum there is a tusk of the *Hippopotamus major*, which was dredged up from the oyster-bank at Happisburgh; it is black and heavy, being penetrated by iron. In the Museum of the Yorkshire Philosophical Society there is a molar tooth of the *Hippopotamus major*, from Overton, near York. In the collection of Mr. Saull, F.G.S., are preserved some fine portions of the under jaw, and several detached teeth of the *Hippopotamus major* from the post-pliocene fresh-water beds at Alconbury, near Huntingdon.

Remains of the extinct Hippopotamus have been found in other limestone caves in England than that at Kirkdale; as, for example, at Kent's Hole, Torquay. Several teeth of the Hippopotamus were found, associated with Mammoth, Rhinoceros, Aurochs, Ox, Hyæna, and Bear, in the cavern at Durdham Down, recently described by Mr. Stutchbury.

With respect to the bones of the extremities of the Hippopotamus, the femur, which equals in size that of the fossil Rhinoceroses, differs most essentially in the absence of the third trochanter, or process from the middle of the outer side. It may be distinguished from the femur of the great Ruminants, as the Aurochs or Giraffe, by the head being more detached from the shaft and more spherical, and by the superior development of the lower extremity, especially the back part of the condyles.

The astragalus is a very characteristic bone: its anterior surface (fig. 150), which, as in other hoofed quadrupeds with toes in even number, is almost equally divided by a low vertical ridge into two articulations, differs from that in the Ruminants and the Hog by the slight concavity of those facets: there is also a well-marked articular surface on the outer side of the bone for the lower end of the fibula, and a similar one on the inner side for the lower end of the tibia or internal malleolus. The anterior view of the astragalus of the Rhinoceros (fig. 149) is placed by the side of that of the Hippopotamus to show the unequal division of the anterior (scapho-cuboid) articular surface, characteristic of the hoofed quadrupeds with toes in uneven number, as the Horse, the Rhinoceros and the Elephant.

The fluviatile accumulations of sand and gravel at Cropthorne, near Evesham, in Worcestershire, in which Mr. Strickland discovered the remains of the Hippopotamus, Bear, Aurochs, and other extinct Mammals, constitute terrace-like hillocks, from one to four miles distant from the present bed of the Avon, above which their summits rise to a height of forty feet. They are very analogous to the deposits on the banks of the Thames, in which the remains of the Hippopotamus were discovered in such abundance by Mr. Trimmer. The value of Mr. Strickland's discovery is greatly enhanced by the care with which the shells of the formation containing the Mammalian fossils were collected and examined by him: of those shells he has determined twenty-four species, five terrestrial, and nineteen fresh-water; of which latter, there appear to be three extinct species. All the others are existing and indigenous to Britain. In reference to this discovery, Mr. Lyell remarks:—"The Hippopotamus is now only met with in rivers where the temperature of the water is warm and nearly uniform; but the great fossil species of the same genus (*H. major*, Cuv.) certainly inhabited England when the testacea of our country were nearly the same as those now existing, and when the climate cannot be supposed to have been very hot."*

We have no evidence that the great fossil Hippopotamus extended so far north as the Mammoth and tichorhine Rhinoceros, with which it is commonly found associated in England and the temperate latitudes of Europe; its remains are not uncommon in the pliocene deposits of Italy, and along the European shore of the Mediterranean. No remains of *Hippopotamus major* have yet been discovered in any part of Asia. The genus is represented in the rich fossiliferous tertiary deposits of the Sivalik Hills by a Hippopotamus with six incisive teeth in the lower jaw, from which difference its discoverers, Capt. Cautley and Dr. Falconer, have proposed for it the sub-generic name of *Hexaprotodon*.

We have no evidence of the Hippopotamus having existed on our planet anterior to the pliocene division of the tertiary epoch: and the ancient extinct, like the recent species, seems to have been confined to the Eastern Hemisphere.

* Principles of Geology, 1837, vol. i. p. 144.

PACHYDERMATA. *CHŒROPOTAMUS.*

Fig. 163.

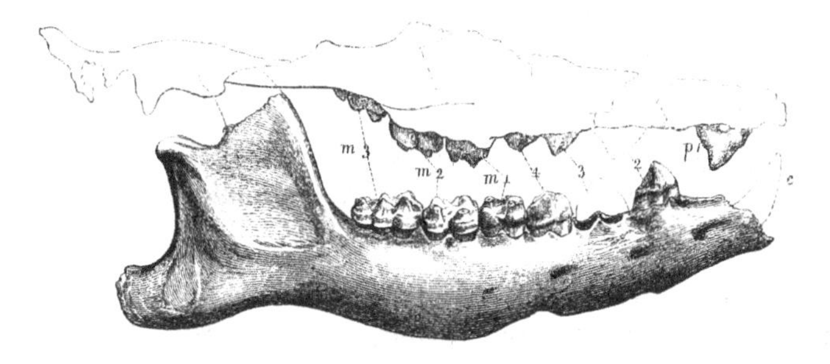

Lower jaw of *Chœropotamus Cuvieri*, $\frac{1}{3}$ nat. size. Eocene marl, Isle of Wight. An outline of upper jaw from Cuvier.

CUVIER'S CHŒROPOTAMUS. Chœropotamus Cuvieri.

Chœropotame, CUVIER, Ossemens Fossiles, 4to. 1822, p. 360, pl. li. fig. 3, A, B, C, pl. lxviii. fig. 1.
Chœropotamus Gypsorum, DESMAREST, Mammalogie, p. 545.
" *Cuvieri*, OWEN, Geological Transactions, second series, vol. vi. p. 41, pl. iv.

SEVERAL interesting forms of Pachyderms with toes in even number, as *Anthracotherium*, (Cuvier,) *Merycopotamus* and *Hippohyus*, (Cautley and Falconer,) which filled up the wide interval that now divides the Hippopotamus from the Hog, formerly existed, and have left their remains in more ancient tertiary deposits than those containing the fossil *Hippopotamus*. Hitherto no remains of these genera have been detected in Britain ; and the nearest link which the fossils of our island afford in the transition from the Hippopotamus to the Hog-tribe, is presented by the *Chœropotamus*. This quadruped must have resembled the Peccari, but was about one third larger : it was the earliest form of the Hog-tribe introduced upon our planet.

Cuvier had recognized amongst the fossil fragments extracted from the gypsum at Montmartre, indications of extinct genera different from the *Palæotheria* and *Anoplotheria*, and to one of the rarest and least satisfactorily represented of these he gave the name of *Chœropotamus*. The fossil figured at the head of the present section not only extends, by its association in the same deposit with *Palæotheria* and *Anoplotheria*, the analogies of the eocene marls of the Isle of Wight with the gypsum beds at Paris, but affords additional information of the osteology and dentition of the extinct genus, which is essential to the determination of its exact affinities.

The fossil in question is the right ramus of the lower jaw, with all the teeth in place, except one premolar, the canine and the incisors. It was discovered by the Rev. D. Fox, in the Seafield quarry, near Ryde, Isle of Wight.

The fragments of the *Chœropotamus* which Cuvier * describes, consist of an incomplete base of the skull with six molar teeth on each side, (fig. 164, A) and a small portion of a ramus of the lower jaw, with the canine (?) and two spurious molars.

The form of the teeth, and the flattened surface of the glenoid cavity, afford sufficient proof of the pachydermal nature of the animal, and its close alliance to the genus *Sus*. But the breadth of the glenoid cavity and the expansion of the zygomatic arches are greater than in any known species of Hog; the Peccari (*Dicotyles*) in these respects, as in the dental details, especially in the proportion and direction of its canine teeth, approaches nearest to the fossil.

Now the points in which the Cuvierian fossils prove that the *Chœropotamus* deviates from the Peccari, are those

* 'Ossemens Fossiles,' ed. 1822, vol. iii. p. 260 ; pls. lxviii. li.

which indicate a nearer approximation in the extinct genus to the carnivorous type; and it is of great interest to find that the ramus of the jaw, so fortunately extracted in an almost entire state from the Isle of Wight strata, exhibits a structure in the prolongation backwards of the angle of the jaw, which has hitherto been found to characterize, almost exclusively, the carnivorous Mammalia. Certain it is that no known pachydermal, or other ungulate species of Mammal presents this conformation. The figure (163) precludes the necessity of a detailed description of this process; it is more compressed and deeper than in the Bear, Dog or Cat tribe, and is not bent inwards in the way which peculiarly characterizes the marsupial jaws, and which so neatly distinguishes the Stonesfield Phascolothere. The condyloid process in the *Chœropotamus* is raised higher above the angle of the jaw than in the true Carnivora, and it is less convex than in the Hog or Peccari. In the size of the coronoid process the Peccari exceeds the true Hogs; and in that respect, as well as in the form and position of its canine teeth, makes a nearer approach to the carnivorous type; but in the *Chœropotamus* the coronoid process is still more developed in correspondence with the greater bulk of the temporal muscle, the size of which is indicated by the span of the zygomatic arches. In the wavy outline of the inferior border of the lower jaw, the Peccari alone, amongst the Hog tribe, resembles the *Chœropotamus*. The two detached molars of the lower jaw described by Cuvier, and which he compares with the third and fourth molars of the Babyroussa, are the fourth and fifth, or penultimate, *m* 2, and antepenultimate, *m* 1, molars, counting backwards, of the *Chœropotamus*, and correspond with the penultimate and antepenultimate grinders of the Peccari. The last molar of the lower jaw, in both the Peccari and Babyroussa, differs from the preceding in having two accessory, smaller and more closely approximated tubercles at the posterior part of the tooth, with a third small tubercle in the middle of the interspace between these and the next pair of tubercles. The Cuvierian fossils did not afford the means of making a

Fig. 164.

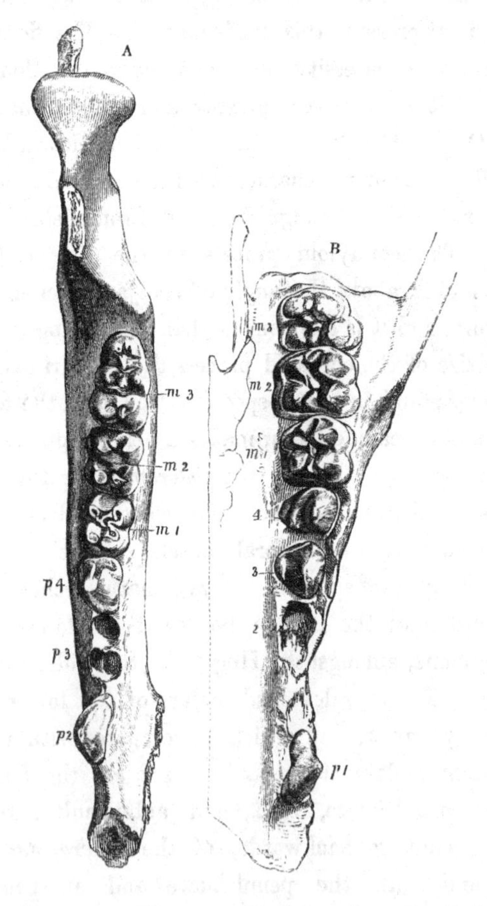

A, Upper jaw, *Chœropotamus*, Montmartre (Cuvier). B, Lower jaw, Seafield, Isle of Wight. ½ nat. size.

comparison between the *Chœropotamus* and these species of the Hog tribe in this particular; but in the present specimen we see that the last molar of the lower jaw (fig. 164, B, *m* 3) presents the same additional posterior tubercles as in the Peccari, and confirms the view taken by Cuvier of the affinities of the ancient Pachyderms.

This tooth offers, also, a miniature resemblance to the corresponding one in the Hippopotamus (see fig. 162, *m* 3).

All the premolars were more simple in comparison with the true molars; the last premolars of the upper jaw (fig. 164, A, 3 and 4,) had each an external large and an internal low and small tubercle, both enclosed by a basal ridge. The true molars are each like two premolars combined, and with the inner tubercles developed to equality with the outer ones; they have also the two small intermediate tubercles and a well-developed cingulum (ib. *m* 1 and *m* 2): the last upper molar (*m* 3) resembles that of the Hyracothere. In the lower jaw the canine had much of the form and proportions of that of a Carnivore. There are three premolars in this jaw; the one which answered to the first premolar above was not developed in the Chœropotamus: the first in place (ib. B, *p* 2) had a compressed pointed crown and a small posterior talon, like that above; the second and third increase in breadth, but are narrower and more simple than those above. The true molars below are also narrower than the upper ones, but are quadricuspid with accessory tubercles, and a largely developed hinder talon in the last molar.

Our fossil jaw fortunately yields a fact essential in characterizing the genus, and which the fragments in Cuvier's possession were too imperfect to afford, viz., the exact number of molar teeth in the lower jaw, which is twelve.

The tooth anterior to the grinders, and which from its shape Cuvier regarded as a canine, is situated closer to the symphysis of the jaw than in any of the existing *Suidæ;* but the Peccari, in this respect also, comes nearest to the *Chæropotamus.* On the outer surface of the jaw, near its anterior extremity, the vascular foramina are as numerous as in the jaws of the Hog tribe.

Nothing as yet is known of the incisors of the *Chæropotamus;* the rest of the dentition closely resembles that of the Peccari, but the premolars are more simple and the canines by their size, shape, and direction, and the lower jaw by the backward prolongation of its angle, alike manifest a marked approximation to the Ferine type. The occasional carnivorous propensities of the common Hog are well known, and they correspond with the minor degree of resemblance, which this existing Pachyderm presents to the same type. The extinct *Chæropotamus,* still better adapted by its dentition for predaceous habits, presents an interesting example of one of those links, completing the chain of affinities, which the revolutions of the earth's surface have interrupted, as it were, and for a time concealed from our view.

It is interesting, also, to perceive that the living subgenus of the Hog tribe which most resembles the *Chæropotamus* should be confined to the South American continent, where the Llama and Tapir, the nearest living analogues of the Anoplotherian and Palæotherian associates of the *Chæropotamus,* now exist, and which was formerly inhabited by a genus — *Macrauchenia,* which connects the Llama with the Palæothere.

PACHYDERMATA. *HYRACOTHERIUM.*

Fig. 165.

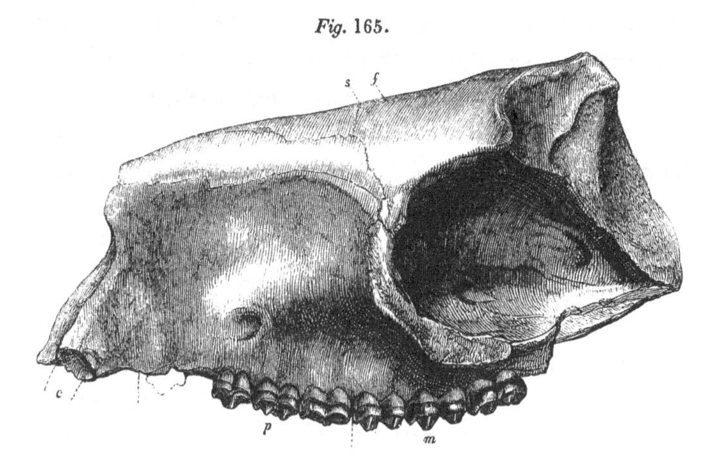

Skull of *Hyracotherium leporinum*, nat. size. Eocene clay.

THE LEPORINE HYRACOTHERE. Hyracotherium leporinum.

Hyracotherium leporinum, OWEN, Geological Transactions, second series, vol. vi. p. 203.

THE fresh-water eocene marls of the Isle of Wight appear to be much richer in mammalian remains than the contemporaneous formation called the London clay; here, however, certain genera, as *Lophiodon* and *Palæotherium*, have been found which exist in the eocene gypsum in France, and the remains of which also occur in the fresh-water marls of the Isle of Wight ; and the interesting fossil to be described in the present section, although it indicates a genus not hitherto found in the older tertiary beds on the Continent, demonstrates the extinct quadruped of which it formed part to have been as distinct, generically, as the *Anoplotherium* or *Palæotherium*, from any living Mammalia, and to have had the nearest affinity to the *Chœropotamus*.

The fossil in question consists of a mutilated cranium (fig. 165) rather larger than that of a hare, containing the molar teeth of the upper jaw nearly perfect and the sockets of the canines. It was discovered in the London clay forming the cliffs at Studd Hill, about a mile to the west of Herne Bay, by William Richardson, Esq., F.G.S., who kindly gave me the opportunity of describing it in the Geological Transactions for 1839.*

The molars are fourteen in number in the upper jaw, and resemble more nearly those of the Chœropotamus than the molars of any other known genus of existing or extinct Mammalia. They consist of four premolars and three true molars on each side. The first and second premolars, counting from before backwards, have simple sub-compressed crowns, surmounted by a single median conical cusp, with a small anterior and posterior tubercle at the outer side, and a ridge along the inner side of its base: they are separated from each other by an interspace nearly equal to the antero-posterior diameter of the first premolar, which measures two lines and a half. The second and the rest of the series are in close juxtaposition (fig. 166). The third and fourth premolars present a sudden increase of size and of complexity of the grinding surface, with a corresponding change of form: their grinding surface supports three principal tubercles or cusps, two on the outer and one on the inner side: there are two smaller elevations, with a depression on the summit of each, situated in the middle of the crown, and the whole is surrounded by a ridge, which is developed into a small cusp at the anterior and external angle of the tooth. These teeth form the

* Geol. Trans., Second Series, vol. vi. p. 203.

principal difference between the dentition of the present genus and that of the Chœropotamus, in which the corresponding false molars are relatively smaller and of a simpler construction, having only a single external pyramidal cusp, with an internal transverse ridge or talon at its base. The true molars, three in number on each side, closely correspond in structure with those of the Chœropotamus. They present four principal conical tubercles, situated near the four angles of the quadrilateral grinding surface. Each transverse pair of tubercles is connected at the anterior part of their base by a ridge, which is raised midway into a smaller conical tubercle with an excavated apex. The crown of the tooth is surrounded by a well-marked ridge, which is developed, as in the third and fourth false molars, into a sharp-pointed cusp at the anterior and external angle of the tooth. The hindmost molar is more contracted posteriorly, and its quadrilateral figure less regular than the two preceding molar.

The sockets of the canines or tusks (figs. 165 and 166, *c*) indicate that these teeth were relatively as large as in the Peccari, and that they were directed downwards. The temporal muscles were as well developed as in the Peccari, the depressed surface for their attachment (figs. 165 and 167, *t*) extending on each side of the cranium as far as the sagittal suture. The frontal bones (ib. *f*) are divided by a continuation of the sagittal suture. The nasal suture, *s*, runs transversely across the cranium parallel with the anterior boundary of the orbits. The lachrymal bone, *l*, extends a very little way upon the face. The external angle of the base of the nasal bone, which is of considerable breadth, joins the lachrymal, and separates the superior maxillary from the frontal bone. The anterior margin of the malar bone encroaches a little way upon the face at the anterior boundary of the orbit. The under surface of the palatal processes of the maxillary bones is rugose, as in the Peccari; the portion of the skull, including the inter-

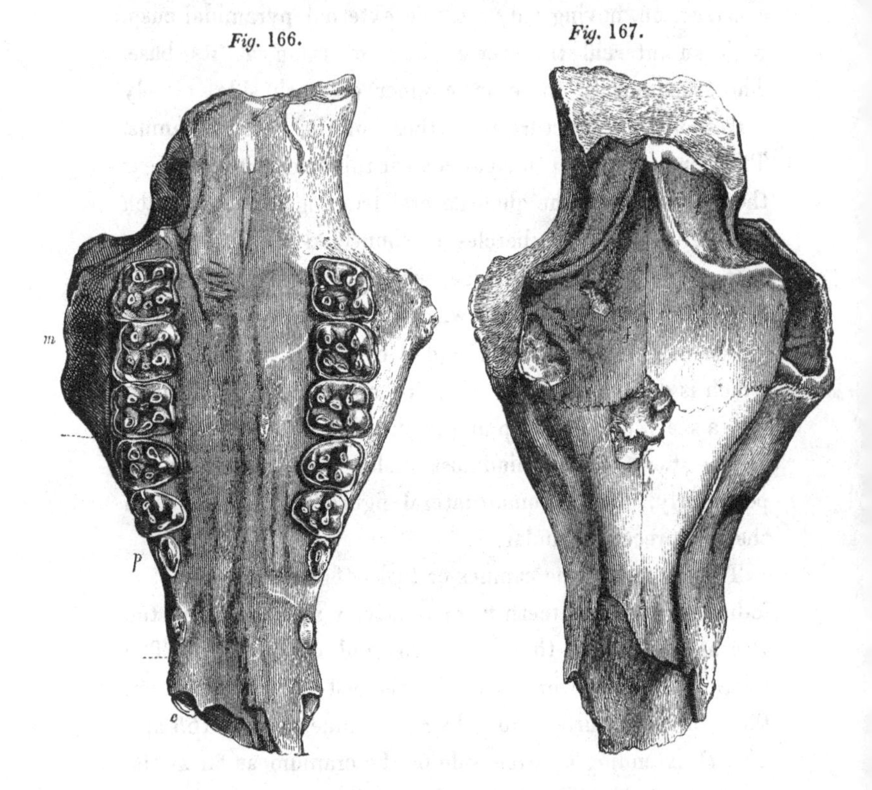

Fig. 166. Under view. *Fig.* 167. Upper view.

Skull of *Hyracotherium*, nat. size. Eocene clay.

maxillary bones and the incisive teeth, is unluckily broken off and lost.

The general form of the skull was probably intermediate in character between that of the Hog and the *Hyrax*. The large size of the eye indicated by the capacity of the orbit, must have given to the physiognomy of the living animal a resemblance to that of the Hare, and other timid Rodentia. Without intending to imply that the present small extinct Pachyderm was more closely allied to the *Hyrax* than as being a member of the same order, and similar in size, I have proposed to call the new genus which it unquestionably indicates, *Hyracotherium*, with the specific name *leporinum*. The form and structure of the molar teeth determine this interesting extinct genus to belong to the same natural family of the Hog tribe, as the Chœropotamus.

From the same deposits at Herne Bay, Mr. Richardson obtained two small dorsal vertebræ, referable by the capacious canal for the spinal marrow (fig. 169) and the articular cavity for the head of the rib, excavated on opposite surfaces of the two vertebræ (fig. 168, *c*), to the Mammiferous class. Their size is that which might be expected in the dorsal vertebræ of the skeleton of the *Hyracotherium leporinum*, and, as there is no character which forbids their reference to a small Pachyderm, allied to the Peccari, they may very probably belong to the same species as the fossil skull which was discovered at the same place.

Fig. 168.

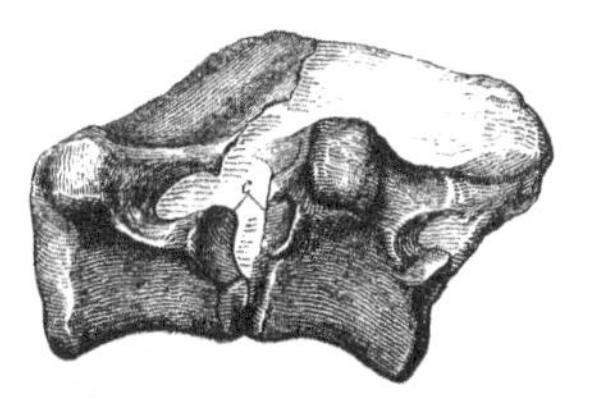

Fig. 169.

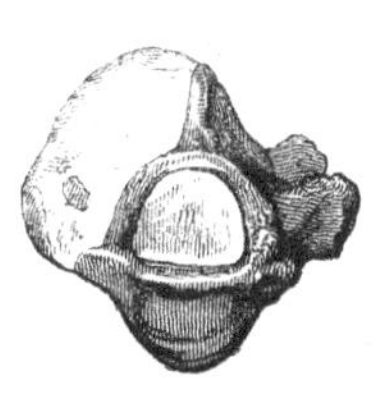

Dorsal vertebræ, nat. size. Eocene clay, Herne Bay.

PACHYDERMATA. *HYRACOTHERIUM.*

Fig. 170. *Fig.* 171.

Last molar. Third premolar.
Nat. size. Eocene sand, Kyson.

THE CUNICULAR, OR RABBIT-LIKE HYRACOTHERE. *Hyracotherium Cuniculus.*

Hyracotherium Cuniculus, OWEN, Annals of Natural History, September 1841. Report of British Association, 1843, p. 227.

IN the eocene sand underlying the red crag at Kingston or Kyson in Suffolk, from which the remains of *Quadrumana*,* *Cheiroptera*,† and *Marsupialia*,‡ have already been obtained, Mr. Colchester has likewise discovered the teeth of other small Mammalian animals, some of which are referable to the small Pachydermal extinct genus *Hyracotherium*, established on the nearly entire cranium from the London clay, described in the preceding section.

The teeth from Kyson are three true molars and one of the false molars, all belonging to the upper jaw. The crowns of the true molars present the same shortness in vertical extent, the same inequilateral, four-sided, transverse section, and nearly the same structure, as in *Hyracotherium leporinum;* the grinding surface also supports four obtuse pyramidal cusps, and is surrounded by a well-developed ridge, produced at the anterior and outer angle of the crown into a fifth small cusp.

These teeth are, however, of smaller size, as will be seen

* Ante, p. 3. † Ante, p. 17. ‡ Ante, p. 71

by comparing figures 170 and 171 with the corresponding molars of the *Hyracotherium leporinum*, (fig. 167). The true molars of these two species further differ in a point not explicable on the supposition of their having belonged to individuals or varieties differing merely in size, for the ridge which passes transversely from the inner to the outer cusp is developed midway into a small crateriform tubercle in the teeth of the *Hyracotherium leporinum*, but preserves its trenchant character in the *Hyrac. Cuniculus*, even in molars which have the larger tubercles worn down.

The premolar, or false molar (fig. 171), in the series of detached teeth from Kyson, which is either the third or fourth, presents the same complication of the crown which distinguishes the *Hyracotherium* from the *Chœropotamus*, but with the same minor modification which distinguishes the true molars of the Kyson species from those of the *Hyrac. leporinum* of Herne Bay; *i. e.*, the two ridges which converge from the two outer tubercles towards the internal tubercle are not developed midway into the small excavated tubercle, as in the *Hyrac. leporinum*, but are simple. The disparity of size between the true and false molars appears to be greater in the *Hyrac. Cuniculus* than in the *Hyrac. leporinum*.

This discovery of a second species of the genus *Hyracotherium*, associated with fossil vertebræ of a Serpent, in the Kyson sand, tends to place beyond doubt the equivalency of that formation with the eocene deposits at the estuary of the Thames, and corroborates the inference deducible from the mammalian, ornithic, and ophidian remains of the London clay, that it was deposited in the near neighbourhood of dry land.

PACHYDERMATA. *SUS.*

Fig. 172.

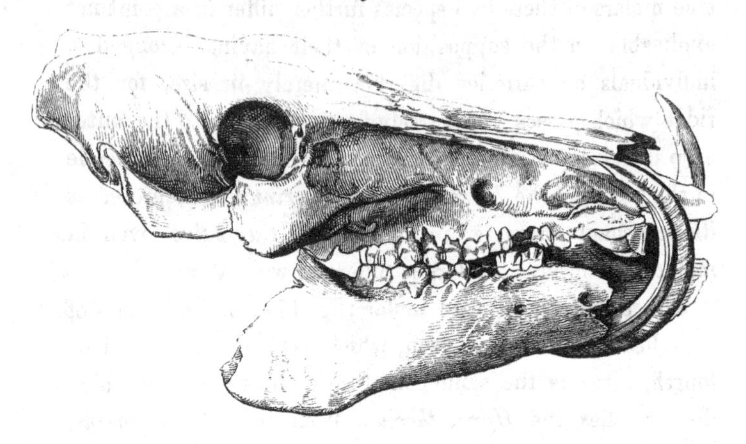

Fossil Skull of Wild Boar from drift, Isle of Portland, ¼ nat. size.

WILD HOG. Sus Scrofa.

Cochon fossile,	Cuvier, Annales du Muséum, tom. xiv. p. 39.
Sus Scrofa fossilis,	H. von Meyer, Palæologica, p. 80.
Sus priscus,	Goldfuss, Nova Acta Acad. Nat. Cuv. t. xi. pt. 2. p. 482.
Sus Arvernensis, (?)	Croizet and Jobert, Ossemens Fossiles du Puy-de-Dome, 4to., p. 157.
Fossil Hog,	Buckland, Reliquiæ Diluvianæ, p. 59.
Sus Scrofa,	Owen, Report of British Association, 1843, p. 228.

When Cuvier communicated his memoir on the fossil bones of the Hog to the French Academy in 1809, he had met with no specimens from formations less recent than the mosses or turbaries and peat-bogs, and knew not that any had been found in the drift associated with the bones of elephants. He repeats this observation in the edition of the 'Ossemens Fossiles' in 1822; but in the additions to the last volume, published in 1825, Cuvier cites the discovery by M. Bourdet de la Nièvre, of a fossil lower jaw of a *Sus*, on the east bank of the lake of Neufchatel, and a fragment of the upper jaw from the cavern at Sundwich, described by Prof. Goldfuss.

Dr. Buckland* includes the molar teeth and a large tusk of a boar found in the cave of Hutton in the Mendip Hills, with the true fossils of that receptacle, such as the remains of the Mammoth, Spelæan Bear, &c. With respect to cave-bones, however, it is sometimes difficult to produce conviction as to the contemporaneity of extinct and recent species. MM. Croizet and Jobert, in their account of the fossils of Auvergne, give more satisfactory evidence of the coexistence of the genus *Sus* with *Elephas*, *Mastodon*, &c., by describing and figuring well-marked fossils of a species of Hog, which they discovered in the midst of their rich fossiliferous tertiary beds. These observers found, however, that the facial part of their fossil Hog was relatively shorter than in the existing *Sus scrofa*, and they have conceived it to represent a distinct species, which they have called *Aper* (*Sus*) *Arvernensis*. Dr. Kaup has described fossils referable to the genus *Sus* from the miocene Eppelsheim sand, in which they were associated with fossils of the *Mastodon* and *Dinotherium*. The oldest fossils of the genus *Sus* from British strata which I have yet seen, are portions of the external incisor of the lower jaw (fig. 173), from fissures in the red crag (probably miocene) of Newbourne near Woodbridge, Suffolk. They were associated with teeth of an extinct *Felis* about the size of a Leopard, with those of a Bear, and with remains of a large *Cervus*. These mammalian remains were found with the ordinary fossils of the red crag; they had undergone the same process of trituration, and were impregnated with the same colouring matter as the associated bones and teeth of fishes acknowledged to be derived from the regular strata of the

* 'Reliquiæ Diluvianæ,' p. 59.

Fig. 173.

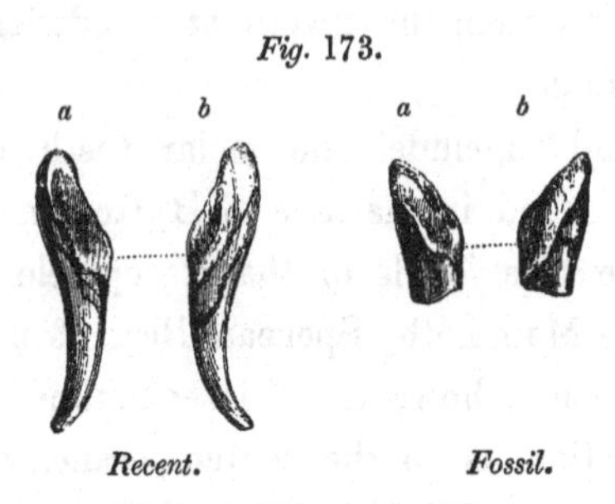

Incisor of Hog. Nat. size.
a a. View of tooth from the inside. *b b.* View of tooth from the outside.

red crag. These mammaliferous beds have been proved by Mr. Lyell to be older than the fluvio-marine or Norwich crag, in which remains of the Mastodon, Rhinoceros and Horse have been discovered; and still older than the freshwater Pleistocene deposits from which the remains of the Mammoth, Rhinoceros, &c., are obtained in such abundance.

I have met with some satisfactory instances of the association of fossil remains of a species of Hog with those of the Mammoth in the newer pliocene freshwater formations of England.

Fig. 174.

Last lower molar, Hog; nat. size.

In the collection of Mr. Wickham Flower there are good specimens of the teeth of the Hog (molars, and a long and sharp tusk), which were taken from the brick-earth at Grays in Essex, twenty feet below the present surface; these teeth were associated with teeth and bones of a Deer, and portions of dark charred wood. Mr. Brown of Stanway has likewise some fossil remains of a young specimen of *Sus* from the freshwater deposits at Grays, which contained remains of the Mammoth and Rhinoceros.

A left upper tusk of a Boar from the newer pliocene beds near Brighton presented a broader longitudinal internal strip of enamel than in those tusks of the Wild Boar of Europe or India which I had for comparison; the longitudinal groove along the unenamelled part was also deeper in the fossil.

The Rev. Mr. Green of Bacton submitted to my inspection the extremity of the tusk of a Wild Boar, and the crown of a tubercular molar of a young Hog, which he had obtained from the blue clay and submerged forest bed at Hasbro' on the Norfolk coast. These remains of the genus *Sus* were in the same fossilized condition as the bones and teeth of the extinct species of Mammalia from the same locality; and I believe them to have been of equal antiquity. These instances of unequivocal fossil remains of the Hog tribe are, however, very rare.

The fine skull of the Wild Boar (fig. 172,) discovered by Capt. Manning in a fissure of the freestone quarries in the Isle of Portland, and described by Dr. Buckland at a late meeting of the Geological Society, has not such decided claims to an equal antiquity with the Mammoth and Trogonthere, and it is unquestionably identical with the existing species of European Wild Boar. I owe to Dr. Buckland's kindness the opportunity of figuring this fossil, which is preserved in Capt. Manning's collection at Portland Castle.

I have received remains of a Hog, associated with bones of a Brown Bear (*Ursus Arctos*) and other existing species of Mammalia, which were obtained by Mr. Whitwell of Kendal, from a limestone cavern at Arnside Knott, near that town.

The anterior part of the left ramus of the jaw of a Hog has been obtained from the drift formation at Kesslingland, Suffolk.

The usual situation of bones of the Hog is that mentioned by Cuvier, viz., in peat-bogs. In the Norwich Museum is preserved the anterior part of the lower jaw of a Hog, which was found four or five feet below the surface in peat-bog upon drift gravel in Norfolk.

A molar tooth with the upper and lower tusks of a Wild Boar have been found associated with remains of the Wolf, Beaver, Goat, Roebuck, and large Red Deer in freshwater marl, underlying a bed of peat ten feet thick, itself covered in some places by the same thickness of shell-marl and alluvium, at Newbury, Berkshire.

In the most recent deposits where the remains of the Hog are usually met with, their identity with the *Sus scrofa* is unequivocal. I have received from Dr. Richardson a collection of bones, not much altered by time, from a gravel-pit in Lincolnshire, near the boundary between the parishes of Croft and Ikeness; among these were remains of the common Hog.

The tusks and molar teeth of a Boar, which were discovered ten feet below the surface of a peat-bog, near Abingdon, Berkshire, were associated with enormous quantities of hazel-nuts in a blackened or charred state, the whole resting on a layer of sand which was traced extending eighteen feet horizontally.

These specimens are preserved in the Museum of the Royal College of Surgeons; they were presented to John Hunter, by Mr. Jones, a surgeon at Abingdon; and the following letter from that gentleman to Hunter is printed in the 4to 'Catalogue of Fossils,' p. 243.

"Dear Sir,

"The under jaw of a Wild Boar, or some other animal, and the nuts which I have taken the liberty to enclose in the box, were a few days since found about ten feet underground by a labourer as he was digging peat or turf.

"Several single tusks have been found, and they were all worn in the manner you will observe these to be at the extremities, and the quantity of nuts was very considerable, and seemed to lay in a layer of white sand between the strata of peat. From whence could they come? Is it possible they could remain there ever since the Deluge?

(Signed) "Wm. Jones.

"Abingdon, Berks, May 23rd, 1787.

"The layer of sand and nuts extended upwards of eighteen feet horizontally."

"*To Mr. Hunter.*"

PACHYDERMATA. *ANOPLOTHERIIDÆ.*

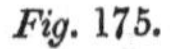

Fig. 175.

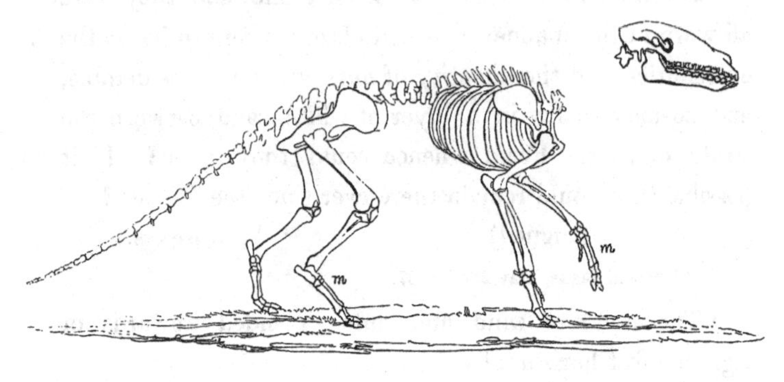

Skeleton of the *Anoplotherium commune*, as restored by Cuvier from fossil remains in eocene tertiary deposits of France. $\frac{1}{38}$ nat. size.

ANOPLOTHERIUM COMMUNE. The Common Anoplothere.

Anoplotherium le plus commun dans les carrières, CUVIER, Annales du Muséum, iii. pp. 370—379, pl. ii., viii., x., xi., xiii.

Anoplotherium commune, „ „ Ossemens Fossiles, tom. iii.

THE ANOPLOTHERIUM appears to have been one of the earliest forms of hoofed quadrupeds introduced upon the surface of this earth; and it is most important, in reference to speculations on the origin of organised species, to bear in mind that this ancient Herbivore presents, in comparison with living species, no indications of an inferior or rudimental character in any known part of its organization; and that, with regard to its dentition, it not only possessed incisors and canines in both jaws, but that those teeth were so equably developed, that they formed one unbroken series with the premolars and true molars, which character is now manifested only in the human species.

Amongst the varied forms of existing Herbivora we find certain teeth disproportionately developed, sometimes to a monstrous size; whilst other teeth are reduced to rudimental minuteness, or are wanting altogether: but the number of the teeth never exceeds, in any hoofed quadruped, that displayed in the dental formula of the Anoplotherium. It is likewise most interesting to find that those species with a comparatively defective dentition, as the horned Ruminants for example, manifest transitorily, in the embryo-state, the germs of upper incisors and canines,* which disappear before birth, but which were retained and functionally developed in the cloven-footed Anoplothere. The dental system of this extinct quadruped realized, in short, that ideally perfect type upon which so many kinds and degrees of variation have been superinduced in the dentition of later and still existing species of hoofed Mammalia.

The outer incisors of the *Anoplotherium commune* have their crowns produced into a low point, and the canine differs only by a slight increase of breadth and thickness of the crown; so that Cuvier, in his original and highly interesting memoir in the 'Annales du Museum,' was induced, in the absence of any evidence of the extent of the intermaxillary bone, to describe this tooth as an incisor, and the canines as being absent in the weaponless pachyderm.† The true canine of the Anoplothere becomes, therefore, from the great breadth and low point of

* Goodsir, in the 'Report of the British Association,' 1838.

† The name *Anoplotherium* (ἀ priv. ὅπλον, weapon, θηρίον, beast), first proposed in this memoir, has reference to the absence of those natural weapons, as tusks, long and sharp canines, horns, or claws, with which other quadrupeds have been supplied.

the crown, very characteristic of the genus, and such a tooth in a perfect state (fig. 176) has been discovered in the eocene freshwater deposits at Binstead in the Isle of Wight.

The first premolar chiefly differs in the increased thickness and greater development of the basal ridge; which, in the three larger succeeding premolars, assumes the

Fig. 176.

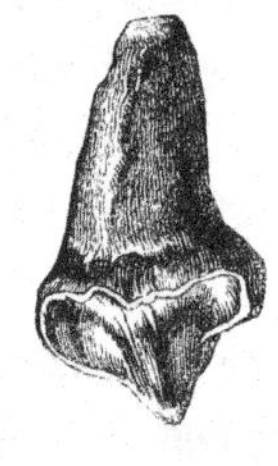

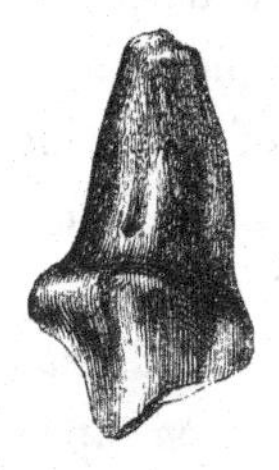

Upper canine, nat. size. *Anoplotherium commune.*

Fig. 177.

First upper premolar, nat size. *Anoplotherium secundarium.* Seafield, Isle of Wight.

character of an inner lobe, and a second lower ridge is developed. When the crown of the anterior premolar is much worn, the enamel lining the valley between it and the basal ridge forms an island, as in the tooth the grinding surface of which is figured in cut 177. This tooth, which is from the freshwater deposits at Seafield quarry, Isle of Wight, indicates by its size the smaller species of Anoplothere, which Cuvier has called *An. secundarium.*

The true molars are three in number, on each side of both jaws of the *Anoplotherium:* those above have large square crowns (fig. 178) divided into an outer and an inner lobe by a valley, *b*, extending from the inner side, two-thirds across, contracting as it penetrates. A second valley crosses its termination at right angles, and forms a curved depression in each lobe, *a b′*, concave towards the outer side of the crown,—this side being impressed by two parallel excavations, *d d*. The peculiar characteristic of the upper molar of the Anoplothere and that by which it may be most readily distinguished from a molar of the Paleotherium, is the large conical tubercle *m* at the wide entry of the valley *b*. The two points of the outer continuous border of the two lobes are first abraded; a double crescentic field of dentine is next exposed, with a detached island on the summit of the internal cone: this, afterwards, from the minor depth of the valley in front of its base, becomes blended with the anterior lobe, *b′*, from which also the crescentic enamel fold is first obliterated, and the pattern of the grinding-surface, which at first resembled that of the Ruminant, is reduced to that of the Palæothere (fig. 110) and Rhinoceros (fig. 122).*

Fig. 178.

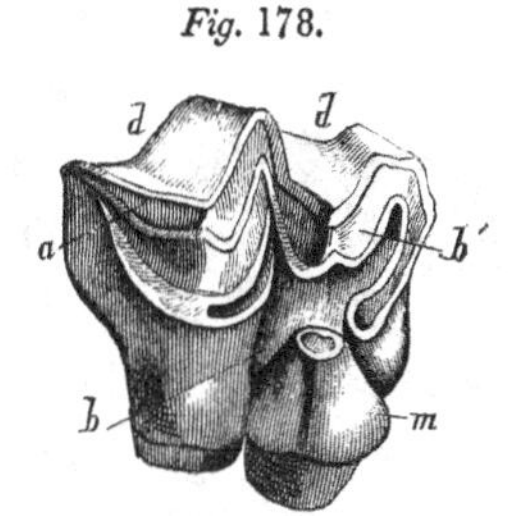

Upper molar, nat. size, of *Anoplotherium commune*. Binstead, Isle of Wight.

The lower incisors and canines much resemble those above. The molar series here, also, consists of four premolars and three true molars; to the latter belongs the tooth discovered by Mr. Thomas Allan of Edinburgh, in the lower freshwater limestone quarry at Binstead, which is figured by Dr. Buckland, in the 'Annals of Philosophy,' vol. x. (1825), p. 361, in a brief memoir containing the first announcement of the remains of the peculiar extinct Pachyderms of the Paris basin in the analogous basin of eocene freshwater deposits in Hampshire. The tooth

* The three principal stages of attrition are well displayed in the fossil upper jaw of the *Anoplotherium commune* from the Montmartre gypsum, figured by Cuvier in the 'Ossemens Fossiles,' 4to., 1822, tom. iii. pl. xlvi. fig. 2.

(fig. 179), which was recognised by Dr. Buckland and Mr. Pentland, as belonging to the *Anoplotherium commune,* is the first of the true molars.

Fig. 179.

Lower molar tooth, nat. size. *Anoplotherium commune.* Binstead, Isle of Wight.

These teeth consist, like those of the Palæotherium, of two semi-cylindrical lobes; but they are more deeply penetrated by narrower enamel folds on their inner side, and are relatively broader transversely, when worn down to the same extent, than those of the Palæotherium, as will be obvious by comparing fig. 179 with fig. 116. The last lower molar tooth has a third small posterior lobe, as in the Ruminants and the Palæotherium.

A general idea of the character of the chief bones of the skeleton may be obtained from the reduced view in cut 175. By comparing it with cut 109, it will be seen that the thigh-bone differs from that (*f*) of the Palæothere in the absence of the third trochanter. The fore-part of the astragalus of the Anoplothere differs from that of the Palæothere in the same way as the astragalus of the Hippopotamus differs from that of the Rhinoceros. The almost equal bipartition of the fore-part of the bone, indicates that the toes of the hind-foot of the Anoplothere were in equal number; and the fossil specimens have shewn them to be two in both fore and hind feet, as in the Ruminants. But the metacarpus and metatarsus, (*m m*,) instead of consisting each of a single 'cannon-bone,' were divided lengthwise, the two primitively separate bones continuing distinct throughout life in the Anoplothere.*

* This condition of the metacarpals and metatarsals has been observed in the exceptional instance of the existing African *Moschus aquaticus*, and in an extinct Ruminant of the Sewalik Hills, by Dr. Falconer, the distinguished elucidator of the Himalayan Fossils.

Whilst the evidence of the Anoplotherium in the eocene strata of the Isle of Wight was the single specimen of a molar tooth in the collection of Mr. Allan, some doubts were entertained of the accuracy of its assigned locality. These were, however, entirely dissipated by the subsequent interesting memoir on the remains of the *Anoplotherium* and *Palæotherium* in the lower freshwater formation of Binstead, near Ryde, by S. P. Pratt, Esq., F.G.S.* I have since received many corroborative instances of different species of both these kinds of ancient Pachyderms, from the eocene deposits in Hampshire, of which the teeth figured in cuts 176, 177, and 178, are examples.†

To the professed naturalist, the following definitions, applied by Cuvier to the extinct Pachyderms of the Paris basin, according to the Linnæan forms in reference to existing animals, must give the most striking evidence of the power of reconstruction of lost species by the application to their fossil remains of the law of organic correlations.

"Genus ANOPLOTHERIUM.

Dentes 44. Serie continuâ.

Primores utrinque 6.

Laniarii primoribus similes, cæteris non longiores.

Molares 28, utrinque 7. Anteriores compressi. Posteriores superiores quadrati, inferiores bilunati.

Palmæ et plantæ didactylæ, ossibus metacarpi et metatarsi discretis, digitis accessoriis in quibusdam.

1. *A. commune.* Statura Asini minoris, caudâ corporis longitudine, crassissima, habitu elongato Lutræ. Verisimiliter natatorius.

* 'Geological Transactions,' 2nd Series, vol. iii. p. 451.

† See 'Geological Transactions,' 2nd Series, vol. vi. p. 41.

2. *A. secundarium.* Similis præcedenti, sed staturâ Suis."

The common Anoplothere was eight feet long, including the tail, and four feet and a half without the tail; the body being about as long as that of a common Ass, but less elevated from the ground; the height to the withers being probably little more than three feet. The long and powerful tail must have formed the chief peculiarity in the living animal's outward form, and must have been of the same service to it in swimming, as the tail of the Coypu and the Otter. Cuvier concludes, therefore, that the extinct aquatic Herbivore swam the ancient lakes of the rising European continent, like the Water Vole and the Hippopotamus, in quest of the succulent roots and stems of aquatic plants;* but we may pause and remark on this conjecture, that the Anoplothere possessed neither the chisel-shaped incisors of the one for gnawing through such roots and stems, nor the great projecting tusks of the other for uprooting and tearing them from the soil; on the contrary, its small, equable and well-opposed upper and lower incisors would indicate that it cropped grass like a horse, and the close resemblance of the molars in the pattern of their grinding surface to those of the Ruminants and horse tribe, strengthens the probability that the Anoplothere came on land to browse or graze.

The existence of many destructive Carnivora at that early period of Mammalian life may partly explain the advantage to the *Anoplotherium commune* of its power of taking shelter in the water, especially as it wanted the means of rapid flight enjoyed by some of its congeners with long and slender limbs—as, for example, the *Anoplotherium*

* "Il allait donc chercher les racines et les tiges succulentes des plantes aquatiques."—Cuv. loc. cit. tom. iii. p. 247.

gracile. It may be more readily conceded that the *Anoplotherium commune*, by virtue of its habits as a swimmer and diver, was either clad with a short close smooth coat of hair like the Capybara and Otter, or was half naked like the Hippopotamus. It is very unlikely, Cuvier well remarks, that the Anoplothere should have been impeded in its swimming by long ears; the auricles were more probably short, as in the *Hippopotamus*, Capybara, and other aquatic quadrupeds.

On the strength of these analogies, and with the proportions demonstrated by the parts of the skeleton, Cuvier has given the subjoined restoration of the outward form of this very remarkable extinct Pachyderm.

Fig. 180.

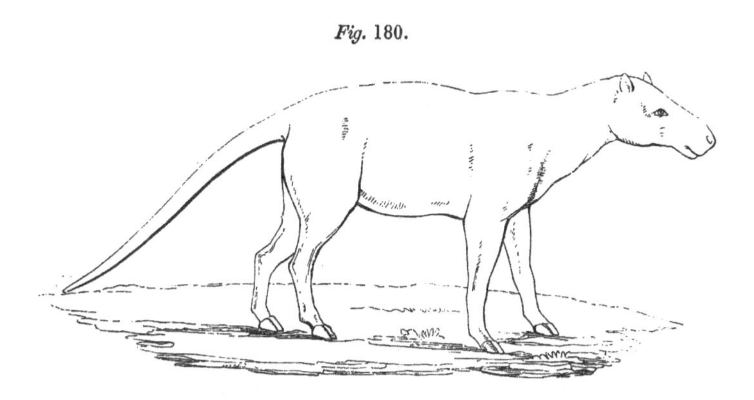

Anoplotherium commune.

PACHYDERMATA. *ANOPLOTHERIIDÆ.*

Fig. 181.

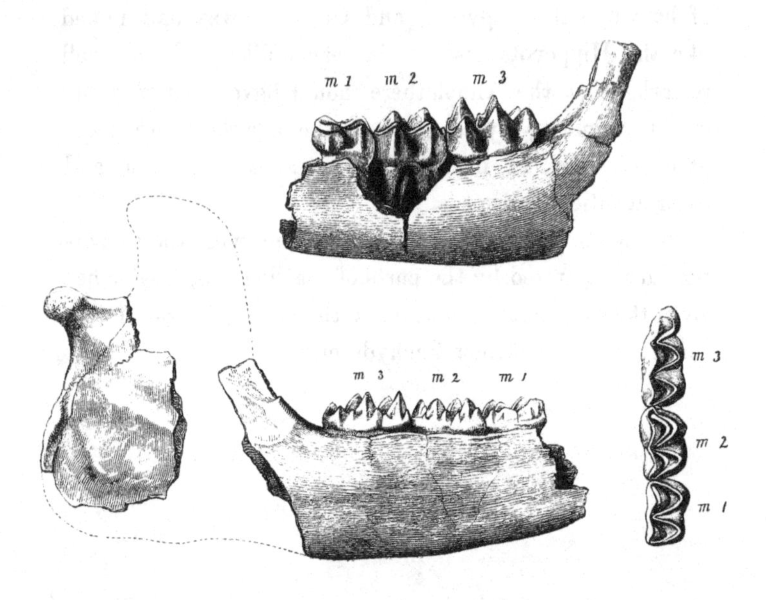

Portion of lower jaw, *Dichobune cervinum*, Eocene marl, Binstead, nat. size.

CERVINE ANOPLOTHERE. DICHOBUNE CERVINUM.

Animal allied to the genus *Moschus*,	PRATT, Geol. Trans., Second Series, vol. iii. p. 451.
Dichobune cervinum,	OWEN, Geol. Trans. Second Series, vol. vi. p. 41. Report of British Association, 1843, p. 225.

WE have seen in the foregoing sections that the extinct Pachyderms of the Paris basin, besides revealing forms which, as in the case of the *Chæropotamus*, tend towards the ferine group of Mammalia, included other genera, which, in the modifications of both their dental and locomotive organs, deviate in another direction, and almost complete the transition from the Pachydermal to the Ruminant order.

Among these genera the *Dichobune* of Cuvier is the most remarkable, inasmuch as the posterior molars (fig. 181, *m* 1, *m*, 2, *m*, 3,) begin to exhibit a double series of cusps, of which the external present the crescentic form; and in one species (*Dichob. murina*, Cuv.,) the crescents are acute and compressed laterally, so that when viewed separately they might be mistaken for the teeth of a true Ruminant.* In the lower jaw of the *Dichobune* the penultimate and antepenultimate grinders present two pairs of cusps, the last grinder three pairs, of which the posterior are small and almost blended together, so that when worn down they appear single. In this respect, as well as in the form of the ascending ramus of the lower jaw, Cuvier, who is not prone to exaggerated expressions, observes that the *Dichobunes* prodigiously resemble the young Musk-deer.†

This resemblance was well appreciated by Mr. Pratt, to whom we owe the discovery of the interesting extinct British quadruped which is the subject of the present section. The species, it is true, is represented by only a single fragment of the skeleton, but this is a characteristic one; it consists of the posterior half of the left ramus of the lower jaw with the three true molar teeth: it was found in the lowest bed of the freshwater marl at Binstead.

* La position et le nombre des pointes y (*Dichob.*) sont les mêmes que dans l'espèce précédente; mais les pointes sont plus aiguës et comprimées latéralement, ce que tend encore davantage à les rapprocher des molaires des Ruminans.—'Ossemens Fossiles,' tom. iii. p. 64.

† Or cette dentition, cette forme de branche montante, cette grandeur même, ressemblent prodigieusement à ce qu'on observe dans les jeunes Chevrotains.—Ibid. p. 65.

In the description which Mr. Pratt has given of this unique fossil, figured in cut 181, he observes, "This jaw appears to be closely allied to the genus *Moschus;* but the loss of the anterior portion renders it difficult to class the fossil correctly, and the greater width of the coronoid process distinguishes it from any described species of that order. This circumstance induced Cuvier (to whom a cast of the specimen had been sent) to suppose it to belong to the genus Anoplotherium, and he had named it *Anoplotherium dichobunes;* but as it was not possible to determine the structure of the fossil from an examination of the cast, I was induced to compare the single tooth above mentioned, with the specimens of the Paris Pachydermata preserved in the Museum of Natural History, and also with the jaws and teeth of all the small Ruminants in the same collection. This was done with the assistance of M. de Blainville, who, after the most careful examination, acknowledged that it was impossible to decide positively without having a more perfect jaw; and he was induced to leave the specimen amongst the Pachydermata, rather because Cuvier had so placed it, than on account of any decisive character. The texture of the tooth approaches, in my opinion, nearer to the Ruminants, while the general form of the jaw gives it the character belonging to the Anoplotherium. It is therefore very desirable to procure more perfect specimens, that this interesting question should be determined, as it is a remarkable circumstance that the teeth of two genera so very different should be so closely allied in form."*

After a close comparison of the original specimen, now in the Museum of the Geological Society, with the corresponding part of the *Moschus moschiferus*, with which it agrees in size, I find that the grinders are relatively

* 'Geological Transactions,' Second Series, vol. iii. p. 453.

broader in the fossil, and that the last molar (*m* 3) has the third or hindmost tubercle distinctly divided by a middle longitudinal fissure, which is not the case in the *Moschus*. The grinding surface is less oblique in the fossil than in the Musk-deer or any other Ruminant; and the shape of the coronoid process differs in a still greater degree from that of the *Moschus* and other Ruminants, and by its superior breadth bespeaks the Pachydermal character of the fossil in question.

These differences forbid its association in the same genus with the Musk-deer. On the other hand, we perceive, both in the structure of the teeth and the form of the jaw, a much closer resemblance between the Isle of Wight fossil in question and the genus *Dichobune*. But besides being somewhat larger than the *Dich. leporinum*, the ascending ramus of the lower jaw differs in form and approaches nearer to that of the true *Anoplotherium*. In this family (*Anoplotheriidæ*), however, Mr. Pratt's interesting fossil indicates a new species, which I have referred to the genus *Dichobune*, under the name of *Dichobune cervinum*.

In cut 181, the upper figure gives a view of the fossil from the outside; the lower figure a view from the inside, with an outline of the impression left by the jaw upon the matrix: to the right are given the grinding surface of the teeth.

Fig. 182.

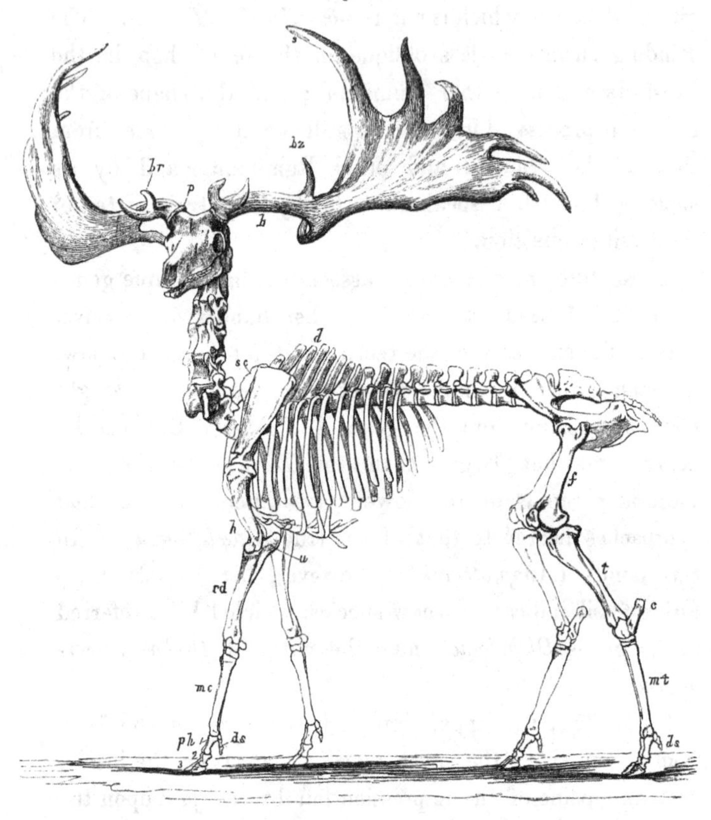

Skeleton of the Gigantic Irish Deer. Height to summit of antlers, 10 feet 4 inches.

GIGANTIC IRISH DEER. Megaceros Hibernicus.

Cervus platyceros altissimus, or Large Irish Deer, MOLYNEUX, Phil. Trans., vol. xix., 1697, p. 485.

" *fossilis,* " " GOLDFUSS, Nova Acta Acad. Nat. Cur. tom. x. pt. ii. p. 455.

Cerf à bois gigantesques,	CUVIER, Ossemens Fossiles, 4to, 1823, tom. iv. p. 70.
Fossil Elk of Ireland,	PARKINSON, Organic Remains, vol. iii. p. 313, pl. xx. fig. ii, (after *Molyneux.*)
Cervus Hibernus,	DESMAREST, Mammalogie, pp. 446, 685.
Cervus megaceros,	HART, A Description of the Skeleton of the Fossil Deer of Ireland, 8vo. 1830.
Fossil Dama of Ireland,	HAMILTON SMITH, Synopsis of the Species of Mammalia, Griffith's Cuvier, 8vo., 1827, p. 306.
Megaceros Hibernicus,	OWEN, Report of British Association, 1843, p. 237.

DR. MOLYNEUX, to whom we owe the first account of the remains of the Gigantic Irish Deer, and by whom they were regarded as a proof that the American Moose was formerly common in Ireland, prefaces his description with the following observation. "That no real species of living creatures is so utterly extinct as to be lost entirely out of the world since it was first created, is the opinion of many naturalists; and it is grounded on so good a principle of Providence taking care in general of all its animal productions, that it deserves our assent."*

The numerous and incontrovertible, though marvellous, results of modern Palæontology, place in a strong light the danger of such a 'petitio principii,' or presumption of the ways in which the benefits of a good Providence are dispensed; and the fallacy of the conclusion founded thereon, in the present instance, is shown both by the now well determined diagnosis of the American Moose, whose dimensions were much exaggerated in the earlier notices of the wild beasts of the North American colonies,† and by the exact comparisons of the osteological characters of the Megaceros with those of all other known Cervine

* Philosophical Transactions, vol. xix. p. 485.

† Molyneux cites 'Jocelyn's New England Rarities' as the source of his ideas regarding the American Moose.

animals. The great extinct Irish Deer surpassed the largest Wapiti, or Elk, in size, and much exceeded them in the dimensions of the antlers. The pair first described and figured in the 'Philosophical Transactions,' measured ten feet ten inches in a straight line from the extreme tip of the right to that of the left antler; the length of each antler from the burr to the extreme tip in a straight line was five feet two inches, and the breadth of the expanded part, or palm, was one foot, ten inches and a half. Dr. Molyneux, after giving the dimensions of the fossil head and its noble attire, says, "Doubtless all the rest of the parts of the body answered these in due proportion;" and he infers the amount of the superiority of bulk of the great Irish Deer over the 'fairest buck' accordingly.

Recent discoveries of the entire skeleton of the Megaceros have, however, shown that the proportions of the trunk and limbs to the vast antlers were not the same with which we are familiar in the existing Deer best provided with these weapons, but that the antlers were both absolutely and relatively larger in the great extinct species: this, in fact, constitutes one of its best characteristics, and involves other differences in the form and proportions of its osseous framework. One of the modifications in the skeleton of the Megaceros, which relates to the vast weight of the head and neck, is the stronger proportions of its limbs; and another and more striking character is the great size of the vertebræ of the neck, which form the column immediately supporting the head and its massive appendages. The extent of these modifications may be appreciated by the following dimensions of the skeleton of the Megaceros, and of that of the great American Moose (*Alces palmata*, var. *Americana*).

	Megaceros.			*Alces.*		
	Ft.	In.	Lin.	Ft.	In.	Lin.
Length of the trunk, from the first rib to the end of the ischium . . .	6	3	3	5	0	0
Height from the ground to the top of the longest dorsal spine . . .	6	0	0	5	6	0
Length of fore leg from the top of the scapula in a straight line . . .	5	7	0	5	4	6
Length of hind leg from the head of the femur in a straight line . .	4	9	3	4	10	9
Circumference of fourth cervical vertebra .	1	10	0	1	0	0
Span of antlers between the extreme tips.	8	0	0	4	0	0

The Elk, or Moose, differs, in fact, from the Megaceros more than any other species of *Cervus*, in the greater proportional length of its limbs,—due chiefly to the peculiar length of the cannon-bones (metacarpi and metatarsi).

The first tolerably perfect skeleton of the *Megaceros* was found in the Isle of Man, and was presented by the Duke of Athol to the Edinburgh Museum; the figure in the 'Ossemens Fossiles,' tom. iv. pl. viii. is taken from an engraving of this skeleton transmitted by Professor Jamieson to Baron Cuvier. Another skeleton was composed and set up by Dr. Hart, in the Museum of the Royal Dublin Society, from a collection of bones found at Rathcannon in Ireland, and this is figured in his 'Description of the Skeleton of the Fossil Deer of Ireland.' A third engraving of a foreshortened view, by Professor Phillips, of the skeleton of the *Megaceros*, from Waterford, in the museum of the Yorkshire Philosophical Society, was published, without description, by Mr. Sunter of York; and this exhibits a more natural collocation of the bones, than do either of the above-cited figures. Three very complete and well-articulated skeletons have since been added to English collections; one of these is in the British Museum, another in the Woodwardian Museum at Cambridge, and a third in the Hunterian Museum at the Royal College of Surgeons in London, from which I have composed the figure engraved in cut 182, and which I believe to convey an exact idea of the port and proportions of the noble extinct animal.

The antlers of the Megaceros spring from the extremities of a strong transverse semicylindrical eminence, which crosses the top of the skull rather nearer the orbits than the occiput; the base of each antler is encircled by the rugged and perforated ridge or ring of bone called the 'burr,' or 'pearl' (*p*), immediately above which the beam sends forward the first branch, or brow-antler (*br*), which is sometimes simple—sometimes expanded and bifurcate at the extremity—rarely divided into three points. The beam or shaft (*b*), is usually subcylindrical, and so continues, gradually enlarging for about one-fourth the length of the entire antler, where it expands into the broad and massive subtriangular plate of bone, called the 'palm,' which sends off from six to nine, but commonly seven branches. The first (*b z*), comes off from the fore-part, is directed forwards, and usually inclines inwards; it answers to the 'bezantler' in the Red-deer. The next branch is sent off, like that in the Fallow-deer, from the back part of the palm a little above or beyond the bezantler; all the remaining branches, usually five in number, are continued from the fore-part and the extremity of the palm. The graceful oblique twist commencing in the beam is continued in the palm, so as to turn its convex surface obliquely forwards and downwards, and its concave surface upwards, backwards, and with a slight inclination towards that of the opposite antler, when the head is carried in the horizontal position. The longest branches are usually the two which come off beyond the bezantler from the fore-part of the palm (*s*); those from the extremity of the palm are generally the shortest, and curve in a direction opposite to the former.

Camper* first recognised the well-marked differences between the Megaceros and the Elk, in the conformation of the skull. The peculiarly developed and prehensile upper lip of the Elk is associated with an unusual elongation of the intermaxillaries and nasal apertures, and a shortening of the nasal bones; but the skull of the Megaceros closely conforms to that of the ordinary deer, and more especially the Rein-deer, as Cuvier† has pointed out. The dentition of the Megaceros displays the ordinary Ruminant type, viz.: $i.\ \frac{0}{8},\ p.\ \frac{3-3}{3-3},\ m.\ \frac{3-3}{3-3}=32$: that is, there are eight incisors in the lower jaw, and six molars on each side of both jaws, the first three being premolars, the last three true molars. There are no canines or their rudiments retained in either sex.‡ The subjoined figure of the first true molar (fourth of the series, counting backwards,) in the upper jaw, well illustrates the peculiar character of the grinding surface of the molar teeth in a Ruminant quadruped; the body of the tooth is divided into two lobes (*a*, *b*,) placed one in front of the other, with the inner side convex, the outer side concave or sinuous from a slight convexity at the middle part. Each lobe is subdivided by a vertical cleft, *e*, lined by enamel and bent, with its convexity turned towards the inner side and its concavity towards the outer

Fig. 183.

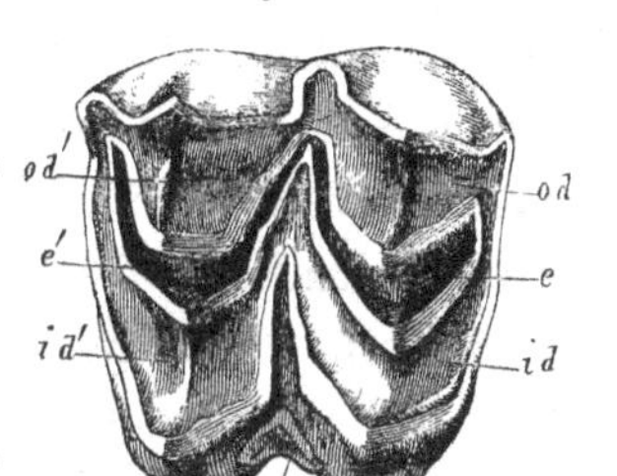

Upper molar of *Megaceros*, nat. size.

* Nova Acta Acad. Petropol. ii. p. 258.

† Op. cit. p. 78.

‡ The two exterior incisors of the under jaw represent the lower canines of the horse, but there are no rudiments of upper canines, as there are in the Red-deer and Rein-deer.

side of the tooth. The concavities and convexities are reversed in the grinders of the lower jaw. The summit of each lobule, or division of the lobe, thus presents a crescentic figure, and, when worn by mastication, exposes a body of dentine (*od*, *id*,) with a raised border of enamel, coated thinly by cement. The crescentic fissures (*e*, *e*) between the lobules, are filled partly by cement, partly, in the recent Ruminant, by masticated food; and when the tooth is much worn, they are divided from each other, and separately inclosed by a crescentic island of enamel: the entire circumference of the complex molar being also invested by a coat of enamel and a thinner layer of cement.

In the Megaceros the inner lobules (*id*) are thicker transversely than in the Aurochs, the crescentic enamel islands are narrower and more simple, and the cemental cavity of each is continued into the other until a later period of attrition. In the Elk, the central crescents intercommunicate for a still longer period, and the crown of the molar is cleft by a crucial incision. There is a small accessory column (*d*) at the internal interspace of the lobes of the tooth in both *Alces* and *Megaceros*, which is not present in the Rein-deer; but it is confined to the base of the fissure, not developed to such a length as in the molars of the Aurochs and other *Bovidæ*. With regard to the premolars, which may be compared to a single lobe of the true molars, the central crescentic island of enamel is more complex than in the Aurochs, the inner border forming a fold near its back part which extends to the outer border. In the lower jaw the first and second premolars are relatively larger and more complex than in the Aurochs. I have been led into these details on account of the close correspondence in size between the teeth of the Megaceros and those of the large fossil Bovine quadrupeds. The differences may seem slight, but they are constant and serve to distinguish the species: they determine, for example, the fossil fragment of the upper jaw with the molar teeth from Kent's Hole, now in the British Museum, which fragment has been reduced to its present form by the teeth of the extinct *Hyæna*, to belong to the *Cervus megaceros*, and thus establish the high antiquity of that extinct species.

The great proportional size of the cervical vertebræ of the Megaceros has been already noticed: the atlas appears like a second occipital bone, but exceeds that cranial vertebra in breadth. The extraordinary development of the muscular part of the neck is indicated by the massive processes, especially of the five vertebræ which follow the axis; and the thick full neck, which is so characteristic a feature in the Stag, must have been still more remarkable in the living Megaceros. The cervical vertebræ of the female were one third smaller than in the male. The dorsal vertebræ are thirteen in number, and the anterior ones are remarkable for the length of the spinous processes (fig. 182, *d*) which gave attachment to the elastic ligaments supporting the head: those of the third, fourth, and fifth dorsals rise to a foot in height.

The six lumbar, the sacral, and the caudal vertebræ, closely agree with those parts in the existing Deer. The sternum consists of seven bones, including the xiphoid; they become broader and flatter to the sixth, which measures five inches across.

The bones of the extremities more resemble those of the Rein-deer than the Elk, but are relatively stronger in proportion to their length than in any existing species of *Cervus*. In fig. 182, *sc* is the scapula or blade-bone; *h*, the humerus or arm-bone; *rd*, the radius or bone of the fore-arm: *u*, the olecranon or process of the ulna, answering to the bone of the elbow, and which in Ruminants is anchylosed to the radius; *mc*, is the metacarpal or cannon-bone of the fore-leg; *ph* 1, 2, 3, the three toe-bones or phalanges of the hoof.

In all Deer, besides the two toes corresponding to the third and fourth in the pentadactyle foot,—the metacarpals of which are blended together to form the cannon-bone, whose bifurcate lower end supports the two hoofs or divisions of the cloven foot,—there are rudiments of the second and fifth toes, which appear externally as the two small posterior supplemental hoofs, *ds*. In the Rein-deer both the upper and the lower ends of the rudimental metacarpals of these abortive toes are present in the skeleton, the intermediate part being absent.* I have recognised the upper end of the metacarpal of the inner or second rudimental toe in a collection of the bones of a Megaceros carefully removed from subturbary shell-marl near Limerick. They are slender pointed styles, about three inches in length; articulated to the bone formed by the confluent trapezoid and os magnum, and to a rough surface on the inner and posterior angle of the upper end of the cannon-bone: a more extensive rough surface on the outer and posterior angle would indicate that the proximal end of the fifth or outer metacarpal had likewise existed, and of larger size, in the perfect skeleton.

The intermediate parts of both rudimental metacarpals are wanting, as in the Rein-deer; but the phalanges supporting the small spurious hoofs have been recovered, and are represented in fig. 182. The middle phalanx is subcompressed, square, about an inch in length; the ungual

* Cuvier, loc. cit. p. 18.

phalanx longer, rough, and rounded at the end: there are two strong sesamoids behind each division of the distal end of the cannon-bone. In the Elk the upper end of the inner supplemental metacarpal is not an inch in length; but the lower end of the metacarpal of each of the spurious hoofs is two-thirds the length of the cannon-bone. In the hind leg of fig. 182, *f* marks the femur or thigh-bone; *t*, the tibia or leg-bone; *c*, the calcaneum, heel-bone or hock; *m t*, the *metatarsus* or hind cannon-bone; *d s*, the spurious hoofs. Both metacarpal and metatarsal cannon-bones are much less deeply indented longitudinally in the Megaceros than in the Rein-deer.

Molyneux, who knew the Moose of North America only by the vague and exaggerated notices of Jocelyn, but who had seen the antlers of the Swedish Elk, accurately points out the difference between them and those of the Megaceros in their much smaller size, in the greatest expansion of the palm being nearest the head, and "the smaller branches not issuing forth from both edges of the horns, as in *ours*, but growing along the upper (anterior) edge only."* To these differences must be added the absence of the brow-antlers in the Elk, and the great breadth and subdivision of the branch answering to the bezantler, which, in the Elk (fig. 192), forms rather a division of the palm.

The antlers of the great Wapiti differ from those of the Megaceros in having no palm, the cylindrical figure prevailing throughout all the ramifications. The Rein-deer differs in the superior length and ramification of the brow-antlers (fig. 197), and in the greater length and different mode of branching of the beam, which is smooth and subcompressed. But the male Rein-deer is that existing species in which

* Loc. cit. p. 503.

the relative size of the antlers to the body comes nearest to the peculiar proportions of those appendages in the Megaceros.

The brow-antler, and the expansion of the beam into a palm, brings the *Megaceros*, as Colonel Hamilton Smith first showed, into that group of the Cervine family to which the Fallow-deer belongs,—this species being, perhaps, the nearest existing representative of the gigantic extinct species; but in the Fallow-deer, (fig. 191) all the branches above the bezantler (*bz*) are sent off from the posterior margin and end of the palm, while in the Megaceros they are all, with one exception, sent off from the anterior and terminal margin. The brow-antler (*br*) in the Fallow-deer is always simple, cylindrical, and pointed; in the Megaceros it is often expanded and sometimes bifurcate at the end, but never so long or so ramified as in the Rein-deer. With justice, therefore, might Cuvier, who had pursued the comparison of the antlers through all the known species of Deer, affirm that "the inspection of the head and antlers alone of the 'Cerf à bois gigantesques' suffices to assure us that it is an extinct animal, like the long-headed Rhinoceros, the little Hippopotamus, the Elephant with long tusk-sockets, and the gigantic Tapir,* which, if they belong to known genera, are not the less unknown, as species, on the actual surface of the earth." †

In fact, the antlers of the great Irish Deer, combining some of the characters of those of the Elk, the Rein-deer, and the Fallow, with others peculiar to themselves, compel the zoologist, guided by the principles so admirably wrought out by Colonel Hamilton Smith ‡ for the subgeneric ar-

* Now known as the still more extraordinary Dinothere, of which not only the species but the genus has passed away.

† Op. cit. p. 82.

‡ Griffith's Cuvier, 8vo. vol. iv. 1827.

rangement of the extensive and diversified species of the great Linnæan genus *Cervus*, to regard the subject of the present section as the type of a distinct subgenus, for which the term *Megaceros*, originally applied by Dr. Hart as the 'nomen triviale' of the extinct species, may be retained, as indicative of the most striking and characteristic feature of the antlers, viz., their great proportional size.

The weight of the skull and antlers of the Megaceros in the Museum of the College of Surgeons in London, is seventy-six pounds avoirdupoise: that of the skull and antlers of the specimen in the Royal Dublin Society is eighty-seven pounds, avoirdupoise. The average weight of the skull, without the horns or lower jaw, is five pounds and a quarter. From the identity of texture of these enormous cranial weapons with those of the Deer-tribe, and from the development of the burr at the base, we may infer that the large bloodvessels, shown by their impressions to have been spread so richly over the surface of the antlers during the period of growth, were ultimately obliterated, and that the antler, then losing its vitality, was undermined by the absorbent process, and cast off. Such shed antlers, showing the characteristic convex surface of the detached base beneath the burr, have been frequently found in Ireland. Dr. Hart has noticed them in his tract above cited, and the base of one in the British Museum is figured in cut 194. It cannot be doubted but that the growth and shedding of the antlers of the Megaceros, obeyed the same periodical law as do those of all existing deer: but, when we reflect that between sixty and seventy pounds' weight of osseous matter was annually thrown out by the carotids in the course of three or four months, we may well exclaim, with Redi, "Maximâ profectò admiratione dignum est tantam molem ramorum tam brevi tempore quotannis renasci et crescere." *

It is true, indeed, that these antlers were subject to the periodical variations of size and form which occur in the existing species of Deer, and which Mr. Bell has illustrated at pp. 400 and 404 of the 'History of British Quadrupeds,' in the instances of the Red-deer and Fallow-deer.

A corresponding suite of antlers of the Megaceros from their first appearance in the young animal, has not yet been recovered, the smaller and simpler specimens probably not attracting the same attention as the larger antlers. It is, however, extremely desirable that such specimens should be collected and preserved whenever they may be met with. The three best-marked varieties which have come under my notice, and which appear to indicate progressive epochs in the age of the animal, are those of which figures are subjoined.

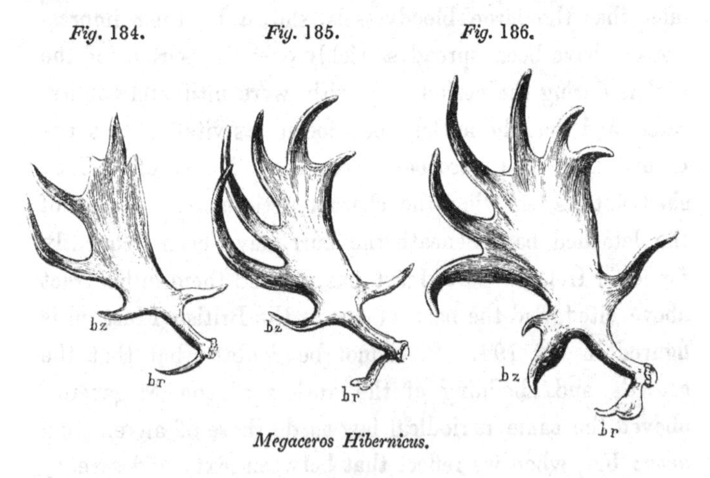

Megaceros Hibernicus.

* 'Experimenta circa res diversas naturales,' 12mo., 1675, p. 156. The wonder of the great Italian physiologist was excited by the comparatively insignificant phenomena in the Red-deer, which, in the course of about ten weeks, developes its antlers, weighing about four and twenty pounds.

The first (fig. 184) which apparently corresponds with the state of the antlers at the fourth year in the Fallow-deer,* is five feet in length, and fourteen inches across the palm: it presents a simple cylindrical and pointed brow-antler (*br*); a short and simple bezantler (*bz*); the hind branch almost straight, and only two long branches from the fore-part of the palm, which terminates in three short straight obtuse points, the middle being the longest. The second figure (fig. 185) shows an expansion and flattening of the brow-antler, an elongation of the bezantler and of the anterior branches of the palm, and the prolongation of the three terminal points into branches: the total number of branches being eight. The length of the antler, following the curve, is six feet; the greatest breadth of the palm fifteen inches. This form of antler corresponds with that at the fifth year in the Fallow-deer. In the third figure, (fig. 186,) the brow-antler is expanded and bifurcate; the bezantler is likewise expanded and divided into two points, but this is a very rare variety. It is shown on the right side in a pair of antlers in the Hunterian Museum, and in both antlers of the remarkably fine skeleton in the Museum of the Royal Dublin Society. The palm is much increased in breadth and sends off six branches besides the posterior one, the number of points in this antler being eleven. The length of the antler following its curve, is seven feet; the breadth of the palm thirty inches. Such an antler would indicate the Megaceros to have reached the prime of its age, like the 'crowned Hart' of the seventh or eighth year. The antlers of the Megaceros, which retain the same expanse of palm with shorter branches, especially the terminal ones, have probably belonged to older animals when the reproductive force was on the decline.

* 'Bell's Quadrupeds,' p. 404, the middle figure.

The dimensions given may be regarded as the average size of antlers having the characters above described: the palm is most subject to variation in regard to its breadth, the two antlers of the same pair rarely agreeing in this respect. In one pair, showing the second character described, I have seen the right antler with a palm twelve inches broad, and the left with one sixteen inches broad. In a pair in the Hunterian Museum, showing the third character, the left palm is nineteen inches across, the right twenty-three inches; but the palm sometimes attains the breadth of three feet. When circumstances have favoured such unusually full development of the antlers, it is sometimes associated with modifications which increase the strength of the supporting beam. There is an example of this kind in the skull and antlers of a noble Megaceros which are fixed over the entry to the hall of the ancient Manor-house of Knowle, near Sevenoaks, Kent: the circumference of the basal ridge or 'burr' is sixteen inches; a strong ridge is developed from the whole under part of the beam, and is continued into a short-pointed snag, above which the beam begins to expand into the palm.

Having noticed the principal varieties of the antlers of the Megaceros depending upon age or individual peculiarities, I may briefly advert to the facts which elucidate the relation of the antlers to the sex of the great extinct Cervine animal. In the existing species of the Deer-tribe, the frontal furniture is peculiar to the males, with the exception of the Rein-deer, and, occasionally, the Elk; naturalists were therefore interested in ascertaining whether the Megaceros pushed its affinity to those large existing species by the development of antlers in the female sex.

Mr. Maunsell, in a letter quoted by Dr. Hart, descriptive of the discovery of numerous remains of the *Megaceros* in shell-marl at Rathcannon, observes that "of eight heads which we found, none were without antlers; the variety in character, also, was such as to induce me to imagine, that possibly the females were not devoid of these appendages." *

Fig. 187.

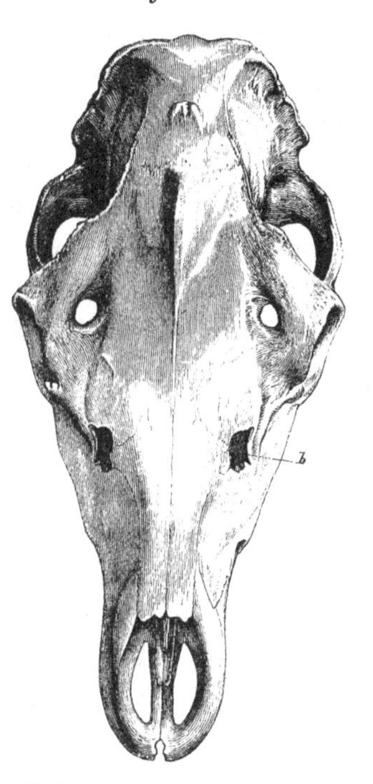

Skull of Female Megaceros, $\frac{1}{8}$th nat. size.

Cuvier was of the same opinion, and he thought that the Megaceros in this respect might resemble the Rein-deer. To this conclusion, also, Dr. Hart is much disposed to subscribe, from having observed that these parts presented differences in size and strength, which appear not to be dependant on difference of age; and he cites an example of the skull of the Megaceros, with teeth much worn down, in which the antlers were less expanded and one-sixth less than those belonging to an evidently younger individual, and which, therefore, he concluded might not unlikely be the male, and the older specimen the female. But in all Deer, the antlers, when the animal has passed its prime, begin to be shorn of those fair proportions that characterise the vigour of life; and the diminished size of the antlers of the aged skull of the Megaceros in Trinity College, shows that their development in

* 'Description of the Skeleton of the Fossil Deer of Ireland, *Cervus megaceros*,' by John Hart, &c. 8vo. 1830, p. 15.

the extinct Deer followed the same laws which govern the succession of the antlers in the existing species.

The first direct evidence on this point was adduced by Professor Phillips, who, in reporting on a donation to the Yorkshire Philosophical Society by G. L. Fox, Esq., of a rich collection of remains of the Megaceros from near Waterford, states; "There is among the specimens the head of a female without horns."* I owe chiefly to the kind interest which the Earl of Enniskillen has been pleased to take in the researches connected with the present work, the opportunity of examining three skulls of the Megaceros, which, evidently mature by their size and state of dentition, and without a trace of the pedestal or place whence antlers could have been shed, must be concluded to have belonged to the female sex. One of these is the subject of cut 187.

They very nearly equal in length the skull of the male; but the occipital bones, and especially the condyles, are smaller, the transverse eminence at the back part of the frontal is wanting, and in its place there rises a longitudinal prominence (fig. 188, *a*,) from the posterior half of the frontal suture, like the median prominence in the skull of the Giraffe. The supraorbital foramen is as large as in the skull of the male;† the preorbital vacuity (fig. 187, *b*,) is somewhat larger than I have usually found it in the

* 'Report of the Yorkshire Philosophical Society,' 1836.

† Dr. Hart, in his description of the skeleton of the Megaceros, says (p. 19) "There is a depression on each side in front of the root of the antler and over the orbit capable of lodging the last joint of the thumb, at the bottom of which is the superciliary hole, large enough to give passage to an artery proportioned to the size of the antler." This foramen, however, is equally large in the female. I have injected the skull of a Fallow-deer with the antlers 'in velvet;' they were supplied by two large branches sent off from the external temporal artery, where it passes behind the orbit: doubtless, the arteries of the antlers had as little connection with the cavity of the orbit, or its superior perforation in the *Megaceros*.

Fig. 188.

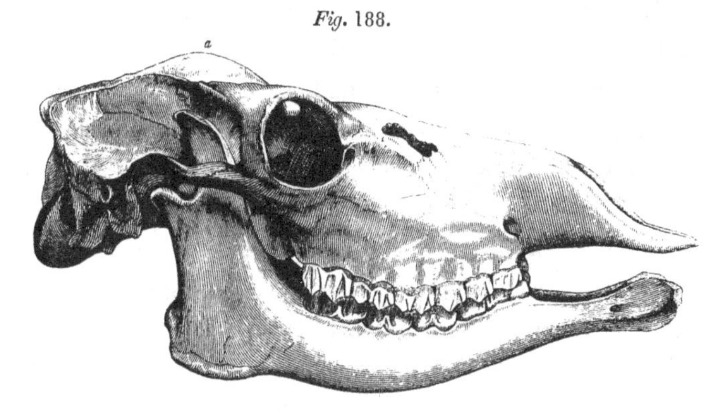

Skull of Female Megaceros, $\frac{1}{8}$th nat. size.

antlered sex. The cervical vertebræ are of nearly equal length with those in the male, but are less by one-third in breadth, and the dorsal spines are one-third shorter; these modifications obviously relate to the non-development of the antlers in the female sex.

Is there any evidence, it may be asked, that the Megaceros coexisted with the human race, or that its extinction was the result of man's hostility? Dr. Molyneux* says that its extinction in Ireland has occurred "so many ages past, as there remains among us not the least record in writing, or any manner of tradition, that makes so much as mention of its name; as that most laborious enquirer into the pretended ancient, but certainly fabulous, history of this country, Mr. Roger O'Flaherty, the author of *Ogygia*, has lately informed me."

The term *shelch* in the romance of the Niebelungen, written in the 13th century, and there applied to one of the beasts slain in a great hunt a few hundred years before that time in Germany, has been cited by Goldfuss, and subsequently by other naturalists, as probably signifying

* Phil. Trans. xix. p. 490.

the *Megaceros*, just as the 'halb-wolf' of the same 'Lied,' has been conjectured to be the *Hyæna*.

The total silence of Cæsar and Tacitus respecting such remarkable animals, renders their existence and subsequent extirpation by the savage natives a matter of the highest improbability; and it has been well observed by Dr. Buckland that "the authority of the same romance would equally establish the actual existence of giants, dwarfs, and pigmies, of magic tarncaps—the using of which would make the wearer become invisible—and of fire-dragons, whose blood rendered the skin of him who bathed in it of a horny consistence, which no sword or other weapon could penetrate."

Some appearances in the bones themselves of the Megaceros, and, perhaps, an undue confidence in the vague statements of their discovery, with remains of the existing deer, hog, and sheep, in peat bogs, have led to the opinion that the Gigantic Deer existed within the time of man. Dr. Hart cites the fact of the discovery of a human body in gravel, under eleven feet of peat, soaked in the bog-water, which was in good preservation, and completely clothed in antique garments of hair, which, it had been conjectured, 'might be that of our fossil animal.' But if any Megaceros had perished, and left its body under the like circumstances, its hide and hair ought equally to have been preserved. Except, however, the solitary instance of fat or adipocere in the shaft of one of the bones discovered by Archdeacon Maunsell, not a particle of the soft parts of the animal seems ever to have been found. Dr. Hart conceives that "more conclusive evidence on this question is derived from the appearance exhibited by a rib, in which he discovered an oval opening near its lower edge, with the margin depressed on the outer, and raised on the inner surface, round

Fig. 189.

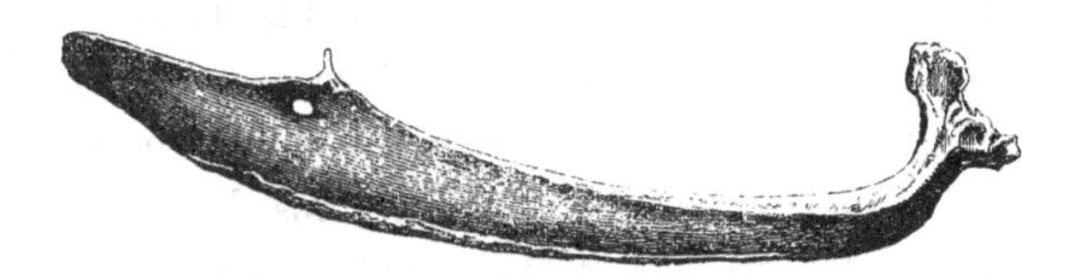

Perforated rib of the Megaceros, from Hart's Memoir.

which there is an irregular effusion of callus."* "This opening," he says, "appears evidently to have been produced by a sharp-pointed instrument, which did not penetrate so deep as to cause the animal's death, but which probably remained fixed in the opening for some length of time afterwards; in fact, such an effect as would be produced by the head of an arrow remaining in a wound after the shaft was broken off."—Op. cit. p. 29.

But a conical arrow-head, with a base one inch in diameter, sticking in a rib with its point in the chest, must have pierced the contiguous viscera, and, rankling there, have excited rapid and fatal inflammation. The evidence of the healing process in the bone, would rather show that the instrument which pierced the rib, had not been left there to impede the operations of the 'vis medicatrix naturæ.' A pointed branch of the formidable antler is as well suited to inflict such a wound, as the hypothetical arrow; and if the combative instincts of the rutting Stag

* Dr. Hart's 'Description,' &c., p. 21. Dr. Hart gives the following analysis by Dr. Stokes of the rib of a Megaceros :—

Animal matter	42·87
Phosphates, with a trace of fluate of lime	43·45
Carbonate of lime	9·14
Oxides	1·02
Silica	1·14
Water and loss	2·38
	100·00

rightly indicate the circumstances under which the wound of the Megaceros was inflicted, they would be those which best accord with the actual evidence of recovery from it.

The Earl of Enniskillen has transmitted to me specimens of carpal and tarsal bones of the Megaceros diseased with exostosis; and there is in his Lordship's collection a lower

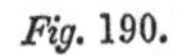

Fig. 190.

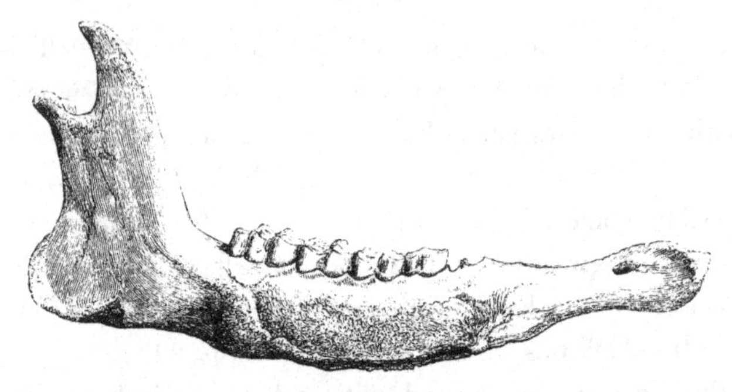

Diseased lower jaw of the *Megaceros*, from shell-marl, Ireland.

jaw of the same extinct species, from which a large part of the outer wall has exfoliated, probably in consequence of a blow received in combat at the rutting-season; a considerable amount of new irregular osseous matter has been formed to replace the lost portion of bone. Cut 190 gives a figure of this interesting example of primæval disease.

To my enquiries as to the places whence the numerous specimens of the Megaceros which I have examined in travelling through both North and South of Ireland, had been obtained, the reply was usually from such or such a bog; but I met with no person who had seen them in the peat itself. In every case where more definite information was afforded by an eye-witness of their discovery, it appeared that the antlers and bones had been dug out of the lacustrine shell-marl beneath the peat or bog earth. The most instructive and precise account of the situation in which the remains of the *Megaceros* have been found in Ireland, is contained in the 'Philosophical Transactions,' vol. xxxiv. p. 122, in a letter from Mr. James Kelly, dated Downpatrick, Dec. 22nd, 1725. He says, "For the first three feet we met with a fuzzy kind of earth, that we call moss, proper to make turf for fuel; then we find a stratum of gravel about half a foot; under which, for about three feet more, we find a more kindly moss, that would make a more excellent fuel; this is all together mixed with timber, but so rotten that the spade cuts it as easily as it does the earth. Under this, for the depth of three inches, we find leaves, for the most part oaken, that appear fair to the eye, but will not bear a touch. This stratum we find sometimes interrupted with heaps of seed, which seem to be broom or furze seed; in other places in the same stratum we find sea-weed, and other things as odd to be at that depth. Under this appears a stratum of blue clay, half a foot thick, fully mixed with shells; then appears the right marl, commonly two, three, or four feet deep, and in some places much deeper, which looks like buried lime, or the lime that tanners throw out of their lime-pits, only that it is fully mixed with shells,—such as the Scots call 'fresh-water wilks.' Among this marl, and often at the bottom of it, we find very great horns, which we, for want of another name, call 'Elk-horns.' We have also found shanks and other bones of these beasts in the same place."

The head and antlers, described and figured by Molyneux in the 'Philosophical Transactions' for 1697, lay about five feet under ground: "the first pitch was of earth, the next two or three feet of turf, and then followed a sort of white marl, where they were found."

Dr. Buckland states, on the authority of Mr. Weaver, that the bones and antlers of the Megaceros which were found in the bog of Kilmegan, near Dundrum, in the county of Down, "lay at the bottom of the peat between it and a bed of shell-marl, resting upon, or being merely impressed in the marl, which is composed of a bed of freshwater shells, from one to five feet thick, and must have been formed while the bog was a shallow lake."

The first specimen of the Megaceros discovered in England consisted of a skull and antlers dug from the depth of six feet out of a peat-moss at Cowthorpe, near North Dreighton, in the county of York.*

Mr. Parkinson refers the beams of two antlers found in the till at Walton in Essex, on account of their large size, to the Great Irish Deer; and I have obtained more satisfactory evidence of the Megaceros from the same newer pliocene stratum, by inspection of the collection of fossils belonging to Mr. Brown of Stanway, in which is preserved, not only the large round beam, but the characteristic brow-antler and part of the palm, as far as where it has expanded to a breadth of ten inches. The length of the brow-antler is five inches and a half, but its extremity is broken off. Mr. Brown has, also, obtained from the same freshwater formation on the Essex coast, the entire lower jaw of the Megaceros.

The base of an antler as large as that of the Megaceros has been dredged up from the oyster-bed at Happisburgh, already referred to as famous for the numerous teeth of the Mammoth which it has yielded.

Remains of the Megaceros found eight feet and a half

* Phil. Trans. 1746, vol. xliv. pl. i. fig. 3.

below the surface of a peat-bog at Hilgay, Norfolk, are preserved in the collection of Mr. Whickham Flower, F.G.S. Antlers of the Megaceros have been disinterred from the marl or gravel beneath peat-bogs in Lancashire.

The formerly unique skeleton of the Megaceros in the Museum of the University of Edinburgh was obtained from a formation in the Isle of Man, which Mr. E. Forbes, Prof. of Botany in King's College, London, informs me is a white marl, with freshwater shells found in detached masses, occupying hollows in the red marl; which red marl, by the proportion of marine shells of the species found in the neighbouring seas, is referable to the newer pliocene period. The cervine fossils have never been met with in the marine or red marls in the Isle of Man, but only in the white marls occupying the freshwater basins of the red marl; and from the position of the beds containing the remains of the Megaceros, Prof. Forbes concludes that this gigantic species must have existed posterior to the elevation of the newer pliocene marl, which is probably continuous with the same formation in Lancashire and at the mouth of the Clyde, forming a great plain, extending from Scotland to Cheshire, and now for the most part covered by the sea. The geographical features of the dry land, the seat of those lakes in which the remains of the Megaceros are most commonly found, would seem, therefore, to have undergone much change since the time of its extinction.

Fragments of the huge antlers and other remains of the Megaceros have been discovered in some of the ossiferous caverns in England. A characteristic specimen, now in the British Museum, was obtained by Mr. M'Enery from Kent's Hole; it consists of part of the upper jaw, with both series of molar teeth; it precisely corresponds with the same parts in the skull of a Megaceros from Ireland.

Thus the evidence of the former existence in England of the gigantic extinct Deer, though less striking and abundant than in Ireland, is complete, and of greater value, inasmuch as it establishes the contemporaneity of that species with the Mammoth, Rhinoceros, and other extinct Mammalia of the period of the formation of the newest tertiary fresh-water fossiliferous strata.

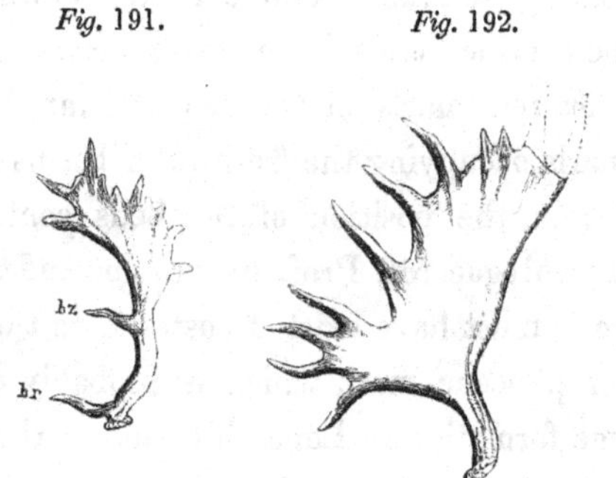

Fig. 191. Antler of Fallow Deer.

Fig. 192. Antler of Elk.

RUMINANTIA. *CERVUS.*

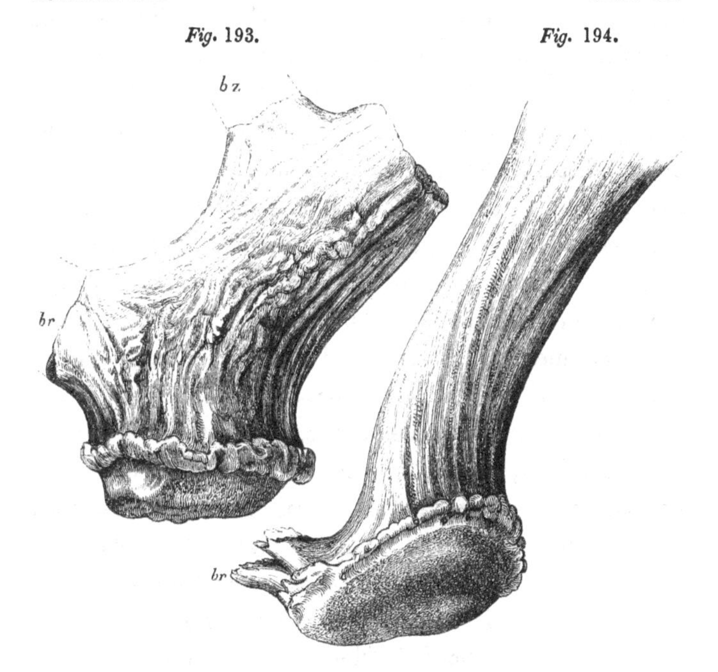

Strongyloceros spelæus, Kent's Hole, $\frac{1}{3}$ nat. size.

Megaceros, shell-marl, $\frac{1}{3}$ nat size.

STRONGYLOCEROS SPELÆUS. Gigantic Round-antlered Deer.

THE base of an antler, equalling in dimensions that of the largest Megaceros, has been found fossil and partly gnawed by Hyænas, in the cavern of Kent's Hole near Torquay. This fragment (fig. 193), fifteen inches in circumference, differs from the antler of the Megaceros (fig. 194) in sending off the bezantler (*bz*) at a shorter distance from the brow-antler (*br*), and in the beam diminishing in size and preserving the cylindrical figure above the origin of the bezantler, by which it may be inferred that the species so represented belonged to the round-antlered section of the Cervine genus, (Elaphine group of Col. H. Smith,) and to which section the subgeneric name *Strongyloceros* may be applied. The existing species in this group which most nearly approaches in size the extinct one indicated by the present fossil, is the great Wapiti Deer of Canada (*Cervus strongyloceros*, Schreber, *Cervus Canadensis*, Brisson); but the fossil differs from those antlers of the Wapiti that have come under my observation in the greater distance between the brow-antler and bezantler. Cuvier, however, figures some specimens which resembled the fossil in this respect.

Such a fragment of an antler as the one from Kent's Hole here described, though it be sufficient to determine the great Deer, of which it once formed part, to have been not only distinct from the Megaceros, but to have belonged to a distinct subdivision of the cervine genus, does not permit a satisfactory determination of its specific distinction from the largest existing species of its own subgenus: but, on the other hand, it affords as little ground for asserting its specific identity with them, and, from analogy, it is more probable that it was a distinct species, which, therefore, I propose to indicate as the *Cervus* (*Strongyloceros*) *spelæus*.

If the trunk and limbs bore the same proportions to the head and antlers as in the Wapiti and Red-deer, as most probably they did, the species indicated by this remarkable fragment of antler must have been the most gigantic of our extinct English Cervine animals.

The fragment of the lower jaw (fig. 195) indicates clearly a Cervine animal with a head larger than that of the Megaceros: this fragment shows a depth of jaw of two inches and a half below the second true molar, but has belonged to an immature animal, which had not shed the last deciduous molar, *a m*, nor had fully acquired the second true molar, *m* 2. Sufficient of the crown of this tooth has risen above the gum to show that it had not the accessory column at the base of the outer interspace of the two lobes, as in the Megaceros and the large Bovine Ruminants; but that it resembled the Wapiti and Red-deer, both in the absence of that column, and in its presence in the first true molar, *m* 1. The last deciduous molar shows the same large proportional size of the third lobe, which characterises this tooth in all Ruminants, and distinguishes it from the last true molars. I conclude, therefore, that this fragment, which is also from Kent's Hole, and has apparently been fractured by the teeth of Hyænas, belonged to another individual of the same great species of Round-antlered Deer, to which I have referred the base of the antler above described.

Whether this species be identical with the fossil *Cervus giganteus* of M. Robert, which he distinguishes from the *Cervus Hibernus*, and discovered associated with the Hyæna and Mammoth in the ferruginous beds at Cussac, Haute Loire, I am unable to say.

Fig. 195.

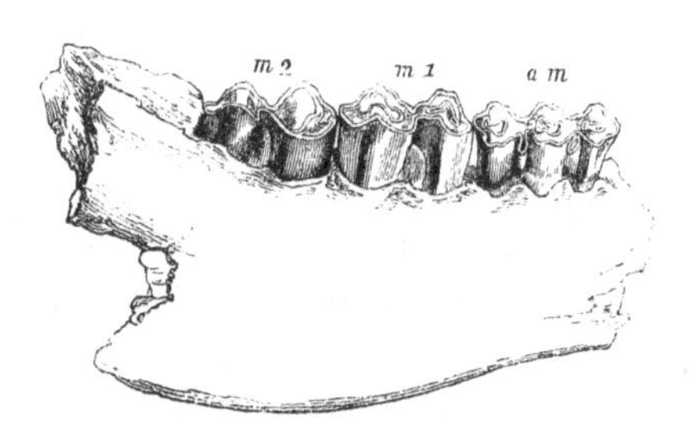

Fragment of under jaw, ⅓ nat. size. (*Strongyloceros spelæus ?*) Kent's Hole.

RUMINANTIA. *CERVUS.*

Fig. 196.

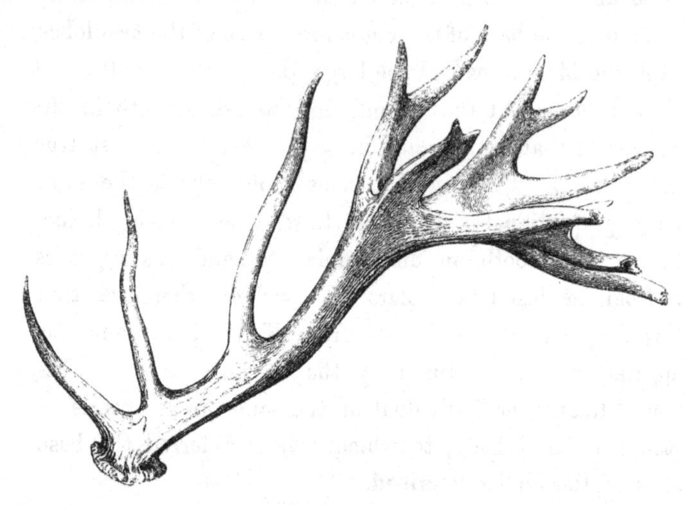

Fossil antler of Red Deer. Alluvium, Ireland.

CERVUS (STRONGYLOCEROS) ELAPHUS.
Red-deer.

Cerf semblable au cerf ordinaire,	CUVIER, Ossemens Fossiles, 4to. 1823, tom. p. 98.
Elaphus fossilis,	H. V. MEYER, Palæologica, 8vo. 1832, p. 91.
Fossil Stag and Deer,	BUCKLAND, Reliquiæ Diluvianæ, *passim.*
Red Deer, Cervus Elaphus,	OWEN, Report of British Association, 1843, p. 236.

THE most common fossil remains of the Deer-tribe are those which cannot be satisfactorily distinguished from the same parts in the species *Cervus Elaphus*, which most abounded in the forests of England until the sixteenth century, and which still enjoys a kind of wild life, by virtue of strict protecting laws, in the mountains of Scotland.

The oldest stratum in Britain yielding evidence of a

Cervus of the size of the Red-deer, is the red-crag at Newbourne. More conclusive evidence of the specific character of this sized Deer is afforded by antlers as well as teeth and bones, and these attest the existence of the *Cervus Elaphus* through intermediate formations, as the newer freshwater pliocene, and the mammoth silt of ossiferous caves, up to the growth of existing turbaries and peat-bogs. I found remains of this round-antlered Deer in all the collections of Mammalian fossils from the fluvio-marine crag, and more recent freshwater and lignite beds in Norfolk, Suffolk, and Essex. Similar remains have been obtained from the lacustrine deposits in Yorkshire; the head with antlers two feet ten inches in length, figured by Knowlton in the 'Philosophical Transactions' for 1746, pl. i. fig. 2, was dug out of a bed of sand in the river Rye, in the East Riding of that county.* Hopkins transmitted the sketch of an antler of a large Red-deer to the Royal Society, which is figured in vol. xxxvii. No. 422, of the 'Philosophical Transactions.' The terminal branches of the crown are broken off, yet the length of the antler is thirty inches; the circumference of the base ten inches, and the length of the brow-antler sixteen inches and three quarters. This was drawn out of Ravensbarrow Hole adjoining Holker Old Park, Lancashire, by the net of a fisherman, in 1727. "The tide flows constantly where it is found, and the land is very high near it."—Ib. p. 257. The antlers attached to the head of the Stag found beneath a peat-moss in the same county, and figured by Leigh in his 'Natural History of Lancashire,' attest an animal of equal size, each antler measuring forty inches in length. Mr. Gale records the discovery of antlers of a Red-deer, with a brow-antler nine inches long, found by the workmen

* See also Young and Bird, 'Geology of Yorkshire,' 4to. pl. 17.

in driving a drain to a lead-mine, about nine yards deep from the surface of the earth, at Lathill Dale, near Bakewell, Derbyshire. The antlers, with other bones, were in "a sort of soft coarse clay, or marl, interspersed with little petrified balls or pellets of the same kind of substance as the tuft."* Mr. Barker subsequently narrates that "some men working in a quarry of that kind of stone which in this part of Derbyshire we call 'tuft,' at about five or six feet below the surface in a very solid part of the rock, met with several fragments of the horns and bones of one or different animals." The antlers, when worked out of the surrounding matrix, proved to be those of a 'crowned Hart,' in which the summit or sur-royal expands and radiates a number of short snags from a funnel-shaped cavity, large enough to contain a thrush's nest, whence the park-keepers, Mr. Barker says, call them 'throstle-nest horns.' The following were the dimensions of the fossil antler as compared with the corresponding one of a recent Red-deer.

	Fossil.		Recent.	
	Ft.	In.	Ft.	In.
Circumference at the insertion into the skull .	0	9⅞	0	7
Length of lowest (brow) antler	1	2	1	0
Length of entire horn	3	3½	2	7½†

The locality in Derbyshire where these remains were found is Alport, in the parish of Youlgreave.

Mr. Okes makes mention of the discovery, at about half a mile eastward of the town of Chatteris, in Cambridgeshire, in a stratum of clay, underlying peat-moss, "of part of the horns of a species of Deer, measuring two feet, and, in circumference at that end by which it is attached to the skull, ten inches." This, Mr. Okes concludes "from its

* Phil. Trans. vol. xliii. p. 266. Tuft is a deposit from calcareous waters on their exposure to air, usually containing portions of plants incrusted with carbonate of lime; it is called by modern geologists, when it is porous, 'tufa,' and when solid 'travertin.'

† Ib. vol. lxxv. p. 353.

magnitude, must evidently belong to the celebrated extinct species found in Ireland," viz. the *Megaceros Hibernicus.* Near the spot where these fossils were found, part of a Mammoth's skull with two grinding-teeth was exhumed from the same stratum of diluvial clay. The dimensions above given do not, however, exceed those of the ancient Red-deer exhumed in Lancashire and Derbyshire; whilst the basal circumference of the antler of the Megaceros is commonly from twelve to sixteen inches. It is more probable, therefore, that the large antlers from Chatteris were remains of the *Cervus Elaphus*, when it existed under circumstances which favoured the full development of its specific characters.

I have been favoured by Jabez Allies, Esq. of Lower Wick, near Worcester, with sketches of nearly equally fine antlers of the Red-deer, of the discovery of which he has given the following account in the *Worcester Journal*, October 3rd, 1844.

"At the southward part of the cutting across the meadow at Diglis, near this city, for the Severn Navigation Lock, several relics of antiquity have been found, namely: —At the depth of about twenty feet in the alluvial soil portions of small trees, bushes, and hazel-nuts, intermingled with fragments of stags-horns and bones; and a little nearer to the river southward, at the depth of about twenty-five feet, the relics of an oak-tree; and still nearer the river, at the depth of about thirty feet, a great number of bones of the deer kind, and of small short-horned cattle and other animals, together with fragments of Roman urns and pans of red earth, and a piece of Samian ware; and a little nearer to the river, at the depth of about thirty feet, the horns and part of the skull of a large Stag; but whether it is of the Elk kind, or of what other species, I have not yet been able to ascertain: and alongside of the latter relic was part of the under jaw of a horse; an antler, probably of the red-deer; and also the greater part of a fine Roman urn of a dark colour.

"It appears to me that there was an ancient dyke at the spot, and that the rill of water which ran into the Severn having in ages past been diverted into another channel, the dyke became gradually filled up by the alluvium which is occasionally deposited upon the plains by the floods of the river, and thereby all the relics were buried at the great depth at which they lay; and in proof of this, the stratum on which they rested was muddy grit, such as we find at the bottom of water-courses. It would have taken an immense time for these relics to have been buried upon the surface of a level plain at the depth they were, for I have shown in my before-mentioned work that the alluvium upon the level plains on the borders of the Severn has only accumulated about four feet since the Roman time."

Mr. Allies informs me that a coin of Marcus Aurelius was found at the depth of about thirty feet, just by where the south gates of the lock stand. The antlers of the large Stag, of which I received figures, have the expanded and branched summits characteristic of the 'crowned Hart:' the breadth of this expansion is not less than eighteen inches; the total length of the antler, in a straight line, is two feet. That of a second antler is two feet seven inches. Mr. Dixon, of Worthing, has found antlers of Red-deer, with Roman and British antiquities, in the superficial deposits at Selsey and Bracklesham. An almost entire skeleton of a large *Cervus Elaphus* has been found in sand several feet beneath the present bed of the Ouse, in the Lewes levels.

Many large antlers of Red-deer were discovered by Mr. Gladdish in the freshwater sandy deposits above the chalk at Gravesend.

Antlers and bones of the Red-deer are found associated with remains of the Mammoth and Rhinoceros in the freshwater deposits at Brentford, Rugby, in the valley of the Thames, and in that of the Severn. Morton has figured such antlers in his 'Natural History of the County of Northampton;' and a very fine fossil antler, wanting the summit, has been acquired by the British Museum, (No. 16,081,) from the collection of Miss Baker of Northampton.

Dr. Buckland, in his account of the fossils from the Hyæna-cave at Kirkdale, says of the fragments of horns of Deer, "One of these resembles the horn of the common Stag or Red-deer, the circumference of the base measuring nine inches and three quarters, which is about the size of our largest stag. A second measures seven inches and three quarters at the same part, and both have two antlers that rise very near the base." "No horns are found entire, but fragments only, and these apparently gnawed to pieces, like the bones; their lower extremity nearest the head is that which has generally escaped destruction; and it is a curious fact, that this portion of all the horns I have seen from the cave shows, by the rounded state of the base, that they had fallen off by absorption or necrosis, and been shed from the head on which they grew, and not broken off by violence."* With respect to the horns so shed, the author afterwards remarks, "It is probable that the hyænas found them thus shed, and dragged them home for the purpose of gnawing them in their den; and to animals so fond of bones, the spongy interior of horns of this kind would not be unacceptable. I found a frag-

* 'Reliquiæ Diluvianæ,' p. 19.

ment of stag's horn in so small a recess of the cave, that it never could have been introduced, unless singly and after separation from the head; and near it was the molar tooth of an elephant."*

Similar fragments of shed antlers of the Red-deer, associated with others referable to the *Megaceros* and the great *Strongyloceros*, have been found in Kent's Hole at Torquay; they all show the effects of gnawing, and indicate that all the three species of Deer co-existed in England with the Hyæna and other extinct carnivora at that remote period.

In Ireland the remains of the *Cervus Elaphus* have been frequently found associated, as in the lacustrine marls of Yorkshire, with the Megaceros; but the most abundant specimens occur in the still more recent turbary and alluvial deposits of that island. The fine crowned antler, one of a pair discovered in the bed of the Boyne at Drogheda, and now preserved in the Museum of Sir Philip Egerton, (fig. 196,) measures thirty inches in length, and sends off not fewer than fifteen snags or branches. Many instances of the discovery of remains of the Red-deer in the morasses and the lacustrine marls beneath peat-mosses of Scotland, have been recorded, and the chain of evidence of the existence of this species of Deer in Britain, from the pliocene tertiary period to the present time, seems to be unbroken. This at least is certain, that a Deer, undistinguishable by the characters of its enduring remains from the *Cervus Elaphus*, co-existed with the Megaceros, the spelæan Hyæna, the tichorhine Rhinoceros, and the Mammoth, and has survived, as a species, those influences which appear to have caused the extinction of its gigantic associates, as well as of some smaller animals, for example the *Trogontherium*, the *Lagomys*, and the still more diminutive *Palæospalax*.

* 'Reliquiæ Diluvianæ,' p. 32.

Fig. 197.

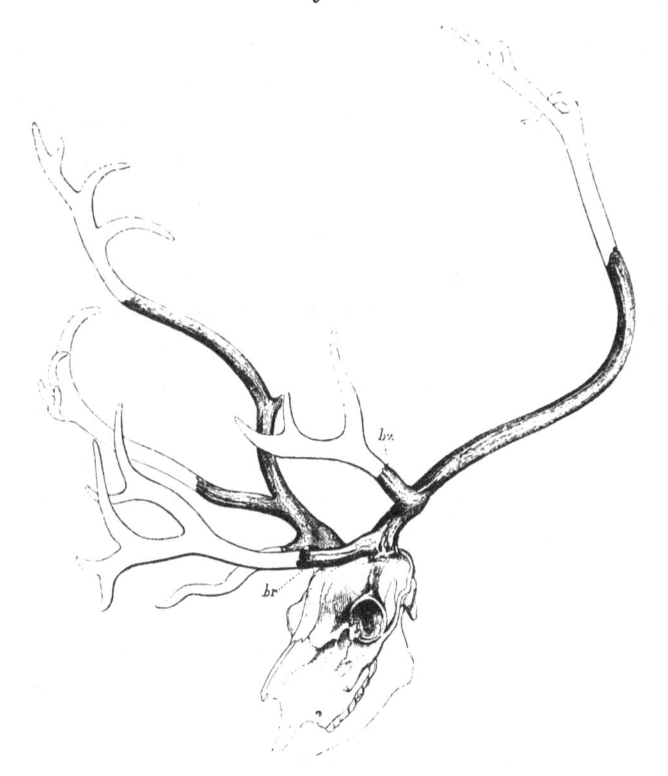

Antlers of Rein-deer, from Bilney Moor, East Dereham.

CERVUS TARANDUS. Rein-deer.

For the first good evidence of remains of the Rein-deer in this island I am indebted to Mr. George Bartlett of Plymouth, who transmitted to me the subjoined account of their discovery, together with the characteristic drawings by the accomplished naturalist and artist therein mentioned.

"The skull and humerus, of which the accompanying

drawings by Colonel Hamilton Smith are representations, were found by me some time since during some researches made with the Rev. H. F. Lyte in a cavern in the lime-rock of Berryhead, Devon. The skull was at a great depth below the original floor of the cave, and was lying in an aluminous silt, and buried beneath a block of limerock many tons in weight, which had no doubt, subsequently to the deposit of the skull, fallen from the roof on it. Not quite so deeply buried, but adhering to the side of the block by a calcareous cement, I found the other bone, the humerus. No bones of any kind were associated with them, and although the lower jaw and horns of the skull were wanting, yet no fragment of bone or organized calcareous matter was near or anywhere around: the tooth fell from the skull on taking it up. They were in a dry situation, and about forty feet perpendicular from the opening of the cave, which is situated in the side of a precipitous hill about seventy feet above the level of the sea."

The fragment of the skull (fig. 198) showed the places from which the antlers had been recently shed, and, by their proximity to the occipital ridge, determined the identity of the fossil with the *Cervus tarandus*. In the Fallow and Red-deer, as in all other recent cervine species corresponding in size with the fossil, the antlers spring from the frontal bones nearer the orbits and further from the occiput. The extinct *Cervus Guettardi* most resembles the Rein-deer in the position of the antlers; but, besides the smaller size of the skull, the antlers rise a little further from the occiput. The precise agreement of the fragment of the skull, of the molar tooth, and of the humerus, in size and form, with those parts in the Rein-deer, verifies the inference from the characteristic position of the antlers, as to the species to which the fossils belong.

More recognisable,— though perhaps not more decisive, evidence of the *Cervus tarandus*, is afforded by the discovery of a fragment of the skull with the antlers attached, beneath a peat-moss in a small moor at East Bilney, near East Dereham, in the county of Norfolk. A drawing of these antlers, transmitted to me by C. B. Rose, Esq., is engraved in cut 197. The characteristic branched brow-antler, though the terminal forks are broken, measured seven inches and a half in length; the length of the beam from the burr to the fractured extremity, was thirty one inches in a straight line; the breadth of the os frontis at the rise of the horns was three inches. These specimens correspond with that variety of the antlers in the Rein-deer which is represented in figs. 13 and 20, pl. iv. tom. iv. of the 'Ossemens Fossiles.'

Fig. 198.

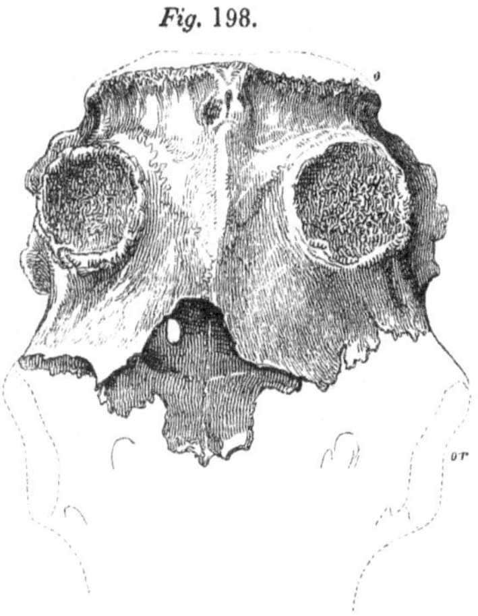

Cranium of Rein-deer, Berry-head Cave, Devon.

A single mutilated antler, retaining thirty-five inches of the beam, with seven inches of the brow-antler, twelve inches of the bezantler, and the commencement of the expansion or palm at the fractured end of the beam, was likewise discovered at the same place. Both these specimens show the smooth subcompressed character of the beam and branches peculiar to the antlers of the Rein-deer amongst the existing species of *Cervus*.

The remains of the quadrupeds found in the lacustrine shell-marls of Scotland, according to Mr. Lyell, all belong to species which now inhabit, or are known to have been indigenous in Scotland. "Several hundreds," he observes, "have been procured within the last century, from five or six small lakes in Forfarshire, where shell-marl has been worked." Those of the stag (*Cervus elaphus*) are the most numerous, and if the others be arranged in the order of their relative abundance, they will follow, according to Mr. Lyell, nearly thus: Ox, Boar, Dog, Hare, Fox, Wolf, and Cat; the Beaver is the rarest. A pair of Deer's horns of large size, and with fine antlers, together with two metacarpal bones, "so deeply grooved as to appear like double bones," were dug up out of a marl-pit beneath five or six feet of peat-moss, on the margins of the Loch of Marlee. In the same place were found the remains of the Beaver noticed at p. 194. Mr. Neill, who has recorded both these discoveries, says, with regard to the deeply-grooved leg-bones, "It has been suggested to me by Dr. Barclay, that they were probably the metatarsal bones of the great species of Deer, which appears to have been contemporary with the Beaver, and to have become extinct much about the same period with that animal."* If the *Megaceros Hibernicus* be the species here referred to, the character of the deep-grooved metacarpal bone will not at all apply to it, since the median longitudinal groove is wider and shallower on both the fore and back part of the metacarpals and metatarsals in the *Megaceros* than in any other species of Deer; the Reindeer is most remarkable for the depth of the grooves, especially the posterior one of the metatarsus. I will not venture to pronounce, in the absence of the specimens, and of any know-

Fig. 199.

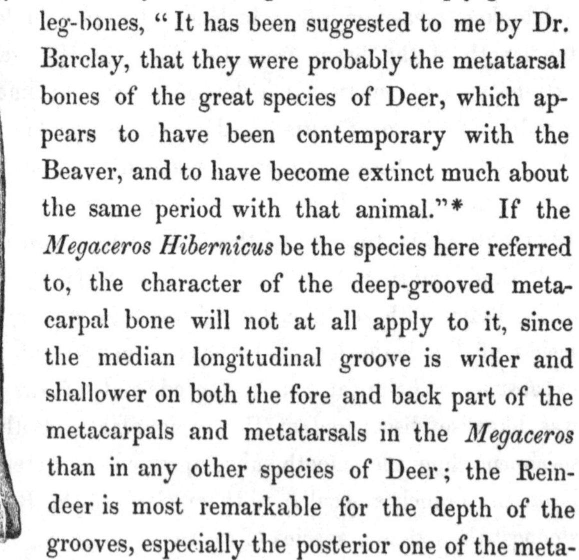

Metatarsal of Rein-deer, $\frac{1}{4}$ nat. size. Fens.

* 'Jamieson's Edinburgh Philosophical Journal,' vol. i. 1819, p. 183.

ledge of the characters of the associated horns, whether the cervine remains referred to by Mr. Neill, belonged to the Rein-deer; but I subjoin a figure of a metatarsal bone, precisely corresponding with that of the existing Rein-deer, which bone was found at the depth of five feet in the fens of Cambridgeshire.

Dr. Fleming* cites a pair of Deer's horns found in a marl-pit at Marlee, which, from their superior size and palmed form, were supposed to be the horns of the Elk-deer; and he refers to a donation to the Royal Society of Edinburgh "by the Hon. Lord Dunsinnan, of a painting in oils of the head and horns of an Elk found in a marl-pit, Forfarshire," and adds: "Whether these two examples from marl-beds should be referred to the Fallow-deer, or the Irish Elk, may admit of some doubt, though it is probable that they belong to the former." The superior size of the palmed antlers militates against their reference to the ordinary Fallow-deer; and the observation of the deeply-grooved metacarpal or metatarsal bones, from the same marl deposit, renders it desirable to compare the specimens and the oil-painting with the large palmed varieties of the antlers of the Rein-deer figured by Cuvier in the fourth volume of the 'Ossemens Fossiles,' 4to. 1823, pl. iv. figs. 11, 18, and 16.

CERVUS DAMA. Fallow-deer.

Of this species as an aboriginal one, coeval with the Red-deer and Megaceros in Great Britain, I have no decisive evidence from actual observation of characteristic fossil or semifossil remains. The portions of palmated antlers and teeth from the peat-moss at Newbury, noticed

* 'History of British Animals,' 8vo., 1828, p. 26.

in my Report on British Fossil Mammalia, accord in size with the Fallow-deer; but more perfect specimens and decisive evidence of this species are desirable, even from that comparatively recent deposit.

In the large cave of Paviland, on the Glamorganshire coast, Dr. Buckland found, with remains of Mammoth, Rhinoceros, Hyæna, &c., "deer of two or three species," and "fragments of various horns, some small, others a little palmated."* The same doubt as to whether the latter are referable to Rein-deer or Fallow arises, as in the case of the palmated fragments from Newbury.

Of the teeth of deer found fossil in the cave at Kirkdale, Dr. Buckland† specifies the smallest as being nearly of the size and form of those of a Fallow-deer.

Portions of jaws and teeth occur in Mr. Green's collection of fossils from the blue clay and lignite beds at Bacton, which accord in size and figure with those of the Fallow-deer: but such specimens are far from yielding satisfactory grounds of identification. Dr. Fleming,‡ however, considers that the evidence on which the claims of the Fallow-deer to be regarded as an indigenous animal are founded is far from doubtful. He quotes Lesly, (De Or. Scot. p. 5,) who mentions, among the objects which the huntsman pursued with dogs, "Cervum, damam, aut capream." And he adds that:—"In the Statistical Account of Ardchatten, Argyleshire (vol. vi. p. 175), it is said, that Fallow-deer run wild in the woods of a much superior size and flavour to any of their species that are confined in parks."

The "damam" of Leslie may mean the hind, or female of the Red-deer. The real wild Fallow-deer has only been recognised by modern Naturalists from the south of Tunis.

* 'Reliquiæ Diluvianæ,' pp. 83, 85. † Ib. p. 18.
‡ 'History of British Animals,' 8vo., 1828, p. 26.

Fig. 200.

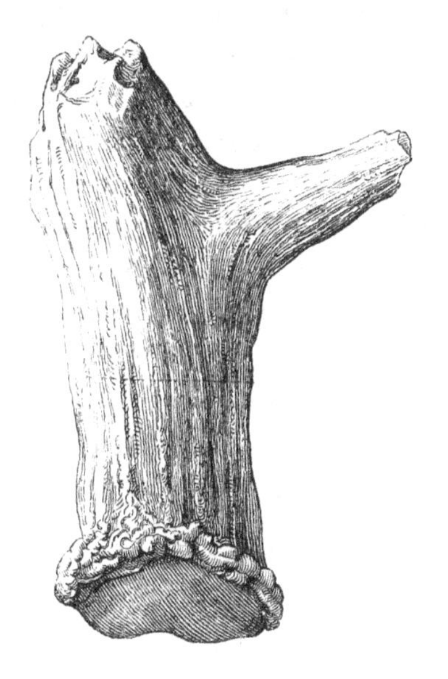

Base of antler, ½ nat. size. Kirkdale Cave.

CERVUS BUCKLANDI. Buckland's Fossil-Deer.

Smaller species of Deer of Kirkdale, BUCKLAND, Reliquiæ Diluvianæ, pp. 19, 264, pl. 9, fig. 5.

OF the former existence of a species of Deer, about the size of the Rein-deer, but differing from all known existing species in Europe, the fossil fragment of the antler above figured affords good evidence. Dr. Buckland, who gave the first description and figure of this specimen, which is from the cave of Kirkdale, especially distinguishes it from the rest as "having the lowest antler at the distance of three inches and a half from the lower extremity, or base, the circumference of which is eight inches."

Such a position of the first branch may be observed amongst existing Deer, in the great Rusa, or *Hippelaphus* of India, and in the Mazama or *Cervus furcatus* of S. America; but it is always directed more obliquely upwards than in the fossil. The *Cervus Guettardi*, amongst fossil species, shows the same relative position as well as direction of the first branch; but this species is smaller than in the Kirkdale fossil, being intermediate between the Rein-deer and the Roe; and the beam is smoother and less cylindrical. The British extinct species would seem, however, to be more nearly allied to the fossil of Etampes * than to any known existing Deer; but it is distinct, and I propose to dedicate it to the distinguished Geologist by whom its chief characteristic was first pointed out.

* Cuvier, tom. cit. p. 89.

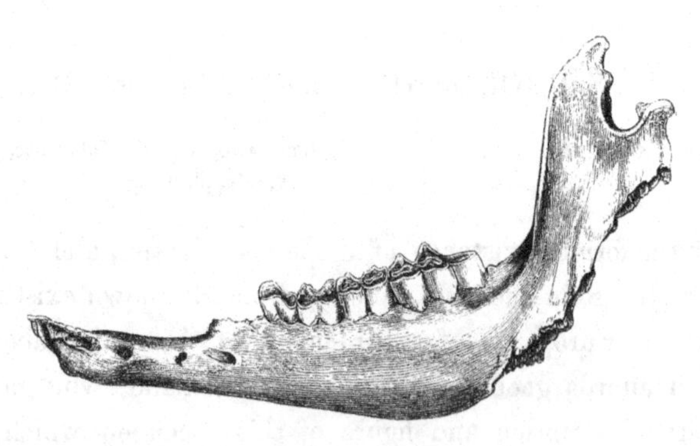

Left ramus, lower jaw of Roebuck, ½ nat. size. Subturbary marl, Newbury, Berks.

Fig. 202. *Fig.* 203.

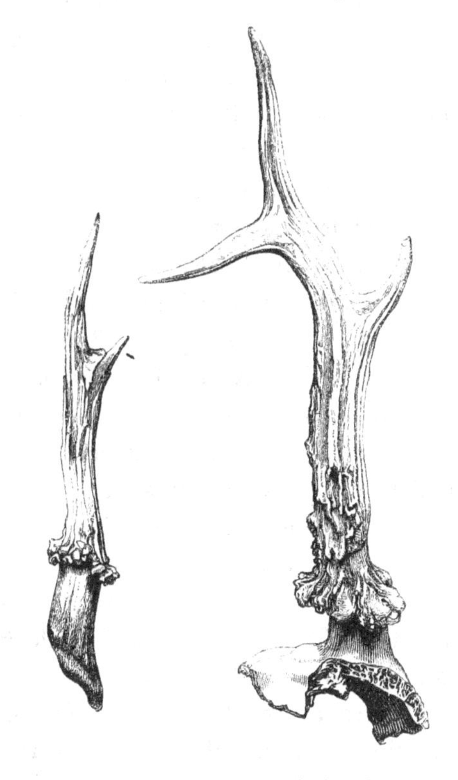

Antler 3rd year. Antler 6th year.

Fossil Roebuck, Fens, Cambridgeshire.

CERVUS CAPREOLUS. Roebuck.

Chevreuil des tourbières,	CUVIER, Ossemens Fossiles 4to. tom. iv. 1823, p. 105.
Capreolus fossilis,	H. v. MEYER, Palæologica, 8vo. 1832, p. 94.
Cervus capreolus,	OWEN, Report of British Association, 1843, p. 238.

THE ROEBUCK is now confined, in Britain, to the district of Scotland north of the Forth, but numerous remains attest a former distribution of the species as extensive as that of the Red-deer. Dr. Buckland specifies, amongst the cervine remains of the Cave of Paviland, an antler, "approaching to that of the Roe." I have received characteristic remains of the *Cervus capreolus* from the ossiferous caves in Pembrokeshire, by favour of Charles Stokes Esq.; and from a fissure of a limestone rock in Caldy Island, off Tenby, Glamorganshire, where the Capreoline antlers were discovered associated with remains of the *Rhinoceros tichorhinus*, by the Rev. R. Greaves. I have also been favoured with fossil antlers and bones of the Roebuck from the limestone caverns in the neighbourhood of Stoke-upon-Trent, by Robert Garner Esq., the author of the History of Staffordshire.

Almost the entire skeleton of a small Ruminant, agreeing in size and general characters with the female Roe, has been discovered in the lacustrine formation at Bacton, with the remains of the Trogontherium, Mammoth, &c. This specimen is preserved in the Norfolk and Norwich Museum. The antlers figured above, the one (fig. 202) of a young Roe of the third year, the other (fig. 203), at the sixth year, were discovered ten feet deep below the fen-land of Cambridgeshire.

In the collection of British fossils belonging to Mr. Purdue of Islington, there is an almost entire left ramus of the lower jaw of a small Ruminant, identical in size and conformation with that of the Roebuck (fig. 201). It was found in a lacustrine deposit of marl, with freshwater shells, below the bed of peat, at Newbury in Berkshire, where skulls and antlers of the Roebuck are not uncommon.

RUMINANTIA. *CAPRA.*

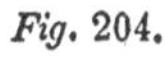

Portion of fossil skull and horn-cores of a Goat, ½ nat. size. Newer fresh-water pliocene, Walton, Essex.

CAPRA HIRCUS. Goat.

Evidence of the smaller hollow-horned ruminating animals is afforded by fossil jaws, teeth, and detached bones of the skeleton, and in a few cases by the characteristic appendages of the skull, which then serve to identify the species or the genus of such fossils.

A fragment of a lower jaw, containing one of the lateral series of six molar teeth, with a part of the skull having the perfect cores of the horns attached, was discovered by Mr. Brown, of Stanway, in the newer pliocene deposits at Walton in Essex: these fossils were in the same condition as the bones of the large extinct Mammalia from the same formation. The jaw and teeth agreed in size and configuration with the same parts in the common

Goat, and also in the Sheep; and the highly interesting question, which of these had existed contemporaneously with the Mammoth and Rhinoceros, was satisfactorily determined by the cranial fragment. In its shape and size, and especially in the character of the cores of the horns, which were two inches in length, subcompressed, pointed, and directed upwards, with a slight bend outwards and backwards, it closely agreed with the common Goat (*Capra hircus*), and with the short-horned female of the Wild Goat (*Capra Ægagrus*). In the Sheep, the greatest diameter of the horn is across the longitudinal axis of the head; in the Goat, it runs almost parallel with it,—a character well shown in the present fossil.

Whether the *Capra Ægagrus* or the *Capra Ibex* should be regarded as the stock of the domesticated Goat of Europe has long been a question amongst Naturalists; the weighty argument which may be drawn from the character of the wild species, which was contemporary with the *Bos primigenius* and *Bos longifrons* in England, is shown by the present fossil to be in favour of *Capra Ægagrus*.

I have been favoured with some remains of the Goat, from a bog in Fermanagh, by the Earl of Enniskillen; but as the evidence of their having been obtained from the subjacent marl was not conclusive, they may have belonged to a comparatively recent period. Remains of the Goat, associated with those of the Ox, Red-deer, Hog, Horse, and Dog, were found in the bed of the Avon, in sinking the foundations of a bridge over that river, near the town of Chippenham. Bones of a Goat or Sheep, similarly associated, have been transmitted to me by Dr. Richardson, from a gravel-pit in Lincolnshire.

RUMINANTIA. *BOVIDÆ.*

Fig. 205.

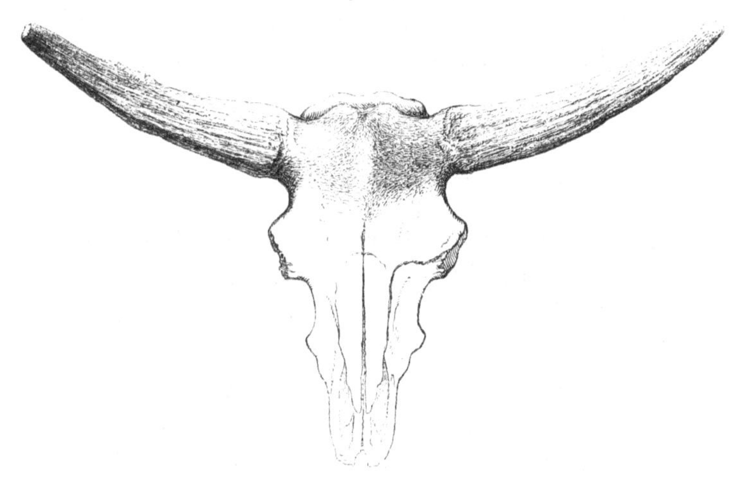

Fossil frontlet and horn-cores of Aurochs, pliocene clay, Woolwich.
(The rest of the skull restored in outline.)

BISON PRISCUS, Great Fossil Aurochs.

Aurochs fossile,	CUVIER, Ann. du Mus. xii. pp. 379, 386, tab. xxxiv. figs. 1, 2, 4, 5. Ossem. Foss. iv. tab. xi. fig. 5, tab. xii. figs. 1, 2, 6, 7.
Bos (*Bison*) *priscus,*	BOJANUS, Nov. Acta. Acad. Nat. Cur. t. xiii.
Bos priscus,	H. v. MEYER, *ib.* t. xvii. p. 1.
Urus priscus,	OWEN, Report of British Association, 1343, p. 232

WHEN the forests of Germany and Belgium were first penetrated by the Roman armies, there were found, amongst other beasts of chase, two large species of wild oxen; the one called "Bison" distinguished by its shaggy coat, the other called "Urus" by the great size of its horns. Both these species continued to exist to later periods of the Empire, and were occasionally captured and exhibited alive in the shows of the amphitheatre.

Cuvier, who has collected, with his usual research, all the notices in the poets and historians of those periods, cites amongst others the following passage from Seneca, which briefly and clearly defines the characteristics of the two species:

> "Tibi dant variæ pectora tigres,
> Tibi villosi terga bisontes
> Latisque feri cornibus uri."

Pliny characterises the Bison by its mane, and distinguishes the "*jubatos bisontes*" from the "*excellentique vi et velocitate uros,*"—the Uri remarkable for their strength and speed.

One of the species of the great primitive European wild cattle, now known as the Lithuanian Aurochs, still survives by virtue of strict protective laws, in extensive forests, which form part of the Russian Empire; and it is distinguished from all the breeds of domestic cattle of Europe, and from the Chillingham wild oxen, by the thicker clothing of hair, which, in the male Aurochs, is developed at the fore-part of the body into a curly felted mane, justifying the distinctive epithets of "villous" and "maned," applied by the Romans to the wild Bison of their period.

The Aurochs, (*Bison*) differs, moreover, from all the species or varieties of ordinary Ox (*Bos*) by more important characters, deducible, fortunately, from those enduring parts of its body which serve to reveal its existence in Europe, at periods more remote than the conquests of Cæsar. The differences observable in the skull, for example, of the *Bos* and *Bison* are thus accurately and distinctly defined by Cuvier:

"The forehead of the Ox (*Bos*) is flat, and even slightly concave; that of the Aurochs (*Bison*) is convex (*bombé*), though somewhat less so than in the Buffalo: it is quadrate in the Ox, its height, taking the base between the orbits, being equal to its breadth; in the Aurochs, measured at the same place, the breadth greatly exceeds the height, in the proportion of three to two: the horns are attached in the Ox to the extremity of the highest salient line of the head, that which separates the forehead from the occiput; in the Aurochs this line is two inches behind the root of the horns: the plane of the occiput forms an acute angle with the forehead in the Ox; the angle is obtuse in the Aurochs: finally, that plane of the occiput is quadrangular in the Ox, but semicircular in the Aurochs."* The ribs never exceed in number thirteen pairs in any species of *Bos* proper; the European Bison or Aurochs has fourteen, and the American Bison fifteen pairs of ribs.

The fossil cranium with horn-cores, described and figured by Klein in the thirty-seventh volume of the 'Philosophical Transactions,' No. ccccxxvi. figs. 1, 2, and 3, and which is now in the British Museum, well illustrates the characters which distinguish the Aurochs: the specimen was dug up near the city of Dantzig.

Faujas,† Cuvier,‡ and H. v. Meyer,§ have added abundant illustrations of the remains of the same species from the superficial deposits of various parts of Europe, some of which carry the antiquity of the Aurochs as far back as the period of the extinct Pachyderms of the newer pliocene deposits. The remains of the ancient European Bisons attest their larger size, and longer and somewhat less bent horns than are manifested by the individuals of the present race, but no satisfactory specific distinction has been detected in the fossils compared with the bones of the Lithuanian Aurochs.

* Menagerie du Museum d'Histoire Nat. Art. *Zebu.*

† 'Essai de Géologie,' tom. i. pl. xvii.

‡ Loc. cit.

§ 'Uber Fossile Reste von Ochsen,' 4to., 1832, tab. viii.—xi.

The former existence of the great Aurochs (*Bison priscus*) in this island is unequivocally established by fossil remains of the cranium and horn-cores from various newer tertiary freshwater deposits, especially in Kent and Essex, and along the valley of the Thames.

One of these specimens (fig. 205) was dug out of a stratum of dark-coloured clay beneath layers of brick-earth and gravel, thirty feet below the surface, at Woolwich; it presents the broad convex forehead, the advanced position of the horns, which rise three inches anterior to the upper occipital ridge, and the obtuse-angled junction of the occipital with the coronal or frontal surface of the skull,—all which characters distinguish that part of the skeleton of the continental fossil and recent Aurochs. The bony cores of the horns extend outwards, with a slight curvature upwards, but are relatively longer than in the Lithuanian Aurochs: the tips of the horn-cores in the fossil are four feet six inches apart; the distance from the mid-line between their bases to the extremity of the core, in a straight line, is two feet five inches.

A characteristic cranium with horn-cores of the *Bison priscus*, obtained by Mr. Warburton from the fresh-water newer pliocene deposits at Walton in Essex, is suspended in the Hall of the Geological Society of London.

Another specimen of the fossil cranium of *Bison priscus*, dug out of a brick-field at Ilford in Essex, presents, with the same essential characters as the preceding, relatively thicker, shorter, and more curved horn-cores. This fossil differs by its shorter horns from the preceding, and more resembles the existing Lithuanian Aurochs: it may indicate the female of the more ancient Aurochs.

A broken skull with perfect horn-cores of the *Bison priscus*, discovered by Mr. Strickland in the fresh-water drift at Cropthorne, Worcestershire, yields the following dimensions: from tip to tip of the horn-cores, following the anterior curves, three feet eight inches; the same in a straight line, three feet four inches.

Hitherto, no fossil skeleton of the same individual has been discovered in a state of such completeness as to enable the anatomist to ascertain the number of the ribs,—a fact which would be of importance in determining the relations of the ancient European Aurochs with the existing Lithuanian Aurochs and the Bison of North America. Cuvier regrets that he had not sufficiently precise knowledge of the formations containing remains of the great fossil Aurochs; but that which M. v. Meyer cites appears to give the required proof of the high antiquity of the *Bison priscus*.*

The brick-earth of Woolwich and Ilford, from which two of the specimens of fossil Aurochs above cited were found, underlies a layer of sand, with pebbles and concretions, containing shells of *Unio* and *Cyclas*; and the remains of both Mammoth and Rhinoceros are unquestionably associated with those of the Aurochs in this formation. The other localities which may be cited, from the less certain character of the proportion of the metacarpal and metatarsal bones—those of the slenderest proportions being referred to the Aurochs,—are Brentford, Kew, Kensington, Wickham, Erith, Grays, Whitstable, Gravesend, Copford, and Clacton.

Professor Phillips has recorded the discovery of the skull with the cores of the horns and the teeth of the great Aurochs at Beilbecks in his 'Geology of Yorkshire,' vol. i. 2nd edition, accompanied by land and fresh-water

* The skull of the Aurochs, No. 10 of v. Meyer's Monograph, forms part of the collection at Darmstadt, and bears the following ticket, "Ochsenkopf aus dem Rhein, bei Erfelden mit dem Rhinoceroskopf in Rhein gefunden." V. Meyer states that the skull of the Rhinoceros belongs to the extinct species *tichorhinus*. Op. cit. p. 34.

shells, and by remains of the Mammoth, Rhinoceros, Felis, large Horse, large Deer, Wolf, &c.

To determine to which subgenus of *Bovidæ* belong the detached teeth, vertebræ, ribs, and other bones of the skeleton—often mutilated and gnawed, is still attended with much difficulty. Such remains, however, sufficiently attest that species as large as the *Bison priscus* and *Bos primigenius* were very extensively distributed throughout England: they have been found in almost all the ossiferous caves which have yielded the fossil remains of *Elephas*, *Rhinoceros*, *Hyæna*, and *Ursus*.

Cuvier* affirms, as the result of his numerous comparisons of the recent and fossil bones of the Bovine animals, that the detached bones resemble each other too much to yield certain specific characters, and that it is necessary to have skulls in order to determine the species.

The fossil metacarpal bones of the gigantic *Bovidæ* found in England, indicate two species by their different proportions; one kind being thicker than the other. The metatarsal bones show a corresponding difference; and the proportions of a metacarpal found associated with the skull of the *Bos primigenius*, to be described in the next section, indicate the more slender bones to belong to the Aurochs, (*Bison priscus*).

This difference is shown by the subjoined admeasurements, and may be more readily appreciated by comparing fig. 207 with fig. 209.

	BISON. Metacarpal. Grays.		BOS. Metacarpal. Grays.		BISON. Metatarsal. Clacton.		BOS. Metatarsal. Grays.	
	In.	Lines.	In.	Lines.	In.	Lines.	In.	Lines.
Length	11	0	10	3	11	3	11	6
Circumference at the middle . .	6	3	6	3	5	3	6	6

* 'Ossemens Fossiles,' vol. iv. p. 140.

The metatarsal bone of the *Bison priscus* from Clacton (fig. 207) shows an unusual prominence of the inner border of the anterior groove for the extensor tendon which traversed the middle of that surface of the metatarsal bone. I have seen the same character in a metatarsal bone of corresponding dimensions found in the brick earth at Kensington; and in two metatarsals of a smaller species or variety of Bison (fig. 206), from the cavernous fissures at Oreston: it may be due, however, to accidental ossific inflammation.

Fig. 206.

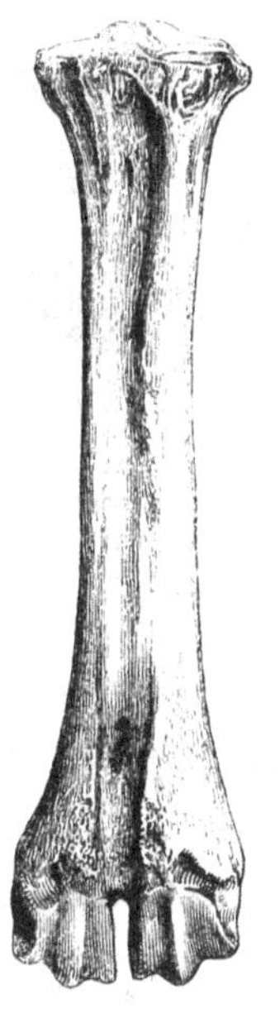

Metatarsal, *Bison minor*, Oreston.

Fig. 207.

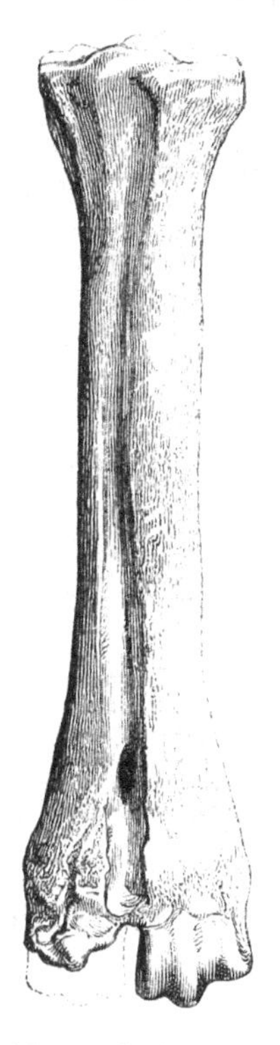

Metatarsal, *Bison priscus*, Clacton.

RUMINANTIA. *BOVIDÆ.*

Fig. 208.

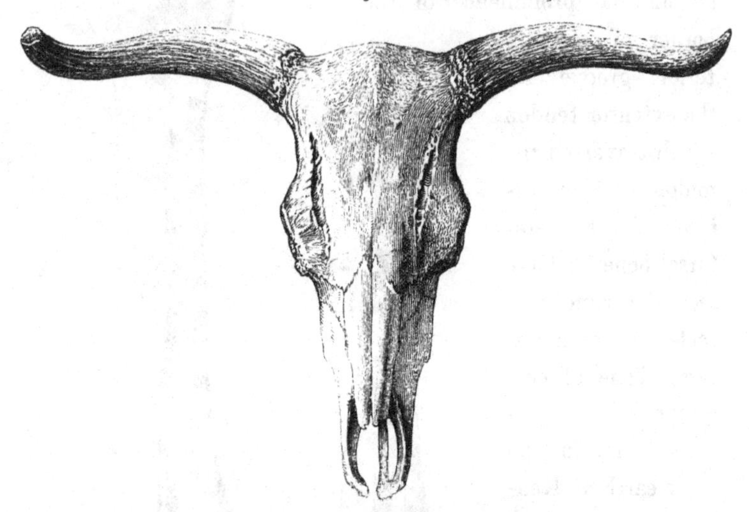

Skull of great extinct Ox, Scotland.

BOS PRIMIGENIUS. Great Fossil Ox.

Bos primigenius, BOJANUS, Nova Acta Acad. Nat. Cur. tom. xiii. pl. II. p. 422.
Bœuf fossile, CUVIER, Ossemens Fossiles, 4to., 1823, tom. iv. p. 150, pl. XI. figs. 1—4.
Fossil Ox, WOODS, Description of the fossil skull of an Ox, 4to. 1839.
Bos primigenius, OWEN, Report of British Association, 1843, p. 233.

OF the two kinds of gigantic oxen which the Romans discovered when they first penetrated the wilds and forests of uncivilized Europe, that which they distinguished by its shaggy coat and mane may be recognised in the still untamed Aurochs of Lithuania; and the remains of the large Bovine animal described in the foregoing section, which belong to the same subgenus (*Bison*), and which show no clear specific distinction from the Lithuanian Aurochs, prove that such an animal was once distributed over Britain, where, its antiquity, as a species, equalled that of the Mammoth, the tichorhine Rhinoceros, the Spelæan Hyæna, and other extinct mammals of the newer tertiary period.

We have next to inquire into the evidence of the *Urus*, that second kind of aboriginal wild ox, which Cæsar describes * as being not much inferior to the elephant in size, and, though resembling the common bull in colour, form, and general aspect, yet as differing from all the domestic cattle in its gigantic size, and especially in the superior expanse and strength of its horns.

Of this species we have the same examples, short of the still-preserved living animal, as of the Bison; and it is most satisfactory to find such proof of the general accuracy of the brief but most interesting indications of the primitive mammalian fauna of those regions of Europe which may be supposed to have presented to the Roman cohorts the same aspect as America did to the first colonists of New England.

In the same deposits and localities which have yielded remains of the Aurochs (*Bison priscus*) there have been found the remains of another bovine animal, its equal or superior in size, but differing from the Aurochs precisely as the Roman poets and historians have indicated, by the greater length of its horns. The persistent bony supports or cores of the horns likewise demonstrate, by their place of origin and curvature, the subgeneric distinction of the great Urus, from the Bison, and its nearer affinity to the

* "Tertium est genus eorum qui *Uri* appellantur. Hi sunt magnitudine paulo infra elephantos, specie et colore et figurâ tauri. Magna vis eorum et magna velocitas: neque homini neque feræ, quam conspexerint, parcunt.—Amplitudo cornuum et figura et species multum à nostrorum boum cornibus differt."—*Cæsar de Bello Gallico*, lib. vi. cap. 29. Valpy's Delphin Classics, 8vo. 1819, p. 254.

domestic ox; whence we may infer that it resembled the ox in the close nature of its hairy covering, which would make the shaggy coat and the mane of the Aurochs more remarkable by comparison.

It is much to be regretted, for the interests of Zoology, that the great Hercynian *Uri* have been less favoured than their contemporary *Bisontes jubati* in the progress of human civilization, and that no individuals now remain for study and comparison, like the Aurochs of Lithuania.

My esteemed friend Professor Bell, who has written the History of existing British Quadrupeds, is disposed to believe, with Cuvier and most other naturalists, that our domestic cattle are the degenerate descendants of the great Urus.* But it seems to me more probable that the herds of the newly conquered regions would be derived from the already domesticated cattle of the Roman colonists, of those "boves nostri," for example, by comparison with which Cæsar endeavoured to convey to his countrymen an idea of the stupendous and formidable *Uri* of the Hercynian forests.

The taming of such a species would be a much more difficult and less certain mode of supplying the exigencies of the agriculturist, than the importation of the breeds of oxen already domesticated and in use by the founders of the new colonies. And, that the latter was the chief, if not sole, source of the herds of England, when its soil began to be cultivated under the Roman sway, is strongly indicated by the analogy of modern colonies. The domestic cattle, for example, of the Anglo-Americans have not been derived from tamed descendants of the original wild

* "I cannot but consider it extremely probable that these fossil remains belonged to the original wild condition of our domestic Ox."—*Bell's British Quadrupeds*, p. 414.

cattle of North America: there, on the contrary, the Bison is fast disappearing before the advance of the agricultural settlers, just as the Aurochs, and its contemporary the Urus, have given way before a similar progress in Europe. With regard to the great Urus, I believe that this progress has caused its utter extirpation, and that our knowledge of it is now limited to deductions from its fossil or semi-fossil remains.

The discovery of the skull and horn-cores of this species, the *Bos primigenius* of Bojanus, in the alluvial beds of rivers, in sub-turbary lacustrine marls, and in the newer tertiary deposits of this country, demonstrates its claim to rank with the British Fossil Mammalia, and at the same time determines its equal antiquity with the Aurochs.

The characters of the *Bos primigenius*, as contrasted with the *Bison priscus*, may be advantageously studied in the magnificent specimen of an entire skull (fig. 208) from near Athol, Perthshire, now in the British Museum. The concave forehead with its slight median longitudinal ridge; the origin of the horns at the extremities of the sharp ridge which divides the frontal from the occipital regions; the acute angle, at which these two surfaces of the cranium meet to form the above ridge (fig. 210), all identify this specimen with the *Bos primigenius* described by Cuvier,* Bojanus,† and Fremery.‡ The cores of the horns bend at first slightly backward and upward, then downward and forward, and finally inward and upward, describing a graceful double curvature: they are tuberculate at the base, moderately impressed by longitudinal grooves, and irregularly perforated. The skull is one yard in length and the span of the horn-cores is three feet six inches; but

* 'Ossem. Foss.' iv. p. 150. † 'Nova Acta Acad. Nat. Cur.' xiii. pt. 2.

‡ 'N. Verh. Koninkl.-Nederlandsch Instituut, Derde Deel,' 1831.

other British specimens of the *Bos primigenius* have shown superior dimensions of the bony supports of the horns. The breadth of the forehead between the horns is ten inches and a half; from the middle of the occipital ridge to the back part of the orbit it measures thirteen inches; the length of the series of upper molar teeth is six inches and a half; the breadth of the occipital condyles is six inches.

In the manuscript catalogue of the British Museum this fine specimen is ascribed to "the Caledonian Ox, Bos taurus, var. *gigantea*." But the wild white variety with black muzzles, ears, and horns, the "boves sylvestres" of Leslie,* which are identical with the cattle preserved at Chillingham, are of very inferior dimensions, and differ particularly in the smaller proportional size, and finer and more tapering figure of the horns. The Kyloes of the mountainous regions of Scotland, which are more likely to have been derived from an indigenous wild race than the cattle of the Lowlands, differ still more from the *Bos primigenius* than does the Chillingham breed in their diminutive size, and very short horns.

"Many of the skulls which occur in marl-pits in Scotland," says Dr. Fleming, "exhibit dimensions superior to those of the largest domesticated breed. A skull in my possession measures twenty-seven inches and a half in length, nine inches between the horns, and eleven inches and a half across the orbits."† These doubtless were of the same species as the skull from Perthshire, in the British Museum; and, from the very recent character of the osseous substances, it may be concluded that the *Bos primigenius* maintained its ground longest in Scotland before its final extinction.

* 'De origine, moribus, et rebus gestis Scotorum, Rome, 1678,' p. 10.

† 'History of British Animals,' p. 24.

It is remarkable that the two kinds of great wild oxen recorded in the 'Niebelungen Lied,' of the twelfth century, as having been slain with other beasts of chase in the great hunt of the Forest of Worms, are mentioned under the same names which they received from the Romans:

> "Dar nach schluch er schiere, einen *Wisent* und einen Elch,
> Starcher *Ure* vier, und einen grimmen Schelch:"

> "After this he straightway slew a *Bison* and an Elk
> Of the strong *Uri* four, and a single fierce Schelch."

The image of the great Urus in the full vigour of life is powerfully depicted in a later poem, destined, perhaps, to be as immortal as the 'Niebelungen:'

> "Mightiest of all the beasts of chase
> That roam in woody Caledon,
> Crushing the forest in his race
> The Mountain Bull comes thundering on."

But the following stanza shows that Scott drew his picture from the Chillingham wild-cattle:

> "Fierce, on the hunter's quiver'd hand
> He rolls his eye of swarthy glow;
> Spurns, with black hoofs and horns, the sand,
> And tosses high his mane of snow."
> SCOTT, *Ballad of Cadgow Castle.*

Mr. Woods cites the fact of the discovery of the skull and horns of the great Urus in a tumulus of the Wiltshire Downs, as evidence that a "very large race of genuine *taurine* oxen originally existed in this country, although most probably entirely destroyed by the aboriginal inhabitants before the invasion of Britain by Cæsar, since they are not mentioned as natives of Britain by him." *

The span of the horn-cores, in the instance cited by Mr. Woods, was thirty-three inches, and the circumference of each at the base fifteen inches and a half. "Many bones

* Op. cit. p. 26.

of deer, boars, &c., were discovered at the same time; also several fragments of pottery of ancient British manufacture." Mr. Woods also conjectures that the conflicts of the first settlers with the Uri might have given rise to the traditionary legends of the great Dun Cow of Guy, Earl of Warwick.

Fig. 209.

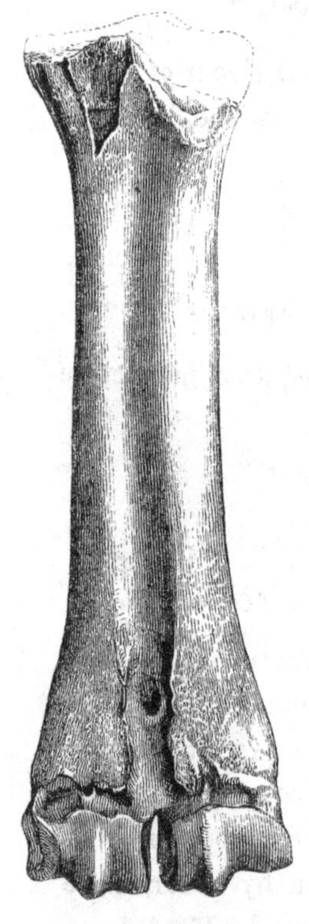

Metatarsus. *Bos primigenius.* Grays.

To return, however, to more positive testimony of the great *Urus* as a British quadruped; I may first mention that the most complete series of the bones of one and the same skeleton of this extinct species is preserved in the private collection of Mr. Wickham Flower; they were obtained from the drift overlying the London clay at Herne Bay. The skull is not so entire as that in the British Museum, but it presents larger dimensions.

The length of each horn-core along the outer curve is three feet three inches; the circumference of the core at its base eighteen inches ten lines; the longest diameter of the base six inches and a half; the chord of the arc described by the core is seven inches and a quarter; from the middle line of the forehead to the tip of the core is two feet two inches. The length of the lower jaw of this specimen is one foot eight inches; that of the series of molar teeth is seven inches. All the true vertebræ, except the atlas, appear to have been recovered, and they include the six remaining cervical vertebræ, thirteen dorsal and six lumbar vertebræ; thus yielding another important character by which this great primeval Ox agrees with the domestic species of the present day. One of the dorsal vertebræ, which retains its spinous process, measures one foot seven inches in height,—a development not greater than might have been expected for the support of the head and horns. One of the scapulæ shows a diseased external surface, ossific inflammation having extended from two depressions in the bone, probably inflicted by the horns of another bull in conflict. The metacarpal and metatarsal bones (fig. 209) give additional exemplifications of the true Bovine character of the present extinct species by their stronger proportions, as compared with those of the Aurochs (fig. 207).

Mr. Brown, of Stanway, has recorded his discovery, in a mass of drift-sand overlying the London clay at Clacton on the Essex coast, of the frontal part of the cranium, with the cores of the horns of a large Bovine animal, which, from the origin, direction, and degree, of curvature of the horns, agrees with the fossil *Bos primigenius.** Each core measured three feet along the outer curve from the base to the tip, the chord of the arc of such curve being eight inches; the diameter of the base was six inches in one direction and five inches in the other. With these parts of the *Bos primigenius* was found a perfect Mammoth's molar tooth, eleven inches in length, eight inches in depth, and three inches across the grinding surface.

Mr. H. Woods, A.L.S., has published a good description and figures of the cranial part of the skull and horn-cores of the *Bos primigenius*, which were discovered in 1838 in the bed of the Avon, about two hundred yards below

* 'Magazine of Natural History,' New Series, 1838. p. 163.

the bridge at Melksham, in a hole sunk in the gravel, and nearly filled with soft black mud. This situation he states to correspond with the places in the neighbourhood of Bath, particularly near Lark-hall, at Tiverton and Newton St. Loe, where such remains have also within a few years been discovered, "mingled with those of the extinct Elephant or Mammoth, Rhinoceros, Bear, Boar, and Horse."[*] The cores of the horns measured in their widest expansion four feet within half an inch, and from tip to tip three feet three inches. The length of each horn-core, following the curvature, was three feet; and these weapons must have been greatly increased when the cores were invested with the horny sheaths in the living animal. The breadth of the forehead between the horns was ten inches, and the breadth across the orbits thirteen inches and aquarter.

Cuvier states, with regard to fossil remains of the *Bos primigenius*, "Il s'en trouve en Angleterre," apparently on the authority of drawings transmitted to him by Mr. Crow. Mr. Parkinson[†] refers his specimens of Bovine fossils, dug up in Dumfriesshire, to the *Bos primigenius*, but without assigning the grounds for this choice.

Cuvier devotes a distinct section to the detached fossil bones of the trunk and extremities of the Bovine tribe, expressing his regret at the numerous sources of uncertainty and difficulty attending their determination when unassociated with the skull; whilst he acknowledges the great importance of ascertaining the species of *Bovidæ* to which the bones from each stratum belonged; whether, for example, an Aurochs, an Ox, or a Buffalo had been the companion of the Elephants, Rhinoceroses, &c. which formerly lived in climates of Europe. At the period of the publication of the second edition of the 'Ossemens

* Op. cit. p. 17.

† 'Organic Remains,' vol. iii. p. 325.

Fossiles' (1823), no authentic example had been recorded of a cranium of either *Bison priscus* or *Bos primigenius* in strata containing bones of the Mammoth and Rhinoceros; and this statement is repeated in the posthumous edition of the 'Ossemens Fossiles,' 8vo, 1835. The skull of the Aurochs in the Darmstadt collection, cited by M. v. Meyer, and the examples of the *Bison priscus* from newer pliocene freshwater deposits in Kent and Essex, described in the foregoing section, leave no reasonable doubt that a large Aurochs was the associate of the gigantic *Pachyderms*, whose representatives at the present day have the Buffaloes for their companions in the tropical swamps and forests. It is true that species of true *Bos* are found wild in the warmer parts of Asia; but no true Aurochs has been discovered within the tropics. That the great Aurochs was associated with a species of *Bos* of equal size in England during the newer pliocene period, is equally demonstrated by the fossils which form the subject of the present sections.

Fig. 210.

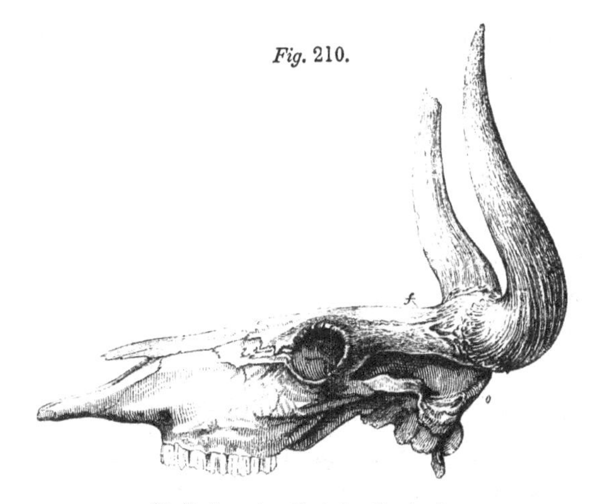

Skull of great extinct Ox, Scotland.

Fig. 211.

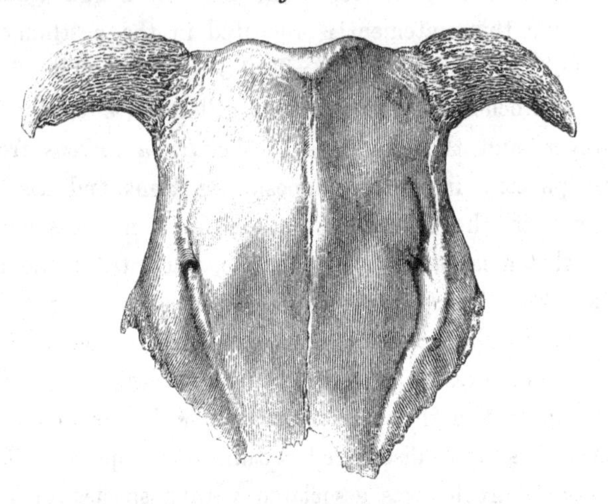

Frontlet and horn-cores or *Bos longifrons*, Bog, Ireland.

BOS LONGIFRONS. Long-fronted, or Small Fossil Ox.

Race of Ox from Irish Bog,	BALL, Proceedings of Royal Irish Academy, January, 1839.
Ancient Bos not exceeding the modern Ox,	WOODS, Description of Fossil Skull of an Ox, 4to., 1839, p. 28.
Bos longifrons,	OWEN, Report of British Association, 1843, p. 235.

THE magnitude of the great Hercynian Urus, and the direction of its horns, have led, as we have just seen, to its being distinguished, by the name of *Bos primigenius*, from the ordinary species of domestic cattle (*Bos taurus*). But " the naturalist well knows," says Cuvier, " that such characters are neither constant nor proper for the distinction of species ;" and it appears, in fact, that the *Bos primigenius*, whilst it retained its great bulk and enormous horns, was subject to some variety in their curvature, unless the *Bos trochoceros* of M. v. Meyer* be actually a distinct and contemporary species. If, however, we admit the justice of Cuvier's remark, and regard the great Urus as a variety of *Bos taurus*, it is not the less an original one, since it was coeval with the Aurochs, and existed long anterior to all records and evidences of domesticated cattle. It was as wild as the Aurochs in the time of Cæsar; and there is as little proof of its having antecedently given origin to the domestic cattle of the Romans as that the Aurochs itself did.

I have already adverted to the high probability that the Roman colonies in Gaul, Belgium, and Britain, derived their domestic cattle from those of the parent State, instead of by the difficult task of subjugating the very formidable species of the fastnesses which those colonists were in progress of reclaiming for the service of civilised life. But, if it should still be contended that the natives of Britain, or any part of them, obtained their cattle by taming a primitive wild race, neither the Bison nor the great Urus are so likely to have furnished the source of their herds as the smaller primitive wild species, or original variety of *Bos*, which is the subject of the present section.

A frontlet and horn-core of this species formed part of the original collection of fossils of JOHN HUNTER, in the manuscript catalogue of which collection it was recorded as having been obtained "from a bog in Ireland." I had entered it, in the catalogue of the museum of the College of Surgeons in 1830, under the name of *Bos brachyceros*, on account of its peculiarly short horns; and, after the imposition of that name by Mr. Gray upon a wild African existing species of *Bos*, I changed the name to *Bos longifrons*, under which the remains of this interesting species

* 'Palæologica,' p. 96.

or variety were described in my 'Report on British Fossil Mammalia.'

Had no other localities for the *Bos longifrons* been known than that of the Hunterian specimen, the species might have been held to be of later date than the *Bos primigenius* and *Bison priscus*, of whose existence, as the contemporaries of the Mammoth and tichorhine Rhinoceros, we have had such satisfactory evidence; I have, however, been so fortunate as to find, in the survey of the collections of Mammalian Fossils in the eastern counties of England, some indubitable specimens of the *Bos longifrons* from freshwater deposits, which are rich in the remains of *Elephas* and *Rhinoceros*.

Mr. Brown of Stanway has obtained the back part of the cranium with the horn-cores from the freshwater newer pliocene deposits at Clacton, and also the frontal part of the skull and horn-cores from similar formations at Walton, both on the Essex coast. Remains of the *Bos longifrons* have also been found in the freshwater drift at Kensington, associated with those of the Mammoth.

This small but ancient species or variety of Ox belongs, like our present cattle, to the subgenus *Bos*, as is shown by the form of the forehead, and by the origin of the horns from the extremities of the occipital ridge (fig. 211); but it differs from the contemporary *Bos primigenius*, not only by its great inferiority of size, being smaller than the ordinary breeds of domestic cattle, but also by the horns being proportionally much smaller and shorter, as well as differently directed, and by the forehead being less concave. It is, indeed, usually flat; and the frontal bones extend further beyond the orbits, before they join the nasal bones, than in the *Bos primigenius*. The horn-cores of the *Bos longifrons* describe a single short curve outwards and forwards in the plane of the forehead, rarely rising above that plane, more rarely sinking below it: the cores have a very rugged exterior, and are usually a little flattened at their upper part.

Remains of this species were described by Robert Ball, Esq., Secretary to the Zoological Society of Dublin, in the 'Proceedings of the Royal Irish Academy for January 1839,' as indicating "a variety or race differing very remarkably from any previously described in works with which the author was acquainted." They consisted principally of parts of the skull with the horn-cores, which had been found at considerable depths in bogs in Westmeath, Tyrone, and Longford.

In the same year Mr. Woods, in his 'Description of the Skull of the *Bos primigenius* from Melksham,' gave a notice of a fragment of a skull, including the cores of the horns, with a wood-cut of the specimen, clearly showing it to belong to the *Bos longifrons*. It was found in a peat-bog in the neighbourhood of Bridgewater. This formation "covers an ancient sea-beach (although now nine miles from the coast), in which such marine genera as *Murex*, *Ostrea*, *Mytilus*, and *Solen* are abundantly embedded, and over these as plentiful a deposit of freshwater species, as *Helix*, *Planorbis*, *Lymnea*, &c., exhibiting the alternate resting of the sea, and a river or lake for very considerable periods." The peat above these deposits is thirty feet in thickness, and the skull of the *Bos longifrons* was deeply embedded in it.

The agreement of the specimens from the more ancient freshwater beds in Essex and Middlesex with those from the later formations of Devonshire and Ireland is ex-

* Op, cit. p. 28.

tremely close in every character, except the accidental ones derived from the difference of the strata.

The following are admeasurements of some of the specimens:

	Hunterian Irish Bog.		Mr. Ball, Bog, Westmeath.	
	In.	Lin.	In.	Lin.
Length from the supra-occipital ridge to the nasal bones	8	0	8	0
Breadth of the skull between the roots of the horns .	5	0	5	0
Breadth of the skull across the middle of the orbits .	6	9	6	6
Circumference of base of horn-core . . .	4	0	3	6
Length following outer curvature , . .	4	0	3	6
Span of horn-cores from tip to tip . . .	12	0	11	0

	Mr. Brown. Fresh-water, Clacton beds.		Mr. Woods* Bog, Bridgwater.	
	In.	Lin.	In.	Lin.
Length from the supra-occipital ridge to the nasal bones	0	0	0	0
Breadth of the skull between the roots of the horns .	5	0	5	0
Breadth of the skull across the middle of the orbits .	0	0	0	0
Circumference of base of horn-core . . .	4	6	0	0
Length following outer curvature . . .	4	0	4	0
Span of horn-cores from tip to tip . . .	12	0	11	3

Additional specimens of the *Bos longifrons* have been transmitted to me by the Earl of Enniskillen from the sub-turbary shell-marl in various localities in Ireland; by Mr. Strickland from the newer pliocene freshwater deposits in the brick-field at Bricklehampton Bank in the valley of the Avon; and by Mr. Allies from the alluvium of the Severn at Diglis.

In the first of those localities the *Bos longifrons* is associated with the *Megaceros Hibernicus*, just as it is in the newer pliocene freshwater deposits in Essex; in the Bricklehampton beds, it occurs with both the *Bison priscus*

* Mr. Woods deemed his specimen to indicate that the races of the genus *Bos* in the ancient time must have been subject to many variations in point of size (op. cit. p. 28); but I have met with none coeval with the *Bos longifrons*, of intermediate size between it and the colossal *Urus* and *Bison*.

and *Bos primigenius;* in the more recent alluvium we find it with the Red Deer and with Roman antiquities.*

In many localities in Ireland the remains of the *Bos longifrons* are found in the peat itself, from which it may be inferred that the species continued to exist after the Megaceros became extinct.

Amongst the numerous specimens of the *Bos longifrons* which have passed through my hands, I have recognised two sizes of the horn-cores, the largest yielding a basal circumference of seven inches, and a length along the outer curve of seven inches; and the smaller size being that which is given in the preceding table of dimensions: the smaller horns may have characterised the female, and the larger horns, which have the same curvature and rugged surface, the males. Mature Bovine metacarpal and metatarsal bones, shorter than those of an ordinary domestic Ox, or not exceeding them in size, but thicker in proportion to their length, have been found fossil in the caves at Kirkdale and Oreston. I suspect these to belong to the *Bos longifrons;* at all events they testify the co-existence of an ordinary-sized *Bos* with the extinct *Carnivora* of that remote period, and one, therefore, more likely to become their prey, than the comparatively gigantic *Bison* and *Urus*.

It has been remarked, in a former section, that the domesticated descendants of a primitive wild race of cattle were more likely to be met with in the mountains than in the lowlands of Britain, because the aborigines, retaining their ground longest in the mountain fastnesses, may be supposed to have driven thither such domestic cattle as they possessed before the foreign invasion, and which we

* The specimens of *Bos longifrons* from Diglis are those referred to at p. 475 as bones "of small short-horned Cattle."

may presume therefore to have been derived from the subjugation of a native species of *Bos*.

In this field of conjecture, the most probable one will be admitted to be that which points to the *Bos longifrons* as the species which would be domesticated by the aborigines of Britain before the Roman invasion. Had the *Bos primigenius* been the source, we might have expected the Highland and Welsh cattle to have retained some of the characteristics of their great progenitors, and to have been distinguished from other domestic breeds by their superior size and the length of their horns. The kyloes and the runts are, on the contrary, remarkable for their small size, and are characterised either by short horns, as in the *Bos longifrons*, or by the entire absence of these weapons.

ADDENDUM TO BOVIDÆ.

The valuable geological services rendered to the Russian Empire by our distinguished countryman, Roderick Impey Murchison, Esq., have been the source of reciprocal benefit to Zoological science at home, the Emperor of Russia having been pleased, at Mr. Murchison's request, to direct the transmission to this country of the prepared skin and the skeleton of the rare Zubr, or existing Wild Aurochs, which species is preserved with so much care in the forests of Lithuania. The specimens have been presented by his Imperial Majesty to the British Museum, where I have had the opportunity of comparing the recent bones with those of the fossil Aurochs since the foregoing sheets went to press.

The metacarpal and metatarsal bones present the same

slender proportions, compared with those of the Ox, which distinguish the fossils. There are fourteen pairs of ribs. The skull shows the same expanse, convexity, and shortness of the frontal region, and the same angle between this and the occipital region, as does the fossil skull of the *Bison priscus;* the horn-cores have the same advanced origin, and the same direction: these, however, are relatively shorter than in most of the fossil skulls, and the general size of the existing Aurochs is less than that of the ancient or fossil specimens. Admitting with Cuvier, that such characters are neither constant nor proper for the distinction of species, we may recognise in the confined sphere of existence to which the Aurochs has been progressively reduced, precisely the conditions calculated to produce a general loss of size and strength, and a special diminution of the weapons of offence and defence. I cannot perceive, therefore, any adequate ground for abandoning the conclusion to which I had arrived from a study of the less perfect materials available to that end, before the arrival of the entire skeleton of the Lithuanian Aurochs, viz.—that this species was contemporary with the Mammoth, the Tichorhine Rhinoceros, and other extinct Mammals of the pliocene period.

Fig. 212.

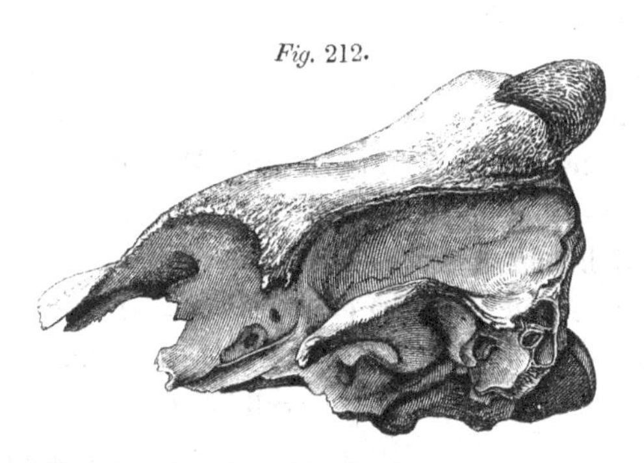

Side view of cranium of *Bos longifrons*, Bog, Ireland.

CETACEA. DELPHINIDÆ.

Fig. 213.

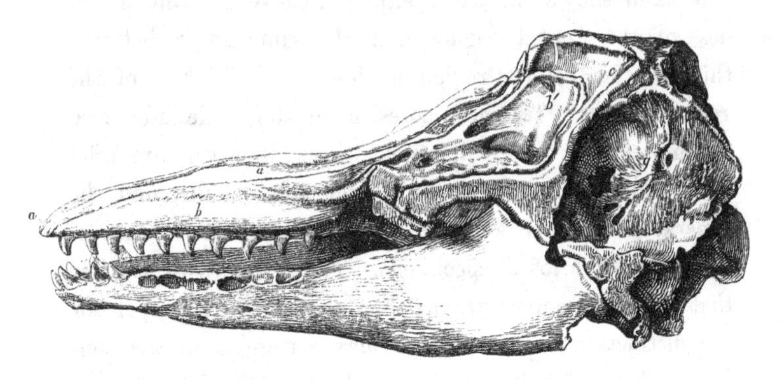

Skull of thick-toothed Grampus. Fen, Lincolnshire.

PHOCÆNA CRASSIDENS. Thick-toothed Grampus.

THE most complete example of the skeleton of a Cetaceous animal which, by the alteration of the osseous texture, and by the peculiar configuration of the bones, claims to rank with the British Fossil Mammalia, is that which was discovered in the year 1843 in the great fen of Lincolnshire beneath the turf, in the neighbourhood of the ancient town of Stamford, and which is now preserved in the Museum of the Stamford Institution.

The skull (fig. 213), which is almost entire, and the teeth, some of which are preserved in the lower jaw, prove the animal to have belonged to the Dolphin tribe (*Delphinidæ*), and to the short-jawed or Porpoise genus (*Phocæna*), and herein to be comparable in point of size with the Round-headed Porpoise (*Phocæna melas*), the Grampus (*Phocæna orca*), and the Beluga (*Phocæna leucas*).

It differs sensibly from the skull of the Beluga by its concave profile, and by its greater breadth in proportion to its length, especially across the maxillary portion: it has shorter temporal fossæ and more numerous teeth. The general resemblance of the fossil to the skulls of the Grampus and Round-headed Porpoise is much closer, and its distinctive character requires more detailed comparison with these for its demonstration. The following are dimensions of the cranium of the fossil and of the closely allied recent species:—

	Crassidens.	*Orca.*	*Melas.*
	In.	In.	In.
1. Length of the skull from the back part of the condyles to the end of the beak	26	36	24
2. Length of the beak from the front end of the malar bones	12	17	13
3. Breadth of the skull across the post-orbital processes of the frontal bones	15	21	15½
4. Height of the skull from the lower part of the condyles to the top of the occipital crest . . .	9½	15	9½
5. Breadth of the beak across its middle part .	7	10	6½
6. Length of the alveolar series, lower jaw, .	9½	15	6½
7. Length of the lower jaw	20½	29	19

These dimensions show the close agreement between the fossil and the skull of the *Phocæna melas* in general size, but the sixth admeasurement demonstrates an important difference in the extent of the dental series: and this difference does not depend on a corresponding difference in the number of the teeth, which might have been merely the effect of age; for the lower jaw of the *Ph. melas*, the subject of comparison, has eleven teeth in the alveolar series of each ramus, whilst the fossil jaw has only ten. The greater extent of the dental series in this jaw depends on the considerably larger size of the teeth in the fossil, as the following dimensions show:—

	Crassidens.	*Orca.*	*Melas.*
	In.	In.	In.
Circumference of the base of the crown of the largest tooth	2	4	1
Length of the crown of do.		12½	⅔

In the relative size of the teeth, their thick conical crowns with slightly recurved and incurved pointed summits, and also in the well-defined coat of enamel, the fossil much more resembles the Grampus. In the skull of a Grampus in the College of Surgeons, the number of teeth is $\frac{12-12}{12-12}$; in the fossil cranium it is $\frac{10-10}{10-10}$? In the latter the enamel has been changed to a light bluish-grey, the dentine to a yellowish-brown. In the fossil lower jaw the number is precisely defined by the actual teeth, or by the distinct sockets: these parts have been restored in the upper jaw of the Stamford specimen.

The *Phocæna crassidens* differs from the *Phocæna melas* in the relatively larger temporal fossæ, by which it resembles the Grampus; and it differs from *Ph. orca*, and resembles the *Ph. melas* in the continuation of the intermaxillary bones backwards to the nasal bones, which they join; but, in the breadth of the intermaxillaries, it is intermediate between the *Ph. orca* and *Ph. melas.* In the latter species, Cuvier correctly states that "the intermaxillaries include nearly two-thirds of the breadth of the beak, whilst in the Grampus they scarcely form one-third:" but, in the *Phocæna crassidens*, the intermaxillary bones form more than half the breadth of the beak. A more definite distinctive character of the fossil skull is the appearance of part of the vomer (fig. 216, *v*,) upon the bony palate, in the same relative position and to the same extent as in the skull of the common Porpoise; for the vomer is not visible on the palate in the Grampus, the Round-headed Porpoise (*Ph. melas*), the Beluga, or the *Delphinus griseus.*

By comparing figure 213 with that of the skull of the Bottle-nosed Dolphin (*Delphinus tursio*) in p. 472 of Bell's 'British Quadrupeds, the difference will be readily appreciated between the Stamford fossil and that large British species of *Delphis*, which is so nearly similar in size. The skull, with the numerous minute teeth, of the Porpoise forms the subject of the vignette at p. 476; and the characteristic cranium of the Beluga is figured at p. 491 of the same work.

I have seen no specimens of these existing British *Delphinidæ* meriting to be regarded as fossils; the subject of the present section presents characters by which it differs not only from the known existing *Delphinidæ* of our own coasts, but from all the species that have been so described and figured as to admit of a comparison.

Of the fossil *Delphinidæ*, described in other works, the *Phocæna Cortesii*, which Cuvier* defines as allied to the *Ph. orca* and *Ph. melas*, is readily distinguished from the *Ph. crassidens* by its more numerous and smaller teeth. The fossil *Delphinus*, allied to the common species,† is distinguished by its still smaller teeth; and another extinct species, from the Faluns of the "Département des Landes,"‡ by the long symphysis of the lower jaw; that of the Stamford fossil being as short as in the Grampus. The fossil Dolphin, described by M. Von Meyer under the name of *Arionus servatus*,§ had a mandibular symphysis not shorter than one-third the entire length of the skull. The *Delphinus* from the "calcaire grossier," du Département de Maine-et-Loire, had seventeen teeth in each alveolar series of the upper jaw. Other recorded extinct Cetacea present still wider differences from the Stamford fossil.

Whether the species or variety of the Grampus indicated

* "Un Dauphin voisin de *l'epaulard* et du *globiceps*." 'Ossemens Fossiles,' 4to., 1823, t. v. pt. i. p. 208.

† Ib. p. 316. ‡ Ib. p. 312.

§ Leonhard and Bronn, Jahrbuch für Mineralogie, 1841, p. 315.

by this fossil may still exist in our seas, remains to be proved: until then, it may be regarded as an extinct species of *Delphinidæ*, for which I propose the name of *Phocæna crassidens*.

Remains of *Delphinidæ* have been found in silt several feet below the surface in the "Beeding levels," and at the mouth of Cuckmeer. The most completely petrified specimen referable to this family of Cetacea, is the anchyloid mass of cervical vertebræ in Professor Sedgwick's museum in the University of Cambridge. Respecting this specimen, which has belonged to a Cetacean as large as the Grampus or Narwhal, the Professor writes to me:— "It was found in the brown clay (*alias* till) near Ely; but I have not the shadow of doubt that it was washed out of the Kimmeridge (or Oxford) clay, for both clays are near at hand. In condition, it is exactly like the bones of those clays, and is utterly unlike the true gravel bones, whether in the dry gravel or the till."

Subjoined is the figure of the anchylosed cervical vertebræ of the *Phocæna crassidens*.

Fig. 214.

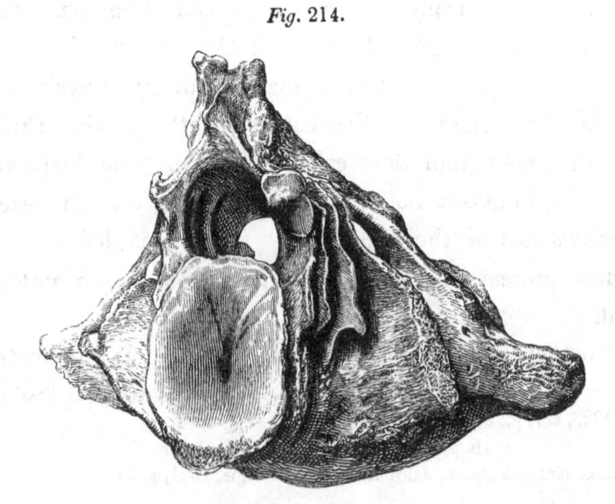

Cervical vertebræ of thick-toothed Grampus. Fens, Lincolnshire.

CETACEA. *MONODON.*

Fig. 215.

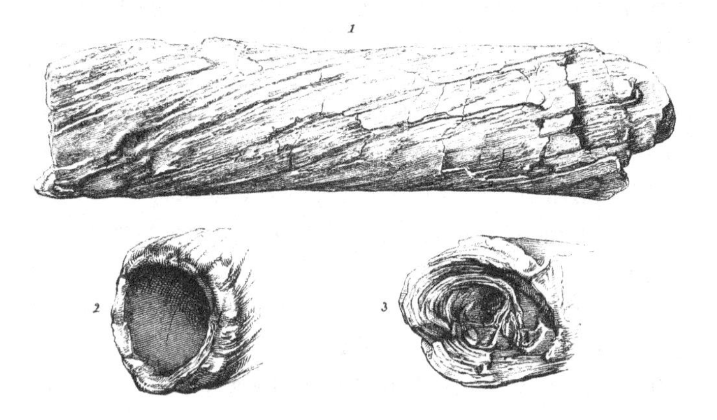

Fossil fragment of the tush of a Narwhal, $\frac{1}{3}$ nat. size.

MONODON MONOCEROS. Narwhal.

Narwhal,	PARKINSON, Organic Remains. 4to., 1811, vol. iii. p. 309.
Narval fossile,	CUVIER, Ossemens Fossiles, 4to., 1825, vol. v. pt. i. p. 349.
Monodon fossilis,	H. v. MEYER, Palæologica, 8vo., 1832, p. 99.
Monodon monoceros,	OWEN, Report of British Association, 1842.

THE following is the evidence of the existence, during the deposition of our tertiary strata, of the very remarkable species of Cetaceous animal, whose spiral tusk so long perplexed the older naturalists, and still figures in heraldry as the horn of the fabulous Unicorn.

Mr. Parkinson, in the work above cited, states that two fossil fragments of the long-projecting and spirally-twisted tooth of the Narwhal were formerly in the museum of Sir Ashton Lever; and he adds, "One of these I now possess, and strongly suspect it to have been found on the Essex coast."

This specimen was purchased at the sale of Mr. Parkinson's collection by the Earl of Enniskillen, who most kindly transmitted it to me to have the figures taken from it which are placed at the head of the present section. The specimen has lost much of the original animal matter, is absorbent, rather friable, and partially decomposed, so that the layers of the basal substance of the dentine might be easily separated. In what length of time, simple exposure to the elements on the sea-shore would produce this state of decomposition, I know not; but I have only witnessed such a state in fossils of the age of the post-pliocene extinct Mammals.

The fragment above figured is the basal part of the tusk; it measures ten inches and a half in length, and nine inches in circumference: fig. 215, [2], shows the short and wide conical pulp-cavity at the inserted end, and fig. 215, [3], the opposite fractured end, where the pulp-cavity begins again to expand as it extends into the exserted part of the tusk. The superficial spiral ridges precisely resemble those at the same heavy implanted part of the tusk in the recent Narwhal.

A portion of a fossilized tusk of a Narwhal is preserved in the museum of Comparative Anatomy in University College, and is said to have been obtained from the London clay in the neighbourhood of the metropolis.* A portion of the skull of a *Monodon monoceros* is said to have been found in the marine silt of the marshy plain called Lewes Levels.

Cuvier mentions a fragment of the Narwhal's tusk, considerably altered in texture, which is preserved in the museum of Natural History at Lyons, and cites a notice of similar fossils found in Siberia.

* Professor Grant, in Thomson's British Annual, 1839, p. 269.

In the Hunterian Collection of Fossils in the Royal College of Surgeons, there is a small fragment of the tusk of a Narwhal, No. 1439, partly decomposed and absorbent from the loss of animal matter, which is stated to be "from Baumann's Cave." This cave is situated on the north-east border of the Hartz, in Blankenburg, and is described by Leibnitz in his 'Protogæa.'*

* See my catalogue of the Fossil Mammalia and Aves in the Museum of the Royal College of Surgeons of England, 4to., p. 286.

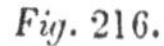

Fig. 216.

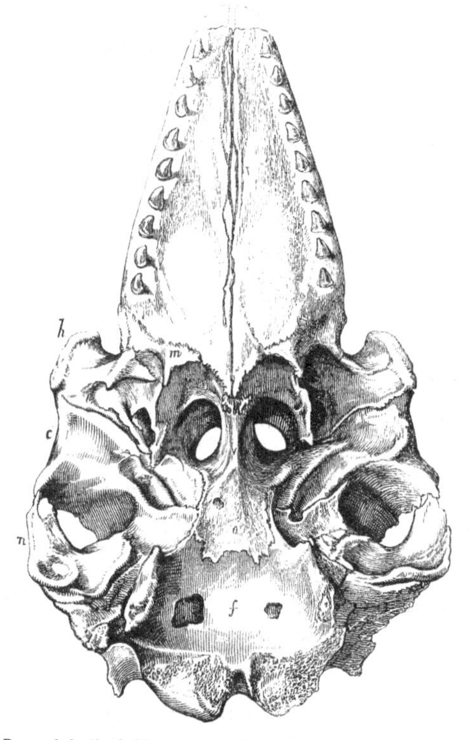

Base of skull of thick-toothed Grampus. Fen, Lincolnshire.

CETACEA. *PHYSETERIDÆ.*

Fig. 217.

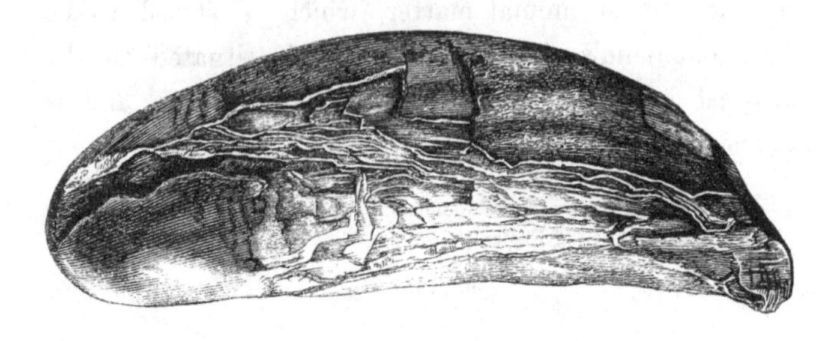

Fossil tooth, ½ nat. size, of Cachalot, newer pliocene, Essex.

PHYSETER MACROCEPHALUS. Cachalot, or Sperm Whale.

Cachalot, OWEN, Report of British Association, 1842, p. 18.

THE evidence of the existence of the Great Sperm Whale, or Cachalot, in European seas at the period when the mammoth and other now extinct mammalia trod the adjoining shores, is precisely of the same nature as that previously adduced for the contemporaneous existence of the Narwhal, viz., the discovery of a fossil tooth, absorbent from the loss of animal matter, and with its substance separating into concentric layers, in the superficial deposits near the coast of Essex. Fig. 217 gives a side-view of this tooth, which is now in my possession, and fig. 218 is a view of a longitudinal section of a recent Cachalot's tooth, to show the characteristic proportions of *c*, the cement; *d*, the dentine; and *o*, the osteodentine, which substances enter into the composition of the Cachalot's tooth.* Satisfactory evidence of the existence of the Cachalot in the present seas of Britain, is given by Professor Bell, in his 'History of British Quadrupeds and Whales,' p. 503.

* Odontography, p. 353, pls. 89, 90.

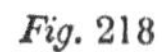

Fig. 218

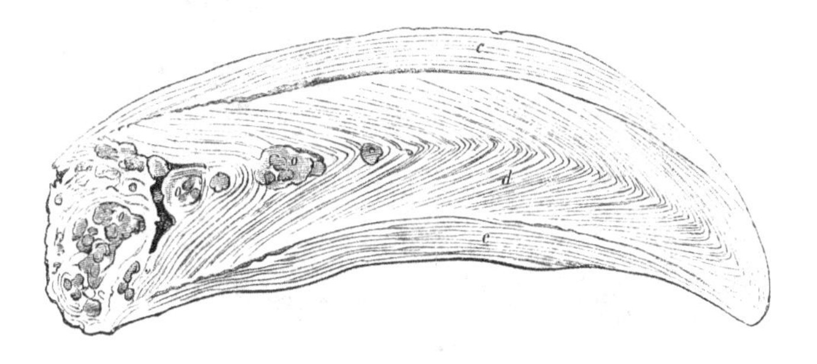

Longitudinal section of a recent tooth of the Cachalot, $\frac{1}{2}$ nat. size.

Fig. 219.

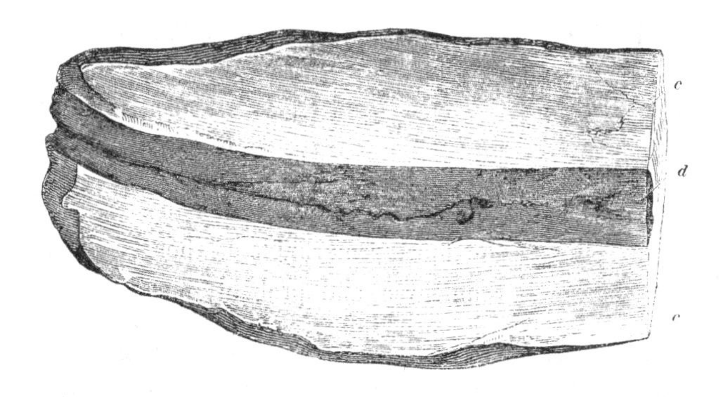

Longitudinal section of a fossil physeteroid tooth from the Red Crag, Felixstow.

CETACEA. *BALÆNIDÆ.*

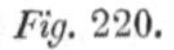

Fig. 220.

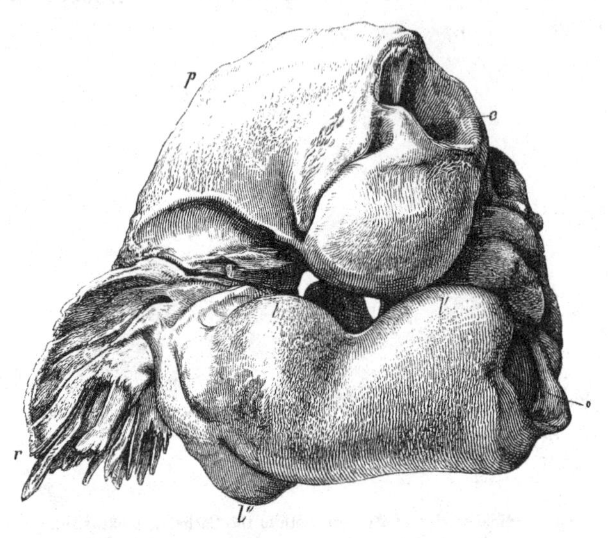

Petro-tympanic, or Ear-bone of the recent Cachalot. ½ nat. size.

CETOTOLITES. Fossil Ear-bones of Whales.

THE Rev. Professor Henslow, in his Paper entitled "On Concretions in the Red-crag at Felixstow, Suffolk,"* in which the attention of geologists was first called to their animal and cetaceous origin, has appended my determinations of some of the most remarkable of those bodies, by which not fewer than three, and probably four, species of whales have been added to the Catalogue of British Fossil Mammalia. I have proposed to call the bodies in question "cetotolites,"† as they consist of portions of the petro-tympanics, or ear-bones of large Cetacea.

* Proceedings of the Geological Society, Dec. 13, 1843, pp. 281, 283.

† κῆτος, whale ; ὦς, ear ; λίθος, stone.

An entire specimen of this compound bone has not yet been obtained from the Red-crag; nor is it likely that it should be. Almost all the fossils of this formation show the action of surf-waves or breakers; those under consideration appear to have been dislodged from a subjacent eocene deposit; and as the massive petrous bone and the tympanic bone of the Cetacea adhere to each other naturally by only two small surfaces, they would hardly escape being broken asunder under the operation of such disturbing forces.

All the specimens which were submitted to me by Professor Henslow consisted of the tympanic portion only, and I have as yet seen but one specimen of the rugged petrous bone; it is now in the collection of Mr. Brown, of Stanway. The tympanic bone may be readily recognised by its peculiar conchoidal shape and extremely dense texture; the recent bone breaking with almost as sharp a fracture as the petrified fossils.

None of these tympanic fossils are entire: the thin brittle outer plate which bends over the thick, rounded, and, as it were, involuted part, like the outer lip of such simple univalves as the *Bullæ* and *Leptoconchi*, is broken or worn away in the best specimens, all of which are rolled and waterworn. I was at once led by their size to the largest of the existing Cetacea for the subjects of comparison, as the Grampus, the Hyperoodon, the Cachalots (*Physeter*), and the true Whales (*Balænoptera* and *Balæna*).

Two or three of the specimens were fortunately sufficiently entire to show the form of the tympanic cavity bounded by the overarching plate, with the proportion and direction of its anterior or Eustachian outlet, and most of them had the opposite or hinder extremity entire. They were thereby seen to differ from the tympanic bones of the *Delphinidæ*, including the *Grampus* and *Hyperoodon*, in having the hinder extremity of the bone simple and not bilobed; and some of them differed, also, in having the anterior outlet of the cavity partially enclosed by the extension of the outer plate around that end.

With regard to the Cachalot (*Physeter*), I had not had the opportunity of comparing the Felixstow fossils, when I first gave an account of them, with the tympanic bone in that genus, which I then knew only by the figures given by Camper * in his characteristic but sketchy style. Cuvier, who founds his notice of the tympanic bones of the Cachalot on the same figures, states that they most resemble those of the *Delphinidæ;* but are less elongated and less bilobed posteriorly.† The figures show still more clearly that the tympanic cavity is continued freely forward out of the anterior end of the bone, and terminates by a relatively wider outlet than in the *Delphinidæ*.‡

The idea thus given of the form of the tympanic bone of the Cachalot, being, as I have since had the opportunity of satisfying myself, in the main correct, the comparison of most of the Cetotolites becomes limited to the true whales (*Balænidæ*), in the few known species of which the distinctive characters of the tympanic bones are afforded by their relative size and the shape of their inferior surface.

In *Balænoptera* the tympanic bones, according to Cuvier, are very small in proportion to the head, and are equally convex at their inferior surface.

* 'Anatomie des Cetacés,' Pls. xxiii. xxv.

† 'Ossemens Fossiles,' 4to., v. pt. i. p. 376.

‡ Cuvier, (Leçons d'Anatomie Comparée, ed. 1799, vol. ii. p. 492,) says, generally:—" L'extremité antérieure de la caisse est toute ouverte:" which character, M. Adrien Camper thinks he meant to apply to the Cachalot more particularly. I shall combine the description of the petro-tympanic bone of the Cachalot, which I have recently received, with that of the *Cetotolite* most resembling it.

As none of the fossils in question have been found *in situ*, with any part of the cranium, their size in proportion to that of the animal cannot be judged of; but in the specimens that have been least injured and water-worn, the inferior surface shows the flattened or gently concavo-convex undulation which characterises the tympanic bone in true Balænæ.

In regard to the differences which are observable in the tympanic bones of the two known species of Balæna (*Bal. mysticetus*, and *Bal. australis*, *capensis*, or *antarctica*,) Cuvier merely observes that "though slight, they add to the motives which led him to believe the Arctic whale and that of the Cape to be specifically distinct." This remark at least encourages us to regard the characters derivable from the tympanic bone as sufficiently determinate to be a guide in the discrimination of species; and with this conviction I have proceeded to compare the fossils in question with the recent tympanic bones of the two above-cited existing species of Balæna.

In these the thick convex involuted portion of the tympanic bone is slightly and unequally raised above the level of the cavity formed by the over-arching wall, but in the *Bal. antarctica* it gradually decreases in thickness to the anterior or Eustachian angle; while in the *Bal. mysticetus* the thicker posterior part is defined by an indentation from the thinner anterior part. In both species the thinner part of the convex border is distinctly continued to the anterior limit of the cavity; in both the extent of the involuted convexity, inwards, is not well defined, but it gradually subsides, and the convexity is exchanged for the concave curve of the overarching wall. The inner surface of this wall is very rugged near the involuted part. I purposely omit the mention of the slight differences in other parts of the tympanic bone of the *Balæna mysticetus* and *Bal. antarctica*, since the condition of the fossils would not admit of the application of those differences in the determination of their affinities.

Fig. 221.

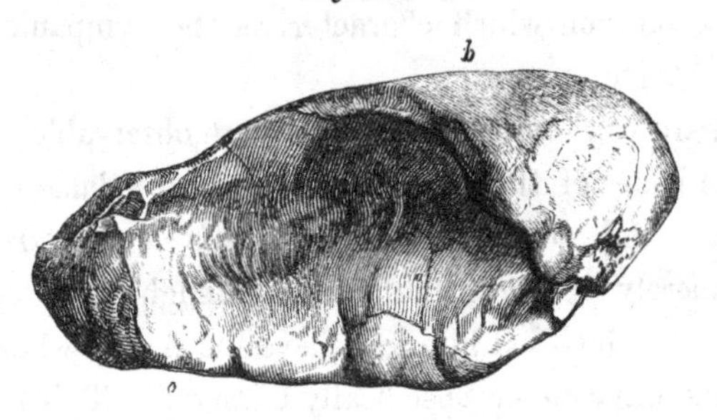

Tympanic bone of *Balæna affinis*, ½ nat. size. Felixstow.

One of the most complete of the fossil tympanic bones, which measures five inches in length, resembles the *Bal. antarctica* in the slight elevation of the outer part of the involuted convexity (*a*), and its gradual diminution to the Eustachian end of the cavity (*o*): it resembles both *Balænæ* in its traceable continuation to that end, and in the gradual continuation of the concave outer wall (*b*) from the involuted convexity; this convexity is indented also, as in both recent Balænæ, by vertical fissures narrower than the marked indentation which distinguishes the *Bal. mysticetus*: these fissures are almost worn out by friction in some of the specimens. The more perfect one under consideration is not, however, identical with the *Bal. antarctica*.

The upper surface of the bone maintains a more equable breadth from the posterior to the anterior end, the outer angle of which, being well marked in the fossil, is rounded off in the recent specimen; the under and outer surfaces of the tympanic bone meet at an acute angle. The above characters are sufficiently marked in the specimens of the fossil tympanic bones, to justify their being regarded as belonging to a species distinct from the known existing *Balænæ*, but nearest allied to the *Bal. antarctica*, and which I propose to call *Balæna affinis*.

Fig. 222.

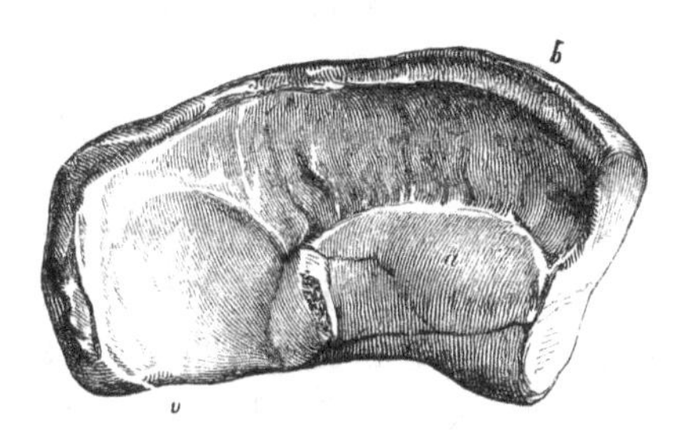

Tympanic bone of *Balæna definita*, ½ nat. size. Felixstow.

A second species is characterised by the distinct definition of the involuted convexity (*a*); and the extent of the slightly concave surface extending from it to the commencement of the overarching wall (*b*) ; the anterior extremity of the involuted convexity is equally well defined, and a wide concavity divides it from the anterior extremity of the Eustachian outlet (*o*). The thickest part of the involuted convexity is not very prominent. The under and outer surfaces of the bone meet at a right angle.

The species indicated by the tympanic bones of this form I have termed *Balæna definita*.

Fig. 223.

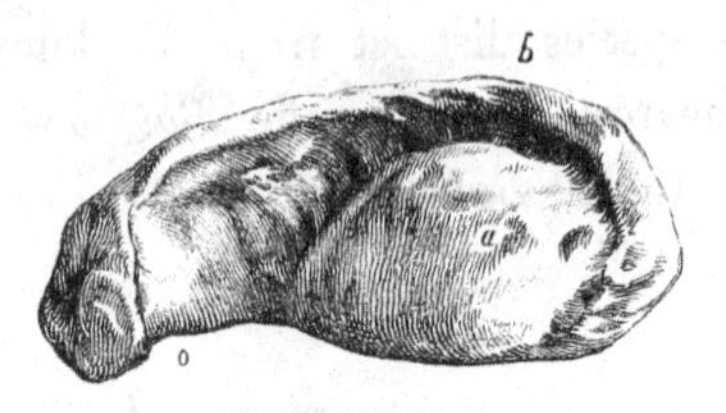

Tympanic bone of *Balæna gibbosa*, ½ nat. size. Felixstow.

A third form of tympanic bone differs from the first in the shorter and more convex form of the involuted part (*a*), the anterior end of which is divided from the anterior end of the cavity by a concave border one inch in extent; the internal border of the involuted convexity is also better defined than in *Bal. affinis;* but the overarching wall (*b*) begins to rise close to it, divided from it only by a deep and narrow rugged fissure, instead of by a broad and gently concave tract, as in *Bal. definita.*

Both the outer and under surfaces of these specimens are more rounded than in the two preceding species; but being more mutilated and water-worn, the differences of the external parts of the bone are of less value. The characters above specified, which are furnished by the involuted convexity, are decisive as to the specific distinction of the present fossil tympanic bone, and the third species of extinct whale, so indicated, I have proposed to call *Balæna gibbosa.*

Fig. 224.

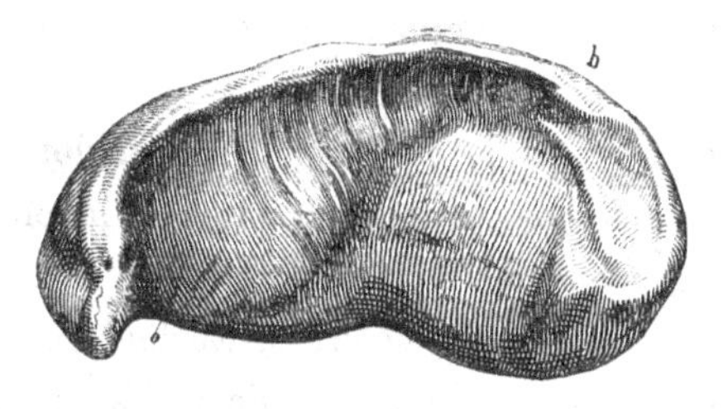

Tympanic bone of Balæna emarginata, $\frac{2}{3}$ nat. size, Felixstow.

There is a fourth form (fig. 224), which differs from the last in the less degree of convexity of the involuted part, but more particularly in its outer border being notched or indented, as in *Balæna mysticetus*, by a vertical angular impression deeper and wider than the smaller vertical fissures.

The comparative shortness of the involuted convexity distinguishes this species from the existing *Balænæ* and the *Bal. affinis*, the notched and less convex involution from *Bal. gibbosa*, and the immediate rising of the overarching

Fig. 225.

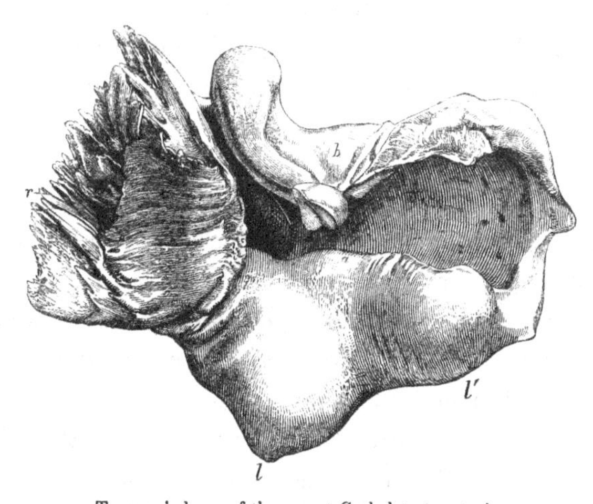

Tympanic bone of the recent Cachalot, $\frac{1}{2}$ nat. size.

wall beyond the inner boundary of the involution from the *Bal. definita.* I have named this species *Balæna emarginata.*

In the petro-tympanic bone of the Cachalot, fig. 220, and the tympanic bone, fig. 225, *a* is the involuted convexity; *b*, the over-arching plate; *l*, outer lobe; *l'*, inner lobe; *l''*, under and outer lobe; *o*, Eustachian outlet; *r*, rough outer process of the tympanic portion; *p*, petrous portion; *e*, its deep excavation.

If we compare the characters of the *Cetotolites*, which have been already described with the petro-tympanic, or its tympanic portion in the Cachalot, we find that that referred to the *Balæna affinis* differs in the continuous, or even, and not bilobed under and back part of the involuted convexity, and in the continuation of the overarching wall around the inner end of the tympanic cavity; it differs, in short, like the tympanic bone of the *Balæna,* in its entire figure from that of the Cachalot.

If we take the *Balæna definita,* we find that besides the absence of the bilobed character of the involuted convexity, it differs by its well-defined anterior border. The *Balæna gibbosa,* in addition to the absence of the bilobation of the involuted convexity, differs from the Cachalot in its limited extent, as well as its greater convexity in the tympanic cavity.

The only Cetotolite which makes any approach to the peculiar characters of that of the Cachalot, is that which I have described under the name of *Balæna emarginata,* in consequence of the vertical notch, or posterior emargination of the involuted convexity, which gives it a slightly bilobed character. It differs in a marked degree from the Cachalot, however, by the very inferior development of the lobe corresponding to the inner one in the Cachalot; and it also manifests the Balænal character in the extension of the origin of the overarching wall around the inner end, throwing the opening of the tympanic cavity backwards.

In the account of these Cetotolites, read to the Geological Society,* the vertical notch or indentation of the border of the involuted convexity of the tympanic bone of the *Balæna mysticetus* is noticed in connection with that character in the *Balæna emarginata;* but in the *Balæna mysticetus*, the involuted convexity does not swell out into lobes at the back part of the bone.

Professor Henslow observes, "It is not a little remarkable that all these specimens should have been procured within a very narrow compass, for I found none beyond the limits of two contiguous indentations in the cliff, a short distance to the north of Felixstow."† Mr. Rose, F.G.S., has, however, since discovered a fractured fossil tympanic bone of the *Balæna definita* in a patch of crag, high up a rather lofty part of the bank of the river Orwell, a short distance below Holywells, Suffolk. The dense texture of this fossil, as of most of the specimens from Felixstow, presented evidence of ferruginous infiltration in a dark layer surrounding the lighter-coloured central part of the bone. The long diameter of this specimen measured three inches and a quarter, its short diameter two inches.‡

* 'Quarterly Journal of the Geological Society,' No. I. p. 40. † Ibid. p. 36.

‡ Numerous other nodules, or pebbles, some cylindrical, some fusiform, with annular, or spiral, or longitudinal convolutions, occur in the Red Crag at Felixstow, which present external indications of animal origin; and yield, upon analysis, 56 per cent of phosphate of lime. They occasionally contain remains of small crabs and fishes,—sharks' vertebræ, for example, like those of the London clay; and they have been regarded by Professor Henslow, who first called attention to their probable origin, as coprolites.—'*Proceedings of the Geological Society*,' December, 1843, p. 282.

CETACEA. *BALÆNODON.*

Fig. 226.

Portion of fossil tooth of Balænodon, Red-Crag, Felixstow. Nat. size.

BALÆNODON PHYSALOIDES. Cachalot-like Balænodon.*

The fossil above figured was submitted to me by Mr. Bowerbank, in the year 1840, with a request that I would endeavour to ascertain its real nature; and having obtained the permission of its owner, Mr. Brown of Stanway, to take a section from it for microscopical examination, I found by this insight into its intimate structure, that it was a fossil tooth, belonging to a Mammalian animal; and in that class, bearing the nearest resemblance in the thickness and structure of the outer layer of cement immediately surrounding the dentine, to the teeth of the Cachalot (*Physeter macrocephalus*).

The most obvious distinction was, that, great as is the thickness of the outer coat of cement in the Cachalot, (fig. 218, *c*,) it was still greater in the fossil, (fig. 219, *c*,)

* *Ziphius*, generic name of a fossil Cetacean, and ὀδοὺς, a tooth.

being twice as thick in proportion to the diameter of the central axis of dentine (figs. 218, 219, *d*). The second difference was the cylindrical form of the slender axis, and the mere filamentary tract of osteo-dentine in its centre, that substance being invariably and abundantly present in the basal part of the same extent of dentine, prior to its contracting to form the conical crown of the Cachalot's tooth (fig. 218, *o o*). The length and slenderness of the cylindrical axis of dentine in the Felixstow fossil tooth are peculiar to it; and the section of this dark-coloured dentinal pith (fig. 227, *d*), at each end of the fragment, would alone have prevented any one cognizant of the form and structure of the Cachalot's tooth, from mistaking the fossil for a tooth of that recent Cetacean. The microscopic structure of the Felixstow fossil shows its near affinity to the *Physeter*, but at the same time proves its specific distinction.

Fig. 227.

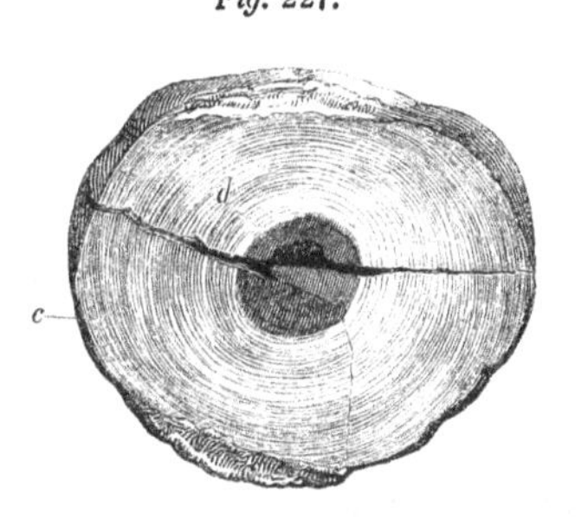

Transverse section of fossil tooth of Balænodon, Red-Crag. Nat size.

The dentinal axis, (fig. 227, *d*,) is finely grooved longitudinally; the margin of its transverse section appearing crenate under a low magnifying power. Viewed under a higher one by light transmitted through a thin slice (fig. 228), its substance is seen to be traversed by dentinal tubes radiating from the centre to the circumference, in a plane more transverse to the axis than in the Cachalot's dentine. The tubes present an average diameter of $\frac{1}{12000}$th of an inch, with interspaces of about twice that diameter: they are thus more closely packed than in the Cachalot. They are minutely undulated, divide and subdivide dichotomously in their course to the periphery of the axis, where they terminate in very fine irregularly bent ramuli; some forming loops, others dilating into minute calcigerous cells. The den-

Fig. 228.

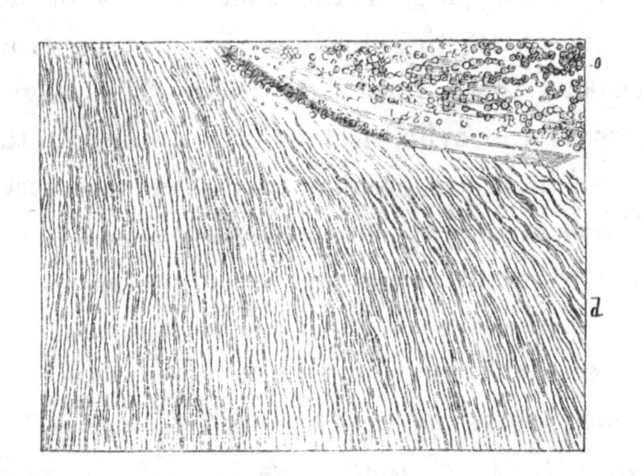

Microscopic structure of dentine; magnified 500 diameters. *Balænodon.*

tinal cells were very conspicuous in many parts of the section, of a sub-circular form, and about $\frac{1}{5000}$th of an inch in diameter.

The clear basis of the cement is almost colourless where it fills the outer indentations of the dentine, and the radiated cells are fewer here: in the rest of the entire thickness of the cement they (fig. 229, *c c*) are very abundant, and arranged at intervals of about thrice their own diameter, which diameter is about $\frac{1}{3000}$th of an inch.

The cetaceous character of the cement is manifested by the numerous and sub-parallel cemental tubes, (fig. 229, *t*,) which run from the outer part of the cement towards the dentine, near which they terminate by fine ramifications, communicating with the radiating tubular prolongations of the cells. The thick cement is also traversed by vas-

Fig. 229.

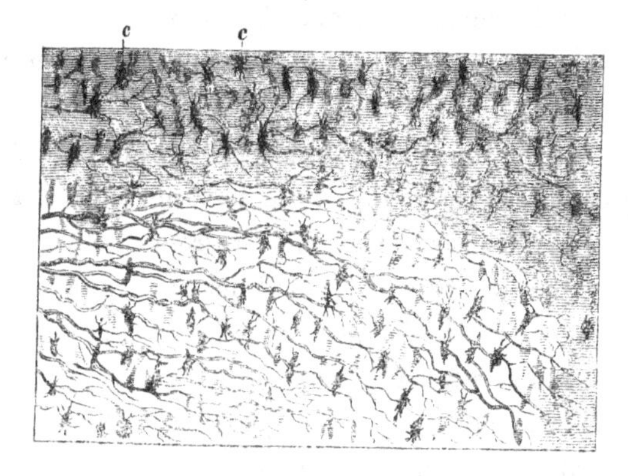

Microscopic structure of cement; magnified 500 diameters. *Balænodon.*

cular or medullary canals, few in number, and irregular as compared with those in the thick cement of the Megatherium's tooth, and more like those in the cement of the Physeter.*

I have recently received from Professor Henslow the conical termination of the crown of two teeth of *Balænodon*, also from the Red Crag at Felixstow, showing the same ferruginous tint, brittle texture, and water-worn surface; but retaining a shape more plainly bespeaking their dental nature. The diameter of the fractured base of the larger conical fragment is two inches; that of the obtuse apex about two-thirds of an inch: the length of the fragment being four inches; the tooth tapers more gradually to its summit, and is less curved than in the Cachalot. This fragment has belonged to part of a larger tooth than the one (fig. 226,) first determined, and the diameter of

* I must refer the Reader to my 'Odontography,' p. 356, pl. lxxxix, *a*, for an account of the microscopic structure of the Cachalot's tooth, which may be compared with the account here given of the tooth of Balænodon.

the axis of the dentine at its base is nine lines; that of the coat of cement is from six to seven lines. The dentine is continued to the apex, forming the obtuse end of the crown, as in the worn teeth of the Cachalot.

The smaller dental fossil transmitted by Professor Henslow, is part of a longer and more tapering cone than the larger one, and thus recedes further from the Cachalot's form of tooth; but the dentine is thicker in proportion to the cement than in the foregoing fossils, and in so far the present resembles more the Cachalot's tooth in structure: where the cone has a diameter of one inch, for example, the dentinal axis, is seven lines across, and the cement three lines in thickness. The exposed surface of the dentine shows the small, close-set, slightly wavy, longitudinal grooves; and the microscopic structure of both fragments agrees with that of the first-described specimen.

The mere difference of form in the fossil teeth might depend, according to the analogy of the Physeter, on a difference of age in the individual, or of place in the jaw, from which such fossils have been derived; but the different relative proportions of dentine and cement in the slender conical tooth, indicates a distinction of species.

We have seen that four species of Cetacea, referrible by the form of the tympanic bones to the Whale family, (*Balænidæ*), but distinct from all known existing species of that family, are more definitely indicated by the remarkable fossils, termed Cetotolites; and it is not improbable that these and the teeth may have been parts of the same Cetaceous animals. We know that the great Whale-bone Whales of the present day, before their jaws acquire the peculiar array of baleen-plates, manifest a true dental system, although the fœtal teeth are transitory and never destined to cut the gum. And as the embryos of existing Ruminants feebly and evanescently manifest in the dark womb, by their upper incisors, their divided cannon bones, and hornless foreheads, the mature and persistent characters of their ancient predecessors the Anoplotheria, so may the equally ancient Whales of the eocene seas have retained and fully developed those maxillary teeth which are transitory and functionless in the existing species.

On this supposition of the relation of the above-described fossil teeth to the tympanic bones in the Crag at Felixstow, the proportions in which they are there found would indicate that the teeth were less numerous in the extinct Cetaceans, from which they have been derived, than they are in the Cachalot. The recent *Ziphius* of the Sechelle Islands has but a single tooth on each side of the lower jaw when full-grown, like the great *Delphinus bidens* of our own seas. But the light of these analogies can give but a dim and distant view of the actual generic characters of creatures, whose former existence is revealed to us by a few fragments of their fossilized skeletons, which have been bruised and worn by ages of elemental turmoil. It may be surprising to many, but not more surprising than gratifying, that the means of investigation at present at our command enable us satisfactorily to determine from such fragments, not only the kingdom of Nature, but the class and the order to which they belonged. They further prove that those ancient Mammals of the deep appertained to the carnivorous section of Cetacea in the Cuvierian system. And if, as is probable, the Whales' form of ear-bone and the Cachalot's character of tooth were combined in the same individual, a distinct family of Carnivorous Cetacea must be established for the eocene fossils, which would form an interesting transitional link between the *Physeteridæ* and *Balænidæ* of the present creation.

That the fossil ear-bones and Cetacean teeth of the Red Crag, have been washed out of the subjacent eocene beds, is probable from the fact of a Cetotolite having been discovered in the London clay itself; and from fragments of other fossil Cetaceous bones having been obtained from the same formation. In the Hunterian collection of fossils, I have determined five considerable fragments of bone to be cetaceous: they were recorded to be "from Harwich Cliff, Essex,"* and were in the same completely petrified condition as the fossil ear-bones from the Red Crag.

BALÆNIDÆ.

The remains of great Whales, referrible to existing genera or species, have been found in Britain, in gravel-beds adjacent to estuaries or large rivers, in marine drift or shingle, as the "Elephant-bed" near Brighton,† and in the newer pliocene clay-beds: but although these depositories belong to very recent periods in Geology, the situations of the cetaceous fossils generally indicate a gain of dry land from the sea. Thus the skeleton of a Balænoptera, seventy-two feet in length, found imbedded in clay on the banks of the Forth, was more than twenty feet above the reach of the highest tide. Several bones of a whale, discovered at Dumore Rock, Stirlingshire, in brick-earth, were nearly forty feet above the present level of the sea. Sir George Mackenzie has re-

* 'Catalogue of Fossil Mammalia, and Birds,' 4to. 1844, Nos. mccccclv, and mccccclix, p. 291.

† See Dr. Mantell's graphic account of his discovery of a fossil jaw of a whale (*Balæna mysticetus*) in this deposit.—'Medals of Creation,' vol. ii. p. 824.

corded the discovery of the vertebra of a whale in a bed of bluish clay near Dingwall: the clay contains many sea-shells, and is evidently a marine deposit; but the spot where the vertebra was found is three miles distant from high-water mark, and twelve feet in height above the present level of the sea. The vertebræ of a whale, discovered by Mr. Richardson in the yellow marl or brick-earth of Herne Bay, in Kent, were situated ten feet above the highest occasional reach of the sea on that coast. A large vertebra of *Balæna mysticetus* was discovered fifteen feet below the surface, in the gravel of the bed of the Thames, by the workmen employed in digging the foundation for the new Temple Church. Dr. Buckland states that "the bones of Whales have been found at Pentuan, in an estuary that is now filled up, on the coast of Cornwall." Mr. Baker, of Bridgewater, possesses the tympanic bone of a Balænoptera, which was dug out of a sand-bank at Huntshill, near that town.

I might add several other instances of the discovery of cetaceous remains in positions to which, in the present condition of the dry land of England, the sea cannot reach; yet the soil in which these remains are embedded is alluvial, or amongst the most recent formation. In most cases the situation indicates the former existence there of an estuary that has been filled up by deposits of the present sea, or the bottom of which has been upheaved.

BRITISH FOSSIL BIRDS.

AVES. *PALMIPEDES.?*

Fig. 230.

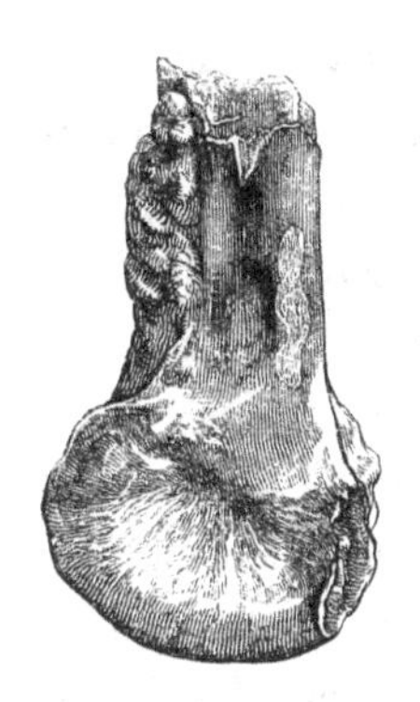
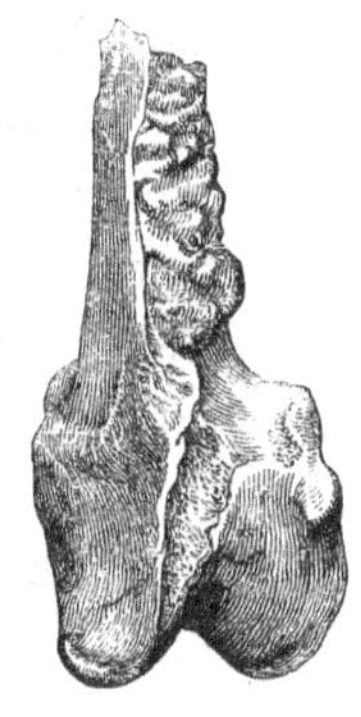

Three views of distal end of tibia, Bird, Chalk. Nat. size.

CIMOLIORNIS* DIOMEDEUS. Long-winged Bird of the Chalk.

Bird allied to Albatross, OWEN, Geological Transactions, 2nd Series, vol. vi. 1840, p. 411, pl. xxxix, figs. 1 and 2.

Osteornis† *diomedeus,* GERVAIS, Thése sur les Oiseaux Fossiles, 8vo. 1844, p. 38.

OF the few actually fossilized remains of birds that have been discovered in England, the most complete and characteristic are those from the London clay. Some fragmentary ornitholites have been discovered in the older pliocene crag, and in the newer pliocene fresh-water deposits and bone-caves. Extremely scanty have hitherto been the recognizable remains of birds from the chalk

* κιμωλία, chalk, ὄρνις, bird.

† This term is applied by M. Gervais, not generically to the fossil in question, but generally to all fossil bones of Birds; and sometimes to bones of other animals, as in the case of his *Osteornis ardeaceus.*

N N

Fig. 231.

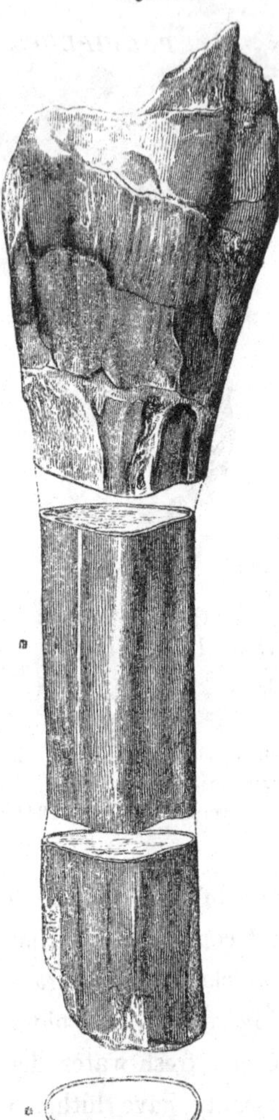

Portions of a wing-bone, nat. thickness; length of the original bone twelve inches.

* Outline of transverse section, at the middle of the bone. Chalk.

formations. The fossil from the Wealden, which I formerly believed, with Cuvier and Dr. Mantell, to belong to a Wading-bird, I have since adduced reasons for referring to the extinct genus of flying reptiles, called Pterodactyle.*

The portions of fossil bone figured in the present section were obtained by the Earl of Enniskillen from the chalk near Maidstone, and were referred by his Lordship and Dr. Buckland to the class of birds: the accuracy of which determination seems to be proved by the smaller fossil (fig. 230), and to be rendered highly probable by the size and general form of the larger fossil (fig. 231). This fossil is the shaft of a long bone, and is twelve inches in length, with one extremity mutilated, but nearly entire, and the other broken off. The shaft of the bone preserves a pretty regular and uniform size, and is slightly bent: it is unequally three-sided, with the sides smooth and flat, and the angles rounded off. The osse-

* See 'Proceedings of the Geological Society,' Dec. 17, 1845.

ous wall is thin and compact, and the cavity large and smooth, like that of the air-bones in birds of flight. It differs from the femur of any known bird in the proportion of its length as compared with its breadth, and from the tibia or metatarsal bone in its trihedral figure and the flatness of the sides, none of which are longitudinally grooved. It resembles most the humerus of the Albatross, both in its form, proportions, and size, but differs therefrom in the more marked angles which bound the three sides. The extremity becomes compressed and expanded, like the distal end of the humerus of the Albatross, but is too much mutilated to allow of the precise degree of similarity or difference to be determined. On the supposition that this fragment of bone is the shaft of the humerus, its length and comparative straightness would prove it to have belonged to one of the longipennate natatorial birds, equalling in size the Albatross.

The trihedral form of the shaft of the bone resembles that of the upper or proximal half of the ulna of the Albatross; but there are no distinct traces of the attachments of the quill-feathers. By the same trihedral form it may be compared with the distal portion of the radius of the Albatross; but this idea can only be entertained by supposing the fossil bird to have been of gigantic dimensions, almost realizing the fabulous 'Roc' of Arabian romance; and the other portions of bone associated with it, and most probably parts of the same bird, render this last supposition still less probable.

The most characteristic of these portions (fig. 230) appears to be the distal end of the tibia, the peculiar trochlear extremity of which, characteristic of the class of Birds, is sufficiently preserved, although crushed. Their relative size to the preceding bone, on the supposition that that bone is the ulna, is nearly the same as in the skeleton of the Albatross.

On comparing the fossil with the lower end of the tibia of a large Eagle, we perceive in this bird a prominence on each side of the condyle, which does not exist in the fossil. In the tibia of the Adjutant Crane, which is more nearly of the size of the fossil, and resembles it more in the sharper margins of the posterior half of the trochlear boundaries, there is a ridge on each lateral surface. In the Albatross there is neither a prominent ridge nor process: the sides of the condyle, especially the outer one, are as even as in the fossil; but the anterior margins of the trochlea are more prominent and thicker in the Albatross.

The only other bone in a bird's skeleton which has a similar trochlear extremity is the metacarpal of the wing, the proximal end of which is formed by the confluent os magnum. In the degree of obliquity, and the extent of the sharp borders of this trochlea, some resemblance to the fossil may be traced; but the fossil, in the greater antero-posterior extent of the articular surface, with which the median groove of the pulley is co-extensive, differs much from the metacarpal joint, and agrees with the tibial trochlea. Moreover, there is no trace of the process which stands out from the radial end of the trochlea in the metacarpus, nor of the smaller process for the attachment of the ligament from its palmar side.

I am at present unacquainted with any bone of the Pterodactyle which presents a deep trochlea, describing three-fourths of a circle, as in the fossil; and I therefore still regard my original view of the nature of these interesting fossils, as the most consonant with known analogies of structure.

AVES. *VULTURIDÆ.?*

Fig. 232.

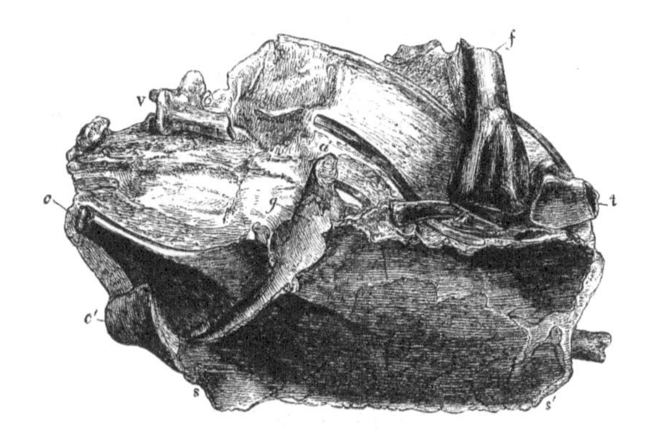

Sternum, &c. of Lithornis, eocene clay, Sheppey. Nat. size.

LITHORNIS VULTURINUS. The Vulture-like Lithornis.

Lithornis vulturinus,* OWEN, Geological Transactions, Second series, vol. vi. p. 206, pl. xxi. figs. 5 and 6.

" " " Catalogue of Fossil Mammalia, and Birds, Museum of the College of Surgeons, 4to, 1845, p. 337.

THE most conclusive evidence of the existence of Birds at the period of the formation of the English eocene tertiary strata, is afforded by the two remarkable ornitholites first described in the above-cited volume of the 'Transactions of the Geological Society of London,' these fossils, including the sternum and the sacrum, two of the most characteristic parts of skeleton of a bird, — both also having been obtained from the London clay at Sheppey. The first of these fossils (fig. 232) forms part of the extensive series of organic remains, which John Hunter had

* λίθος, stone, ὄρνις, bird.

collected before the year 1793, and which was afterwards transferred, with the rest of his anatomical collections, to the Royal College of Surgeons. The second ornitholite (fig. 233) is in the museum of James S. Bowerbank, Esq., F.R.S. I have since been favoured with the view of the sternum of another species of bird from the eocene clay near Primrose Hill, through the kindness of N. T. Wetherell Esq.; and Mr. König has published a figure of the fossil cranium of a bird from Sheppey, in the last part of his '*Icones Fossiles Sectiles*.'

The Hunterian fossil includes, with the mutilated sternum *s s'*, the sternal ends of the two coracoid bones *c c'*, a dorsal vertebra *v*, the lower end of the left femur *f*, the proximal end of the corresponding tibia *t*, portions of two other long bones, and a few fragments of the slender ribs; all of which are cemented together by the grey indurated clay usually attached to Sheppey fossils.

The entire keel, and the posterior and right margins of the sternum, are broken away; but the obvious remains of the origin of the keel, and the length of the sternum, forbid a reference of the fossil to the Struthious or strictly terrestrial order. The lateral extent and convexity of the body of the sternum, the presence and course of the secondary intermuscular ridges, *r*, and the commencement of the keel close to the anterior border of the sternum, remove the fossil from the Brachypterous family of web-footed birds, and lead us to a comparison of the fossil with the corresponding parts of the skeleton in the ordinary birds of flight.

Sufficient of the sternum remains for the rejection of the Gallinacea, and those Grallatorial and Passerine birds which have that bone deeply incised; and the field of comparison is thus restricted to such species as have the sternum either entire or with shallow posterior emarginations. Between the fossil and the corresponding parts of the skeleton of such birds, a close comparison has been instituted in regard to many minor details and modifications, — as, for example, the secondary muscular impressions and ridges on the broad outer convex surface of the sternum; its costal margin and anterior angle, *a*; the form and extent of the coracoid groove, *g*; the conformation of the sternal end of the coracoid bone, *c*; together with the form and relative size of the preserved articular extremities of the femur and tibia. But, without repeating all the details of these comparisons, it may be sufficient to state that, after pursuing them from the Sea-Gull and other aquatic species, upwards through the Grallatorial and Passerine orders, omitting few of the species and none of the genera of these orders, to which belong British birds approaching or resembling the fossil in size, the greatest number of correspondences with the fossil were at length detected in the skeletons of the Accipitrine species.

The resemblance was not, however, sufficiently close to admit of the fossil being referred to any of the existing native genera of Raptorial birds. The breadth of the proximal end of the coracoid removed the fossil from the Owls (*Strigidæ*), and the shaft of the same bone was too slender for the *Falconidæ*; the femur and tibia were, likewise, relatively weaker than in most of our Hawks or Buzzards. But in the small Turkey-Vulture (*Cathartes Aura*), besides the same general form of the bones, so far as they exist in the fossil, there is the same degree of development, and the same direction of the intermuscular ridge on the under surface of the sternum, which divided the origins of the first and second pectoral muscles. The outer angle of the proximal end of the coracoid is produced in the same degree and form, and a similar intermuscular ridge is present on the anterior and towards the outer part of the coracoid. The preserved extremities of the femur and tibia have the same conformation and nearly the same relative size in the fossil as in the existing *Cathartes*. In this genus, nevertheless, there is a deeper depression on the outer surface of the sternum external to the coracoid groove than in the fossil; but this difference is less marked in some of the large *Vulturidæ*. The vertebra, the shaft of the coracoid, and the preserved portions of sternal ribs, are relatively more slender. The fossil, moreover, indicates a smaller species of bird than is known amongst the existing *Vulturidæ*.

The anterior or inner wall of the coracoid groove is broader, the anterior angular process narrower, and the body of the sternum more convex, than in the Heron or Bittern; and the proximal end of the coracoid has a different form in the fossil. In the Sea-Gull the keel rises from a more curved surface of the sternum than in the fossil; the inner wall of the coracoid goove is broader; and the outer angle of the sternal end of the coracoid has a different form and position. I regret that I have not yet had the opportunity of comparing with this interesting specimen the skeleton of the small European Neophron, (*Vultur percnopterus*, Linn.;) but, in the meanwhile, I deem it best to retain the subgeneric distinctive appellation originally proposed for the eocene species of bird represented by the present very remarkable Hunterian fossil.

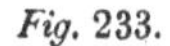

Fig. 233.

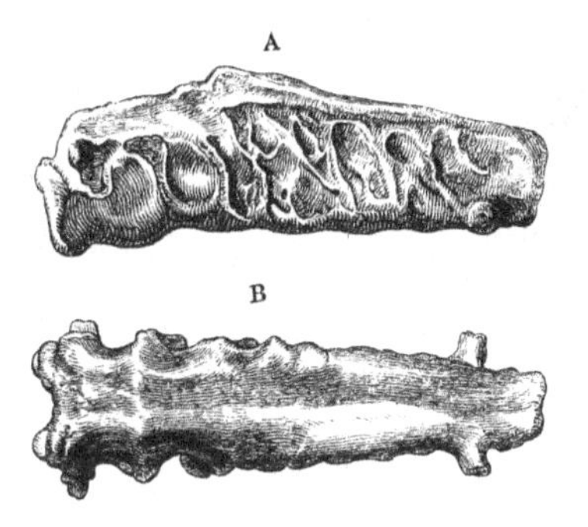

Sacrum of a Bird, eoceue clay, Sheppey. Nat. size.

Figures 233, A and B, give two views of the second ornitholite from Sheppey, alluded to in the preceding section: in the side-view, A, ten vertebræ are anchylosed together, and the under view, B, shows the complete confluence of the vertebral bodies. The long sacrum thus formed is peculiar to birds, and relates to the unfavourable position of a horizontally disposed trunk for support on a single pair of limbs,—compensation being made by the great extent of the axis of the trunk, which is grasped, as it were, by the iliac bones, and the weight transferred by these to the heads of the obliquely-placed thigh-bones. As all birds, whether they have the power of flight or not, are bipeds and prone, the long sacrum is common to all, and its structure does not present such well-marked modifications as to permit any satisfactory deductions from the present mutilated specimen of the precise position in the feathered class of the ancient bird which it represents.

The specimen itself forms part of the choice collection of J. S. Bowerbank, Esq., F.R.S.

AVES. *HALCYONIDÆ.?*

Fig. 234.

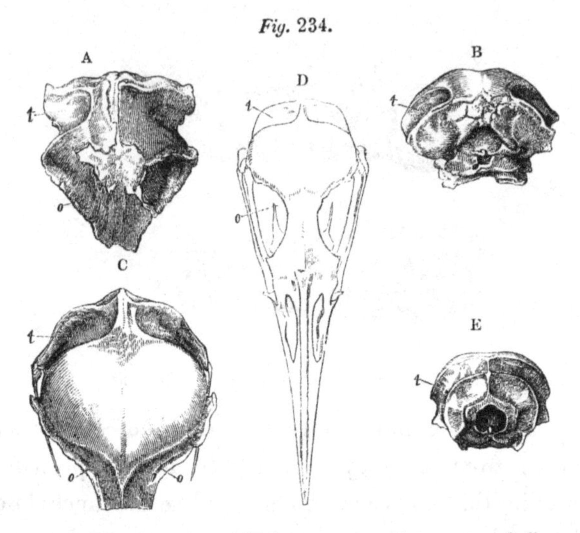

A, B, Fossil cranium of Bird, eocene clay, Sheppey. C, Gull.
D and E, Kingfisher. Nat. size.

HALCYORNIS TOLIAPICUS. Bird probably of the family *Halcyonidæ*.

Larus toliapicus, KÖNIG, Icones Fossiles Sectiles, fig. 193.

THE fossil cranium of a bird, from the London clay at Sheppey, figs. 234, A and B, and 235, forms part of the collection in the British Museum, and is the subject of fig. 193 of the useful work commenced in 1838, by Mr. König, under the title 'Icones Fossiles Sectiles.'* There is much resemblance between this fossil and the corresponding

* M. Pictet, (Paléontologie, 8vo. t. i. 1844, p. 347,) and M. Gervais, in his Geological 'Thèse sur les Oiseaux Fossiles,' 1844, p. 25, have cited No. 91, of Mr. König's 'Icones Fossiles Sectiles,' as that of the skull of a new genus of Palmiped,—for which the name '*Bucklandium*' was proposed; but, on examination of the original specimen in the British Museum, I found it to be the fossil cranium of a fish, allied to the genus *Ephippus*.

part of the skull of the smaller species of Gull; but the absence of the frontal chevron-shaped ridge, defining the excavation (fig. 234, C, *o*) for the supra-orbital glands, which ridge is present in all the species of *Larus*, as it is in most other long-winged marine birds, forbids a reference of the fossil to that genus. The occiput is also relatively broader in the fossil (fig. 234, A).

In the general form of the cranium, I have hitherto found the nearest resemblance to the fossil in the Kingfisher (*ib.* D and E); but the temporal fossæ (*t*) extend further upon the upper surface of the skull. In the subjoined cut (fig. 235), I have restored the skull after the pattern of that of the Kingfisher, in order to render the fossil more intelligible.

Fig. 235.

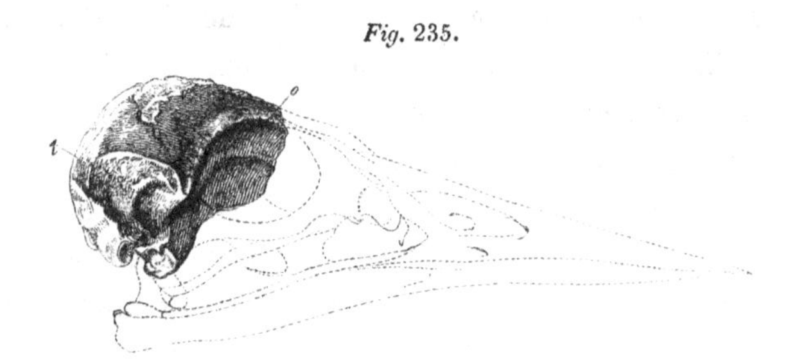

Fossil cranium of Bird, eocene clay, Sheppey. Nat. size.

AVES. *GRALLATORES.*

Fig. 236.

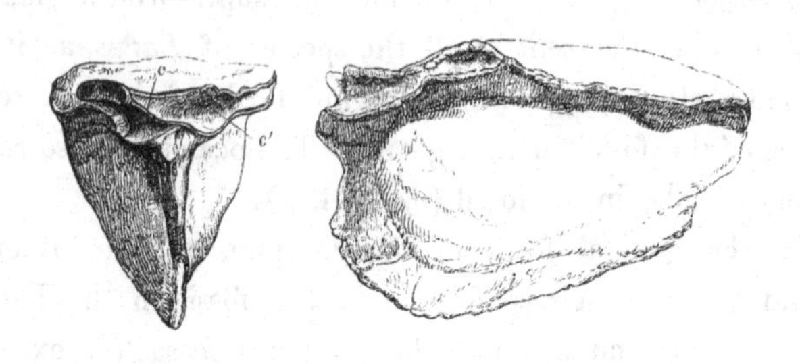

Sternum of a Small Wader, London clay, Primrose Hill. Nat. size.

SMALL WADING BIRD.

A SMALL species of the Grallatorial Order is indicated by the slightly decussating coracoid grooves, *c c*, fig. 236, at the fore part of the fossil sternum above figured. This peculiarity may be seen in the sternum of the existing species of the Heron family, (*Ardeidæ*,) and thus assists in the determination of the present fossil, in the absence of the characters which the posterior margin of the sternum would have afforded if entire.

The fossil was obtained by N. T. Wetherell, Esq., of Highgate, from the London clay near Primrose Hill, during the excavation of the tunnel of the London and Birmingham Railway; and was obligingly transmitted to me by that gentleman for the illustration of the present work.

The existence of fossil remains of Birds in the eocene gypsum of Montmartre, has long been know. Lamanon appears to have described the first in 1782;* and Camper made mention of a second, in a Paper on Fossils, printed in the Philosophical Transactions for 1786. The stimulus

* 'Journal de Physique,' t. xix, p. 175.

which Cuvier gave to the collection of the fossils of this noted locality, brought to light so many examples of eocene ornitholites that they form the subject of a special chapter in the 'Ossemens Fossiles,'* and have been referred, or rather approximated, to the genera *Haliætus*, *Buteo*, and *Strix*, in the Order *Accipitres;* to the genus *Coturnix* in *Gallinacea;* to the genera *Ibis*, *Scolopax*, and *Pelidna*, in *Grallatores*, and to the genus *Pelecanus* amongst the *Palmipedes*. Mr. J. W. Flower possesses some small Ornitholites from the fresh-water eocene deposits at Hordwell, Hants, including a tarso-metatarsal of a Bird, closely resembling that figured in Cuvier's 'Ossemens Fossiles,' t. iii. pl. 72, fig. 2; which is the most common kind in the fresh-water eocene at Montmartre. In one of the last-discovered fossil birds of Montmartre, noticed by Cuvier, the trachea or windpipe, and the little sclerotic bony plates of the eye-ball, were preserved. Fossil eggs of birds have been found in the fresh-water tertiary deposits of Cournon in Auvergne, and fossil feathers in the calcareous beds of Montebolca.

M. Escher of Zurich has obtained from the neocomian schists or lower greensand of the Canton of the Glaris, an ornitholite, which, from the characters of the bones of the wings and feet, M. v. Meyer has referred to the Passerine Order: this ancient bird was about the size of a Lark.

With regard to the pliocene ornitholites of Britain, I have recognized the humerus of a bird of flight, about the size of a Barn-Owl, which was discovered in the same bed of Norwich crag that has yielded remains of the Mastodon: the specimen is now in the collection of Mr. Fitch of Norwich. Extremely rare are the remains of birds in the fresh-water deposits or marine drift of the

* 4to, tom. iii. p. 302, pl. 72, 75.

newer pliocene period, which so abound in mammalian fossils. The light bodies of birds float long on the surface after death; and for one bird that becomes imbedded in the sediment at the bottom, perhaps ninety-nine are devoured before decomposition has sufficiently advanced to allow the skeleton to sink. Dr. Buckland has figured the fossil humerus of a bird, as large as that of a Wild Goose, from the diluvial clay of Lawford.* But most of the British ornitholites of this geological period have been discovered in ossiferous caverns. Dr. Buckland enumerates the following from the Cave at Kirkdale.† "Raven, Pigeon, Lark, a small species of Duck, resembling the 'Summer Duck,' and a bird not ascertained, being about the size of a Thrush." The fossils sanction a reference to the genera *Corvus*, *Columba*, &c., but not a closer determination. The pigeon is represented by a left ulna, and was of a species larger than our wild or domestic kinds. The humerus of a Small Wader has likewise been found. Similar ornitholites have been discovered in the cave at Kent's Hole, and in that at Berry Head, near Torbay. From the latter locality I have recognised the following remains:—the scapula, humerus, ulna, and proximal end of the metacarpus of a Falcon, rather larger than the *Falco peregrinus*. The remains of birds become more common in the fen and turbary deposits, and more easily referable to existing species; but I have restricted myself in the present work to the description of those which are actually fossil.

* 'Reliquiæ Diluvianæ,' p. 27, pl. 13. figs. 9 and 10.

† 'Reliquiæ Diluvianæ,' p. 15.

INDEX.

THE END.

London: Printed by S. & J. Bentley, Wilson, and Fley, Bangor House, Shoe Lane.

Printed in the United States
By Bookmasters